Topics in Applied Physics
Volume 128

Available online at
SpringerLink.com

Topics in Applied Physics is part of the SpringerLink service. For all customers with standing orders for Topics in Applied Physics we offer the full text in electronic form via SpringerLink free of charge. Please contact your librarian who can receive a password for free access to the full articles by registration at:

springerlink.com → Orders

If you do not have a standing order you can nevertheless browse through the table of contents of the volumes and the abstracts of each article at:

springerlink.com → Browse Publications

For further volumes:
http://www.springer.com/series/560

Topics in Applied Physics is a well-established series of review books, each of which presents a comprehensive survey of a selected topic within the broad area of applied physics. Edited and written by leading research scientists in the field concerned, each volume contains review contributions covering the various aspects of the topic. Together these provide an overview of the state of the art in the respective field, extending from an introduction to the subject right up to the frontiers of contemporary research.

Topics in Applied Physics is addressed to all scientists at universities and in industry who wish to obtain an overview and to keep abreast of advances in applied physics. The series also provides easy but comprehensive access to the fields for newcomers starting research.

Contributions are specially commissioned. The Managing Editors are open to any suggestions for topics coming from the community of applied physicists no matter what the field and encourage prospective editors to approach them with ideas.

Managing Editor

Dr. Claus E. Ascheron
Springer-Verlag GmbH
Tiergartenstr. 17
69121 Heidelberg
Germany
claus.ascheron@springer.com

Assistant Editor

Adelheid H. Duhm
Springer-Verlag GmbH
Tiergartenstr. 17
69121 Heidelberg
Germany
adelheid.duhm@springer.com

Subhash L. Shindé • Gyaneshwar P. Srivastava
Editors

Length-Scale Dependent Phonon Interactions

Springer

Editors
Subhash L. Shindé
Sandia National Laboratories
Albuquerque, NM, USA

Gyaneshwar P. Srivastava
Department of Physics
University of Exeter
Exeter, UK

ISSN 0303-4216　　　　　　　ISSN 1437-0859 (electronic)
ISBN 978-1-4614-8650-3　　　ISBN 978-1-4614-8651-0 (eBook)
DOI 10.1007/978-1-4614-8651-0
Springer New York Heidelberg Dordrecht London

Library of Congress Control Number: 2013948233

© Springer Science+Business Media New York 2014
This work is subject to copyright. All rights are reserved by the Publisher, whether the whole or part of the material is concerned, specifically the rights of translation, reprinting, reuse of illustrations, recitation, broadcasting, reproduction on microfilms or in any other physical way, and transmission or information storage and retrieval, electronic adaptation, computer software, or by similar or dissimilar methodology now known or hereafter developed. Exempted from this legal reservation are brief excerpts in connection with reviews or scholarly analysis or material supplied specifically for the purpose of being entered and executed on a computer system, for exclusive use by the purchaser of the work. Duplication of this publication or parts thereof is permitted only under the provisions of the Copyright Law of the Publisher's location, in its current version, and permission for use must always be obtained from Springer. Permissions for use may be obtained through RightsLink at the Copyright Clearance Center. Violations are liable to prosecution under the respective Copyright Law.
The use of general descriptive names, registered names, trademarks, service marks, etc. in this publication does not imply, even in the absence of a specific statement, that such names are exempt from the relevant protective laws and regulations and therefore free for general use.
While the advice and information in this book are believed to be true and accurate at the date of publication, neither the authors nor the editors nor the publisher can accept any legal responsibility for any errors or omissions that may be made. The publisher makes no warranty, express or implied, with respect to the material contained herein.

Printed on acid-free paper

Springer is part of Springer Science+Business Media (www.springer.com)

Preface

The concept of a phonon as the elementary thermal excitation in solids dates back to the start of the twentieth century. Phonons are elementary excitations arising from collective simple harmonic oscillations of atoms about their equilibrium sites in crystalline solids. Phonons manifest themselves practically in all properties of materials. For example, scattering of electrons with acoustic and optical phonons limits electrical conductivity. Optical phonons strongly influence optical properties of semiconductors, while acoustic phonons are dominant heat carriers in insulators and technologically important semiconductors. Phonon–phonon interactions dominate thermal properties of solids at elevated temperatures.

The emergence of techniques for control of semiconductor properties and geometry has enabled engineers to design structures whose functionality is derived from controlling electron interactions. Now, as lithographic techniques have greatly expanded the list of available materials and the range of attainable length scales, similar opportunities for designing devices whose functionality is derived from controlling phonon interactions are becoming available.

Currently, progress in this area is hampered by gaps in our knowledge of phonon transport across and along arbitrary interfaces, the scattering of phonons with crystal defects and delocalized electrons/collective electronic excitations, and anharmonic interactions in structures with small physical dimensions. There is also a need to enhance our understanding of phonon-mediated electron–electron interactions. Closing these gaps will enable the design of structures that provide novel solutions and enhance our scientific knowledge of nanoscale electronics and nanomechanics, including electron transport from nanoclusters to surfaces and internal dissipation in mechanical resonators. This becomes particularly important because of the great potential for use of these materials in energy harvesting systems (e.g., photovoltaics and thermoelectrics), next generation devices, and sensing systems.

This book is aimed at developing a somewhat comprehensive description of phonon interactions in systems with different dimensions and length scales. Chapters are written by acknowledged experts and arranged in a sequence that will enable the researcher to develop a coherent understanding of the fundamental concepts related to phonons in solids. Coverage of their propagation and interactions leads eventually into their behavior in nanostructures, followed by phonon interactions-based practical applications, such as thermal transport. There are nine chapters in all: the first five cover theoretical concepts and developments, the next three are devoted to experimental techniques and measurements, and the last chapter deals with fabrication, measurements, and possible technological applications.

The chapter by Tütüncü and Srivastava (Chap. 1) describes theoretical calculations for phonons in solids, surfaces, and nanostructures. Their calculations utilize the density functional perturbation theory (DFPT) and an adiabatic bond charge method (BCM). The authors introduce and explain the concepts of folded, confined, and gap phonon modes in low-dimensional systems (surfaces and nanostructures). The computed results evolve into the size, dimensionality, and symmetry dependence of phonon modes.

The chapter by Wang et al. (Chap. 2) highlights selected concepts in the theory of acoustic and optical phonons in confined systems. They cover the basic concepts of elastic and dielectric continuum models. Examples of phonon confinement in dimensionally confined structures are provided. These include phonons in single-wall carbon nanotubes, phonons in multi wall nanotubes, graphene sheets, graphene nanoribbons, and graphene quantum dots to name just a few. The chapter also touches upon the mechanisms underlying carrier–phonon scattering processes.

In Chap. 3, Srivastava outlines the theories that are generally employed for phonon transport in solids. This chapter describes in detail the steps in deriving the lattice thermal conductivity expression within the single-mode relaxation-time approximation. Explicit expressions for various phonon scattering rates in bulk and low-dimensional solids are provided. Numerical evaluation of scattering rates and the conductivity expression is presented using both Debye's isotropic continuum scheme and a realistic Brillouin zone summation technique based upon the application of the special phonon wave-vectors scheme. Results of the conductivity are presented for selected bulk, superlattice, and nanostructured systems.

The chapter by Garg et al. (Chap. 4) covers "first-principles determination of phonon lifetimes, mean free paths, and thermal conductivities," in selected crystalline materials. The thermal properties of insulating, crystalline materials are essentially determined by their phonon dispersions, the finite-temperature excitations of their phonon populations treated as a Bose–Einstein gas of harmonic oscillators, and the lifetimes of these excitations. The authors present an extensive case study of phonon dispersions, phonon lifetimes, phonon mean free paths, and thermal conductivities for the case of isotopically pure silicon and germanium.

Mingo et al. (Chap. 5) present ab initio approaches to predict materials properties without the use of any adjustable parameters. This chapter presents some of the recently developed techniques for the ab initio evaluation of the lattice thermal conductivity of crystalline bulk materials and alloys as well as nanoscale materials including embedded nanoparticle composites.

The chapter by Hurley et al. (Chap. 6) focuses on the interaction of thermal phonons with interfaces. They show that phonon interactions with interfaces fall into two broad categories, which are defined by interfaces with two different geometries that form the boundary of nanometer size channels (e.g., grain boundaries, superlattice interfaces, nanowires, and thin films). The authors also demonstrate that the Boltzmann transport equation provides a convenient model for considering boundary scattering in nanochannel structures, while for internal interfaces such as the grain boundaries found in polycrystals, it is more natural to consider transmission and reflection across a single boundary. Also addressed are experimental techniques for measuring phonon transport in nanoscale systems, including experimental results using time-resolved thermal wave microscopy on specimens with grain boundaries having known atomic structure.

Yamaguchi (Chap. 7) reviews time-resolved phonon spectroscopy and phonon transport in nanoscale systems. He touches upon time-resolved acoustic phonon measurements using ultrafast laser spectroscopy, with particular emphasis on the methods that are relevant to transport measurements. The chapter discusses tunable acoustic spectroscopy with a combination of picosecond acoustics and laser pulse shaping. The development of the spectrometer demonstrates direct measurement of the group velocity and mean free path of acoustic phonons at variable frequency up to about 400 GHz.

The chapter by Kent and Beardsley (Chap. 8) delves into "semiconductor superlattice sasers at terahertz frequencies": They discuss the design criteria for the superlattice lasers to be used as the gain medium and acoustic mirrors in saser devices. They elucidate potential applications of sasers in science and technology, viz., nanometer-resolution acoustic probing and imaging of nanoscale structures and devices. They also touch upon more recent developments in the area of THz acousto-electronics, e.g., the conversion of sub-THz acoustic impulses to sub-THz electromagnetic waves using piezoelectric materials.

Lazic et al. (Chap. 9) discuss growth techniques that allow formation of different types of nanostructures, such as quantum wells, wires, and dots on the surface of a single semiconductor crystal for creating surface acoustic wave (SAW) devices. They describe how SAWs propagating on the crystal surface provide an efficient mechanism for the controlled exchange of electrons and holes between these nanostructures. They explore this ability of dynamic SAW fields to demonstrate acoustically driven single-photon sources using coupled quantum wells and dots based on the (Al,Ga)As (311)A material system. Also addressed is the growth of the coupled nanostructures by molecular beam epitaxy, the dynamics of the acoustic carrier transfer between them, and the acoustic control of recombination in quantum dots.

This book has been developed and evolved for more than 2 years and the authors have made every attempt to take into account the latest developments in their fields. We hope that the concepts covered in these chapters will endure and inspire many young scientists to initiate their own research in these exciting and promising fields.

The editors are very thankful to all of the authors for their interest and patience and to the Springer New York staff for their highly professional handling of the production of this volume. They would also like to thank the colleagues who contributed their precious time to reviewing the manuscripts.

Albuquerque, NM, USA Subhash L. Shindé
Exeter, UK Gyaneshwar P. Srivastava

Contents

1. **Lattice Dynamics of Solids, Surfaces, and Nanostructures** 1
   H.M. Tütüncü and G.P. Srivastava

2. **Phonons in Bulk and Low-Dimensional Systems** 41
   Zhiping Wang, Kitt Reinhardt, Mitra Dutta,
   and Michael A. Stroscio

3. **Theories of Phonon Transport in Bulk and Nanostructed Solids** 81
   G.P. Srivastava

4. **First-Principles Determination of Phonon Lifetimes, Mean
   Free Paths, and Thermal Conductivities in Crystalline
   Materials: Pure Silicon and Germanium** 115
   Jivtesh Garg, Nicola Bonini, and Nicola Marzari

5. **Ab Initio Thermal Transport** 137
   N. Mingo, D.A. Stewart, D.A. Broido, L. Lindsay, and W. Li

6. **Interaction of Thermal Phonons
   with Interfaces** 175
   David Hurley, Subhash L. Shindé, and Edward S. Piekos

7. **Time-Resolved Phonon Spectroscopy and Phonon
   Transport in Nanoscale Systems** 207
   Masashi Yamaguchi

8. **Semiconductor Superlattice Sasers
   at Terahertz Frequencies: Design, Fabrication and
   Measurement** 227
   A.J. Kent and R. Beardsley

9. **Acoustic Carrier Transport in GaAs Nanowires** 259
   Snežana Lazić, Rudolf Hey, and Paulo V. Santos

Index 293

Contributors

R. Beardsley School of Physics and Astronomy, University of Nottingham, University Park, Nottingham, UK

Nicola Bonini Department of Physics, King's College London, London, UK

D.A. Broido Department of Physics, Boston College, Chestnut Hill, MA, USA

Mitra Dutta Department of Electrical and Computer Engineering, University of Illinois at Chicago, Chicago, IL, USA

Jivtesh Garg Department of Mechanical Engineering, Massachusetts Institute of Technology, Cambridge, MA, USA

Rudolf Hey Paul-Drude-Institut für Festkörperelektronik, Berlin, Germany

David Hurley Idaho National Laboratory, Idaho Falls, ID, USA

A.J. Kent School of Physics and Astronomy, University of Nottingham, University Park, Nottingham, UK

Snežana Lazić Paul-Drude-Institut für Festkörperelektronik, Berlin, Germany

Departamento de Física de Materiales, Universidad Autónoma de Madrid, Madrid, Spain

W. Li CEA Grenoble, Grenoble, France

L. Lindsay Naval Research Laboratory, Washington, DC, USA

Nicola Marzari Theory and Simulation of Materials (THEOS), École Polytechnique Fédérale de Lausanne, Lausanne, Switzerland

N. Mingo CEA-Grenoble, Grenoble, France

Edward S. Piekos Sandia National Laboratories, Albuquerque, NM, USA

Kitt Reinhardt Air Force Office of Scientific Research, Arlington, VA, USA

Paulo V. Santos Paul-Drude-Institut für Festkörperelektronik, Berlin, Germany

Subhash L. Shindé Sandia National Laboratories, Albuquerque, NM, USA

G.P. Srivastava School of Physics, University of Exeter, Exeter, UK

D.A. Stewart Cornell Nanoscale Facility, Cornell University, Ithaca, NY, USA

Michael A. Stroscio Department of Electrical and Computer Engineering, University of Illinois at Chicago, Chicago, IL, USA

H.M. Tütüncü Sakarya Üniversitesi, Fen-Edebiyat Fakültesi, Fizik Bölümü, Adapazar, Turkey

Zhiping Wang Department of Electrical and Computer Engineering, University of Illinois at Chicago, Chicago, IL, USA

Masashi Yamaguchi Department of Physics, Applied Physics, and Astronomy, Rensselaer Polytechnic Institute, Troy, NY, USA

Chapter 1
Lattice Dynamics of Solids, Surfaces, and Nanostructures

H.M. Tütüncü and G.P. Srivastava

Abstract We present results of lattice dynamical calculations for solids, surfaces, and nanostructures. The calculations have been made by employing two levels of theoretical approaches: the density functional perturbation theory (DFPT) and an adiabatic bond charge method (BCM). The concepts of folded, confined, and gap phonon modes in the low-dimensional systems (surfaces and nanostructures) are explained with examples. The computed results are used to discuss the size, dimensionality, and symmetry dependence of phonon modes.

1.1 Introduction

A phonon is a quantum of crystal vibrational energy. It can be viewed as an elementary excitation arising from relative motion of atoms in a crystal. It can be described as a quasi particle of zero spin which at a given temperature of the crystal is governed by the Bose–Einstein distribution function. Establishment of dispersion relationship between the energy $\hbar\omega$ of a phonon with its momentum $\hbar q$ and polarisation s is known as the topic of lattice dynamics. Phonon dispersion relations are essential ingredients for understanding a large number of properties of three-dimensional solids, surfaces, and nanostructured materials. Theoretical studies on lattice dynamics of solids developed in 1910s with the theoretical works by Born von Kármán and by Debye (see [1]). Measurements of bulk phonon spectrum started in 1950, based on the neutron scattering technique. The concept

---

H.M. Tütüncü (✉)
Sakarya Üniversitesi, Fen-Edebiyat Fakültesi, Fizik Bölümü, 54187, Adapazarı, Turkey
e-mail: Tutuncu@sakarya.edu.tr

G.P. Srivastava
School of Physics, University of Exeter, Stocker Road, Exeter EX4 4QL, UK
e-mail: G.P.Srivastava@exeter.ac.uk

of surface phonons can be traced back to the work in the nineteenth century by Lord Rayleigh [2]. Theoretical studies of surface phonons started in 1940s with the work of Lifshitz and Rosenzweig [3]. Due to small cross sections, the neutron scattering technique is generally not suitable for surface studies. Two main experimental techniques for measurement of surface phonon dispersion curves, started in 1970s and now well advanced, are: (i) inelastic helium atom scattering and (ii) high-resolution electron energy loss spectroscopy (HREELS). Another experimental technique, which provides high resolution to resolve sharp modes due to adsorbates on surfaces, is the surface infrared spectroscopy (SIRS). A good description of the experimental techniques can be found, among others, in [4, 5]. Theoretical studies of surface phonons began in 1950s, and a brief description of early studies can be found in [4]. Fabrication techniques and development of refined experimental techniques have allowed identification of phonon modes in low-dimensional and nanostructured materials since the 1970s (see [6] and references therein). Differences in phonon modes and their dispersion relations in solids (3D), surfaces (2D), and nanostructured materials (2D, 1D, or 0D) can essentially be related to dimensionality and symmetry changes.

Calculations and discussions of phonon dispersion relations have traditionally be attempted by treating the system either as a continuum or as a network of atoms adopting to a well-defined crystal structure. The essential ingredients required for the application of a continuum theory are elastic constants of the medium. Development of a lattice dynamical theory requires knowledge of interatomic force constants. The chapter by Wang et al. in this book reviews the basic concepts and some applications of the elastic continuum dielectric continuum models in bulk and low-dimensional systems. In this chapter we review the basic concepts of the atomic-level theories of lattice dynamics. Using results of detailed calculations for bulk, surfaces, and nanostructured materials, we will discuss the dimensionality and symmetry dependence of phonon modes.

1.2 Elements of Lattice Dynamical Theory

Lattice dynamical studies can be broadly categorised as based on (i) phenomenological models, (ii) first-principles methods, models requiring no adjustable parameters, and (iii) hybrid models, basically phenomenological in nature but including useful parameters obtained from first-principles calculations of electronic structure and atomic geometry. Phenomenological methods usually rely upon fitting interatomic interaction parameters to reproduce available experimental measurements, but are of limited applicability as the parameters are not in general transferable to other situations. Recent developments in theoretical and computational techniques have made first-principles lattice dynamical studies of bulk materials a routine exercise, and of surfaces and nanostructures an affordable reality.

1 Lattice Dynamics of Solids, Surfaces, and Nanostructures

Let $x(lb) = x(l) + x(b)$ denote the position of bth atom of mass M_b in lth unit cell of a periodic system. A trial function for atomic vibrations can be expressed as

$$u_\alpha(lb) = \frac{1}{\sqrt{M_b}} \sum_q U'_\alpha(q;b) \exp[i(q \cdot x(lb) - \omega t)], \qquad (1.1)$$

with U being the vibrational amplitude. Solutions to the lattice dynamical problem can be obtained by solving the following eigenvalue equations

$$\omega^2 e_\alpha(b;qs) = \sum_{b'\beta} C_{\alpha\beta}(bb' \mid q) e_\beta(b';qs). \qquad (1.2)$$

Here $e(b;qs)$ is the eigenvector of atomic displacements and $C_{\alpha\beta}(bb' \mid q)$ is a Fourier component of inter-atomic force constant matrix $\Phi_{\alpha\beta}(0b;lb')$:

$$C_{\alpha\beta}(bb' \mid q) = \frac{1}{\sqrt{M_b M_{b'}}} \sum_l \Phi_{\alpha\beta}(0b;lb') \exp[-iq \cdot (x(0b) - x(lb'))]. \qquad (1.3)$$

The force constant matrix $\{\Phi_{\alpha\beta}(lb;l'b')\}$ is obtained from the harmonic part of the total crystal potential energy, and its consideration is at the heart of the topic of lattice dynamics.

In this section we will provide a brief discussion of lattice dynamical studies using an *ab initio* method, the phenomenological adiabatic bond charge model for tetrahedrally bonded semiconductors, and a hybrid approach employing the adiabatic bond charge model and structural as well as electronic information obtained from the application of the *ab-initio* method.

1.2.1 Structural Modelling and Periodic Boundary Conditions

Lattice dynamical studies of crystalline bulk solids are usually made by imposing the three-dimensional Born–van Kàrman periodic boundary conditions. Lattice dynamics of low-dimensional structures, such as surfaces and nanostructures, can also be studied by adopting to suitable structural modelling and imposing the three-dimensional Born–van Kàrman periodic boundary conditions. For example, a solid surface can be modelled in the form of a slab geometry with an artificial periodicity introduced normal to the surface. This will require consideration of a supercell containing a few atomic layers of the material under study and a reasonably sized vacuum region. The in-plane size of the supercell is chosen according to required surface "reconstruction". The sizes of the atomic slab and the vacuum region are chosen with the consideration that intra as well as inter slab waves experience minimum interference from each other. A nanowire can be modelled in a similar

Fig. 1.1 Arrangement of ions (*open and filled spheres*) and bond charges (*small filled spheres*) in tetrahedrally bonded diamond and zincblende structures

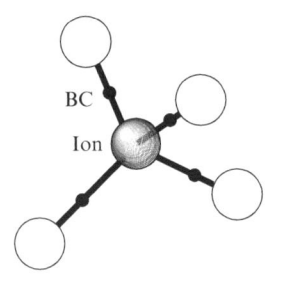

fashion, but with consideration of vacuum regions along the four sides normal to its growth direction. Obviously, modelling a nanodot will require surrounding it with a vacuum region in all directions.

1.2.2 Adiabatic Bond Charge Model

The adiabatic bond charge model was originally developed by Weber [7] for calculating phonon dispersion relations of semiconductors with diamond structure and extended by Rustagi and Weber [8] for zincblende semiconductors. The basic concept of the bond charge model requires that the electronic charge distribution between neighbouring atoms should be clearly identifiable in terms of well-defined maxima around an ion where bond charges (BCs) could be placed. This can be successfully done for semiconductors with either purely covalent bonding (e.g. Si, Ge) or at least with reasonably covalent bonding (such as III–V and II–VI semiconductors). In this model valence charge distribution between two neighbouring ions of charges $Z_1 e$ and $Z_2 e$ is represented, as shown in Fig. 1.1, by a massless point charge (BC) Ze at a distance $r_0(1 \pm f)/2$, where r_0 is the nearest neighbour distance and f is a measure of ionicity, with the charge neutrality constraint $Z_1 = Z_2 = -2Z$. Three types of interactions are considered: (i) ion–ion, ion–BC, and BC–BC Coulomb interaction, (ii) ion–ion and ion–BC central short-range interaction, and (iii) Keating type bond-bending (i.e. non-central) interaction between BCs, as well as between ions and BCs. A total of four (six) adjustable parameters are needed to set up the dynamical equation of motion for materials based on the diamond (zincblende) structure. When dealing with composite structures, such as alloys or superlattices made of two materials, additionally needed hetero-inter-atomic force constants can be constructed as suitable average of the corresponding terms for the constituent materials.

1.2.3 First-Principles Methods

First-principles methods for setting up the dynamical matrix require total energy and atomic force calculations without any attempt towards parametrisation. The most reliable of such methods are based on the application of the Density Functional

Theory (DFT) for electron–electron interaction. One of the most practiced approaches is the use of DFT within the plane wave pseudopotential method for solving the electronic states and total crystal energy. Excellent discussion of these techniques can be found in several books, including [9–11]. With total energy and forces available, phonon calculations can be performed at several levels of sophistication. A brief discussion of these will be provided here.

(a) *Frozen phonon approach*:

This approach is applied to calculate the phonon frequency of a pre-determined atomic displacement pattern, corresponding to a chosen phonon wave vector. As can be imagined, only phonons at a few symmetry points in the Brillouin zone of the chosen crystal can be studied. For a displacement pattern with a "frozen" vibrational mode of amplitude u, the total energy difference δE, or linear atomic force magnitude F, can be computed. The phonon frequency is then computed as $\omega = \sqrt{\Lambda/M}$, where M is the atomic mass and Λ is the force constant determined as $\Lambda = F/u = 2\delta E/u$. For calculating the frequency of the degenerate LO/TO zone-centre optical modes in diamond structure, a relative displacement of the two basis atoms in the primitive unit cell is required along the [111] direction. For computing the modes LA, LO, TA, and TO at the zone boundary X, appropriate displacement patterns using a four-atom unit cell are required so that a phonon wave vector of magnitude $q = |\Gamma - X|$, with wave length equal to the cubic lattice constant, is required [12].

(b) *Restricted dynamical matrix approach*:

In some cases, such as molecular adsorption on surfaces of non-polar semiconductors, zone-centre phonon modes can be studied by constructing a restricted dynamical matrix from total energy differences (harmonic contribution) or forces (linear part) corresponding to $3n$ displacements of n atoms under investigation. Phonon eigensolutions can be obtained by diagonalising such a matrix.

(c) *Planar force constant approach*:

Phonon dispersion relations in non-polar semiconductors along a symmetry direction can also be studied by using the planar force constant approach, developed by Kunc and Martin [13]. In this approach a superlattice is constructed, containing a finite number, say N, of atomic planes normal to phonon wave vector q along the chosen symmetry direction. The atomic plane through a chosen origin is given a displacement u_0 and harmonic interplanar force $F(n)$ on plane n is evaluated. The force constant between the plane through the origin and the nth plane is calculated from the relation $k_n = -F(n)/u_0$. With interplanar force constants determined, the dynamical matrix can easily be set up and phonon dispersion relations obtained.

(d) *Linear response approach*:

An unrestricted calculation of phonon eigensolutions for a periodic system can be made by adopting a linear response approach within the density functional scheme for total energy and forces. This scheme, in the context of the plane wave pseudopotential method and the density functional theory (DFT), has been discussed in detail by Giannozzi et al. [14]. Here we point out the essential ingredient of the scheme.

For a given atomic displacement u the electronic part of total energy is expressed as

$$E_{el}(u) = E_{el}(0) + \delta E_{el}(u, \Delta\rho), \qquad (1.4)$$

where $\Delta\rho$ is the resulting change in the electronic charge density. The electronic contribution to the force constant matrix Φ depends on $\Delta\rho$ and the change ΔV_{KS} in the Kohn–Sham potential. A self-consistent change in ΔV_{KS} is expressed up to terms linear in $\Delta\rho$. Similarly, using first-order perturbation theory, $\Delta\rho$ is expressed in terms of ΔV_{KS}. An iterative solution is obtained for $\Delta\rho$ and ΔV_{KS} by solving the two relationships. For details of calculations of phonon dispersion relations in semiconductors, the reader is encouraged to follow [9, 10, 14].

1.3 Phonons in Bulk Solids

Phonon dispersion relations and density of states of a large number of bulk crystalline solids are available in the literature. However, most results have been obtained by employing phenomenological theories based on parameters to fit experimental measurements at certain high symmetry points in the respective Brillouin zone. In this section we present results of *ab inito* calculations for three selections of crystalline solids: diamond structure group-IV semiconductors, lanthanum metal in the face-centre cubic (fcc) and double hexagonal close packed (dhcp) structures, and MgCNi$_3$ in the ABO$_3$-like cubic perovskite structure.

1.3.1 *Phonons in Group-IV Semiconductors*

Ab initio phonon dispersion and density of states for α-Sn, Ge, Si, and C in the diamond structure are presented in Fig. 1.2. In this figure, filled circles indicate the experimental phonon results [15–18] for these materials. The experimental data well reproduced by *ab initio* calculations. Due to two atoms per unit cell in the diamond structure, there are six vibrational modes for any chosen **q** point. One can see all the six phonon branches along the Γ-K, L-K, and K-W symmetry directions. However, the situation is simplified along high symmetry directions where irreducible representations of phonon wavevectors have higher dimensions. For example, there are only four distinct phonon branches along the Γ-X and Γ-L symmetry directions since transverse branches become degenerate along these directions.

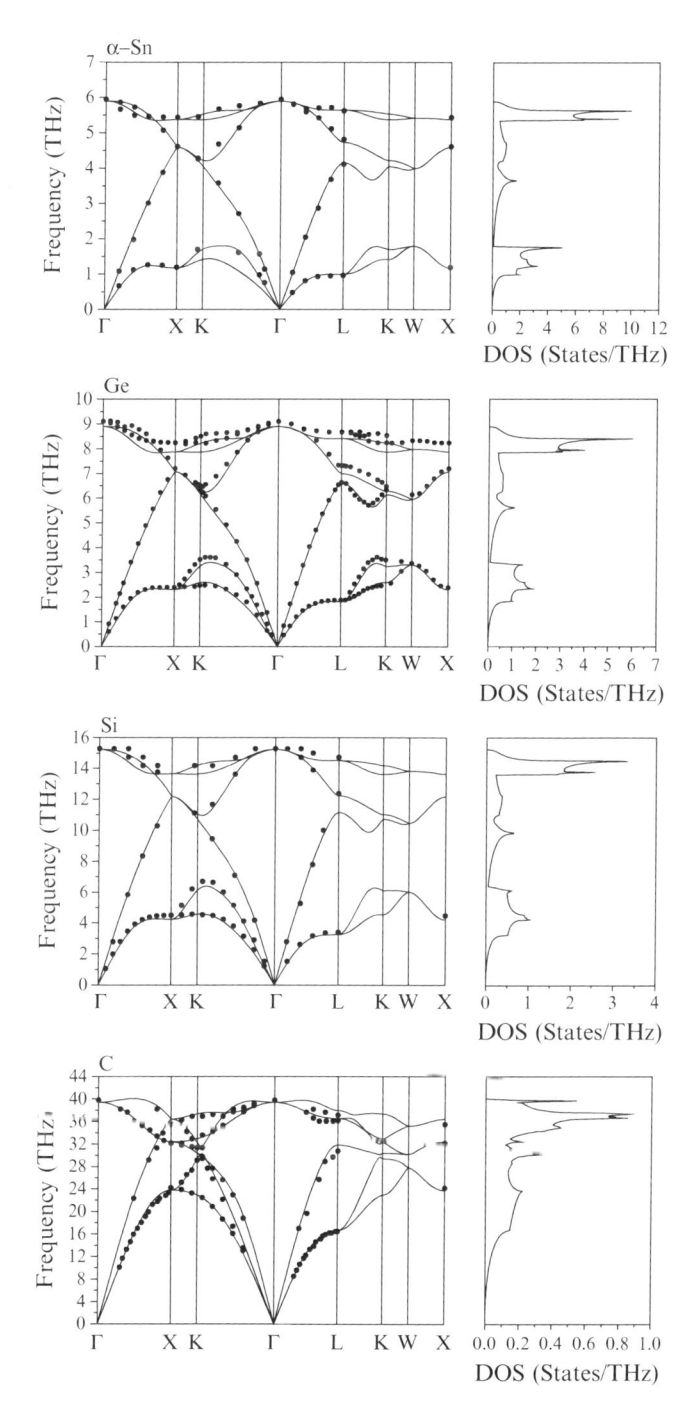

1 Lattice Dynamics of Solids, Surfaces, and Nanostructures   7

Fig. 1.2 *Ab initio* phonon dispersion and density of states for α-Sn, Ge, Si, and C in the diamond structure. Experimental data (*filled circles*) are taken from [15–18]

We first note a couple of features in the dispersion curves and density of states for α-Sn, Ge, and Si. The transverse and longitudinal acoustic (TA and LA) branches behave normally in the long-wave limit with steep slopes. However, towards the zone edges the TA branches flatten out, leading to a clear peak in the density of states. The maximum frequency of the spectrum occurs at the zone centre, where the longitudinal optical (LO) and transverse optical (TO) modes are degenerate. Away from the zone centre the longitudinal optical (LO) branch is quite dispersive along the [100] symmetry direction, and lies below the transverse optical (TO) branch close to the zone edge X.

The lattice dynamical properties of diamond are different from those of other group-IV semiconductors. The typical flatness of the TA branches is completely lacking in diamond. As a result of this, there is no pronounced TA peak in the phonon density of states for diamond. The maximum of the spectrum is not at the zone centre. The LO branch always lies above the TO branch along the [100] direction. Furthermore, the (LA) branch crosses the TO branch along this symmetry direction close to the zone edge X. These differences can be related to several distinctive features of diamond such as its electronic structure, small lattice constant, large bulk modulus, and large cohesive energy.

Having discussed the success of the *ab initio* method in correctly obtaining phonon dispersion curves, we now briefly mention that the application of the phenomenological adiabatic bond charge model (BCM) has also been found to generate highly respectable results for above discussed group-IV solids. This can be clearly seen from Fig. 1.3 which presents the phonon dispersion curves of α-Sn, Ge, Si, and C in the diamond structure. The BCM calculations also indicate the typical flatness of the (TA) phonon branches in α-Sn, Ge, and Si close the boundaries of the Brillouin zone.

We should, however, point out that despite good agreement with *ab inito* results and experimental measurements, being empirical in nature BCM calculations have limited predictive power. The reason for this is that the parameters entering the BCM model are fitted to experimental data. On the other hand, phonon studies using *ab initio* methods use only atomic numbers as input for calculations, but in practice are limited to systems with unit cells containing not more than 100 atoms or so. Thus, BCM method may be a good practical choice for studying phonon calculations in systems requiring a large number of per unit cell, for example when modelling nanostructured materials.

1.3.2 Phonons and Superconductivity in fcc and dhcp Lanthanum

The unique physical and chemical properties of the rare earth metals have attracted interest for decades. Lanthanum (La), the first member of the rare-earth series of elements, can exist in both the double hexagonal closed packed (dhcp) phase and

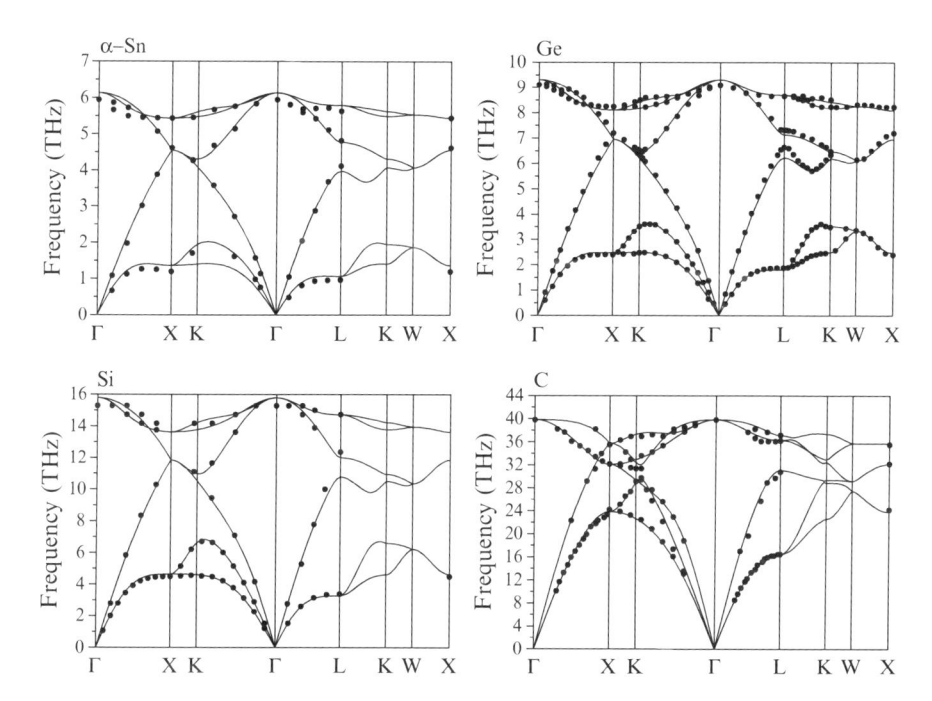

Fig. 1.3 Adiabatic bond-charge model phonon dispersion for α-Sn, Ge, Si, and C in the diamond structure. Experimental data (*filled circles*) are taken from [15–18]

the face-centered cubic (fcc) phase. A normally formed La includes both crystal structure phases. This metal exhibits particularly interesting properties. Due to its high electronic density of states at the Fermi level and specific phonon spectrum, one would expect strong electron–phonon coupling, and therefore a reasonable high superconducting transition temperature. This has led to several experimental and theoretical studies of the phonon and superconducting properties of lanthanum.

The fcc phase with only one La atom per unit cell is characterised with the cubic point-group O_h and space group $Fm\bar{3}m$. The dhcp phase of La has the hexagonal symmetry with point-group D_{6h} and space group $P6_3/mmc$. This structure is different from the well-known hexagonal closed packed (hcp) structure, in that the stacking sequence is ABAC for dhcp instead of ABAB for hcp. This results in doubling the c-axis lattice constant and thus an ideal c/a ratio of $2 \times \sqrt{8/3}=3.2650$. This structure includes four atoms per unit cell, with the atomic positions in lattice coordinates as (0,0,0), (0,0,1/2), (1/3,2/3,1/4), and (2/3,1/3,3/4). The equilibrium lattice parameters (a and c), bulk modulus (B), and the pressure derivative of the bulk modulus (B'), for both structures as determined at zero pressure ($P = 0$ GPa), are given in Table 1.1. The experimental and *ab initio* theoretical results in this table compare very well with each other.

The *ab initio* results [19] for the phonon dispersion curves and the phonon density of states for La in the fcc and dhcp phases are presented in Fig. 1.4. As may be

Table 1.1 Static properties of fcc and dhcp lanthanum. LAPW and LMTO are the full potential linearised augmented plane wave method and the self-consistent linear-muffin-tin-orbital method, respectively. LDA and GGA indicate the local density approximation and the generalised gradient approximation, respectively

Source	a(Å)	c(Å)	B(GPa)	B'
fcc La [19]	5.350		25.70	2.60
LAPW [20]	5.310			
LAPW [21]	5.320		26.10	2.78
LMTO [22]	5.110		24.00	3.00
GGA [23]	5.344		26.59	2.66
Experimental [24]	5.31		24.80	2.80
Experimental [25]			23.10	
dhcp La [19]	3.801	12.262	26.30	2.89
LDA [26]	3.619	11.678	30.00	
GGA [27]			27.50	
GGA [26]	3.784	12.203	24.39	
Experimental [24]	3.773	12.081		
Experimental [28]			24.3	

seen, all phonon frequencies for both structures are real and positive and there are no phonon branches with dispersions that dip towards the zero frequency. This indicates that both phases of La are dynamically stable. The agreement between the *ab initio* and experimental results for fcc La is satisfactory. An interesting feature of the fcc La phonon spectrum is that the lower TA branch becomes very soft along the [111] direction. However, no anomalies occur in the LA branch along any high symmetry direction. The interesting aspect of the phonon spectrum for the dhcp La is the anomalous dispersion exhibited by the lower acoustic branch along the Γ-M direction close to the zone boundary M. Thus, this branch exhibits a dip at the M point. Away from the zone centre, all the optical phonon branches are quite dispersive along all the symmetry directions. Figure 1.4 also shows the calculated phonon density of states for dhcp La. In this phase there are several phonon branches (acoustic and low-lying optical) with frequencies less than 1.5 THz. Higher-lying optical branches are strongly dispersive with no sharp peaks, as usually observed in several metals.

The zone-centre phonon modes are of special theoretical interest because they can be observed by various experimental methods. The optical zone-centre phonon modes in dhcp belong to the following irreducible representations:

$$E_{2u} + E_{2g} + E_{1u} + B_{2g} + B_{1u} + A_{2u}.$$

The A and B modes are non-degenerate and correspond to atomic vibrations along the c-axis (or the z-axis), while the E modes are twofold degenerate and correspond to atomic vibrations in the x-y plane (normal to the c-axis). Thus, we have a total of six distinct optical phonon modes. These have been calculated at frequencies

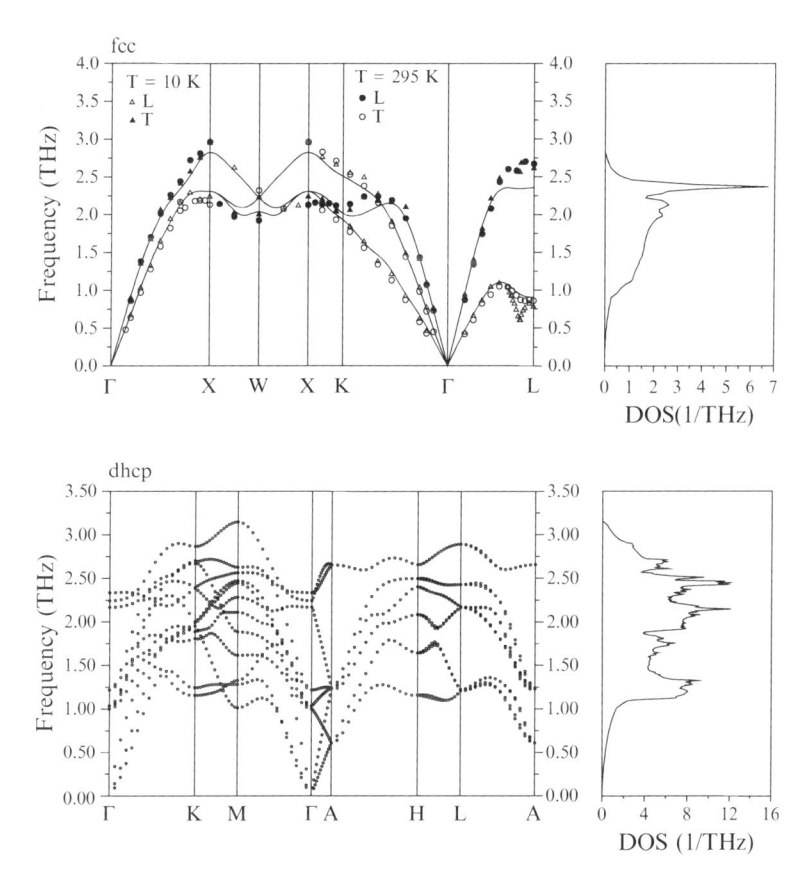

1 Lattice Dynamics of Solids, Surfaces, and Nanostructures

Fig. 1.4 The phonon spectrum and vibrational density of states for fcc and dhcp La. For fcc La, *open and closed circles* are experimental data at T=295 K, while *open and closed triangles* indicate experimental data at T=10 K [29]. Taken from [19]

0.99 THz (E_{2u}), 1.03 THz (E_{2g}), 1.22 THz (E_{1u}), 2.16 THz (B_{2g}), 2.24 THz (B_{1u}), and 2.33 THz (A_{2u}). In addition to their frequencies, the eigendisplacements of these phonon modes are displayed in Fig. 1.5. The lowest one (E_{2u}) results from opposing vibrations of intermediate hexagonal La atoms against other hexagonal La atoms (i.e. the triangularly arranged La atoms do not move for this mode). For the E_{2g} mode the triangularly arranged La atoms vibrate against each other while all the hexagonal La atoms are stationary. The E_{1u} mode includes atomic vibrations from all the La atoms in the unit cell. This phonon mode is characterised by opposing vibrations of triangle and hexagonal La atoms. The B_{2g} phonon mode is due to opposing vibrations of triangle La atoms in the [001] direction while all the hexagonal La atoms remain stationary. For the B_{1u} phonon mode, the intermediate hexagonal La atoms move against other hexagonal atoms in the [001] direction. Finally, all the La atoms in the unit cell vibrate for the A_{2u} mode.

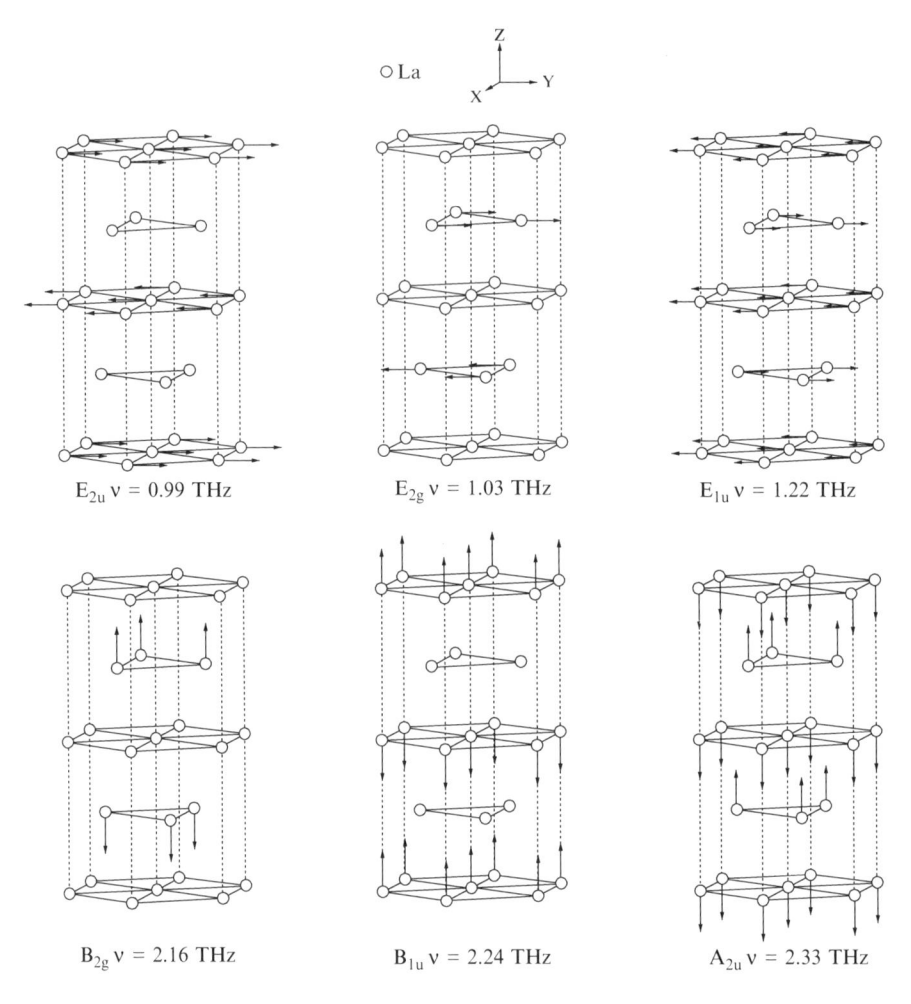

Fig. 1.5 Eigenvector representation of zone-centre phonon modes in double-hexagonal closed packed lanthanum. Taken from [19]

A quantity relevant to discussion of Bardeen–Cooper–Schrieffer (BCS) type superconductivity is the electron–phonon spectral function $\alpha^2 F(\omega)$, which depends on the electronic density of states at Fermi level $N(E_F)$ and the phonon line-width γ_{qs} [30]

$$\alpha^2 F(\omega) = \frac{1}{2\pi N(E_F)} \sum_{qs} \frac{\gamma_{qs}}{\hbar \omega_{qs}} \delta(\omega - \omega_{qs}). \quad (1.5)$$

Figure 1.6 presents the calculated electron–phonon spectral function $\alpha^2 F(\omega)$ and the phonon density of states for fcc and dhcp La. One can see that the shape of

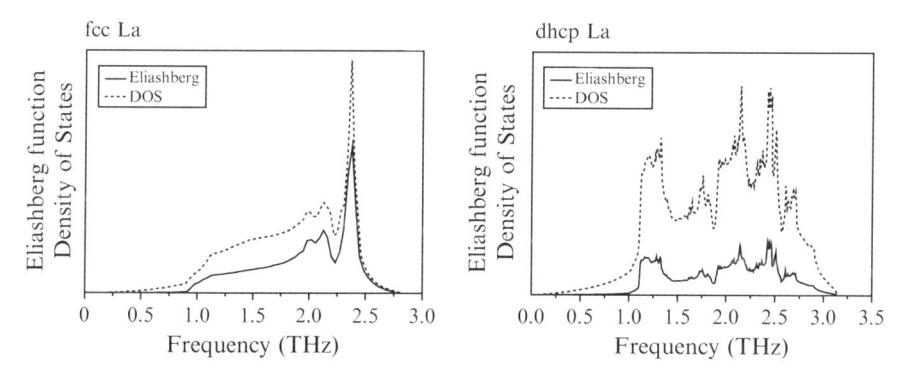

1 Lattice Dynamics of Solids, Surfaces, and Nanostructures

Fig. 1.6 A comparison of the phonon density of states (*dashed lines*) and Eliashberg spectral function $\alpha^2 F(\omega)$ (*solid lines*) for both phases of lanthanum. Taken from [19]

Table 1.2 Comparison of theoretical superconducting state parameters for dhcp La and fcc La. Taken from [19]

Phase	λ	T_C	$N(E_F)$(States/eV)	ω_{ln}(K)	$<\omega>$(K)	$\left(<\omega^2>\right)^{1/2}$ (K)
fcc	1.06	5.88	1.63	85	92	93
dhcp	0.97	4.87	1.49	82	88	91

the electron–phonon spectral function is similar to that of the phonon density of states for both phases in the entire frequency range. This observation indicates that phonons of all frequencies contribute to the electron–phonon coupling for both structures. From the electron–phonon spectral function, the total electron–phonon coupling parameter is found to be 1.06 for fcc La and 0.97 for dhcp La. The calculated electron–phonon coupling parameter for dhcp La can be compared with its experimental value [31] of 0.85. A comparison of the superconducting parameters for dhcp La and fcc La is listed in Table 1.2. The values of superconducting parameters for dhcp La are all lower than their corresponding values for fcc La. This can be related to the difference between their electronic density of states at the Fermi level ($N(E_F)$). The calculations suggest that the $N(E_F)$ value of 1.49 states/eV for dhcp La is smaller than the corresponding value of 1.63 states/eV for fcc La.

1.3.3 Phonons and Electron–Phonon Interaction in MgCNi₃

In recent years, considerable effort has been spent to investigate ground state properties of MgCNi$_3$ because of a recent discovery of superconductivity in this material [32]. Experimental works [33] have shown that the Bardeen–Cooper–Schrieffer (BCS) electron–phonon interaction mechanism can be used to explain the transition temperature (T_C) in this material. A more recent experimental investigation [34] of the specific heat suggests that this material is characterised

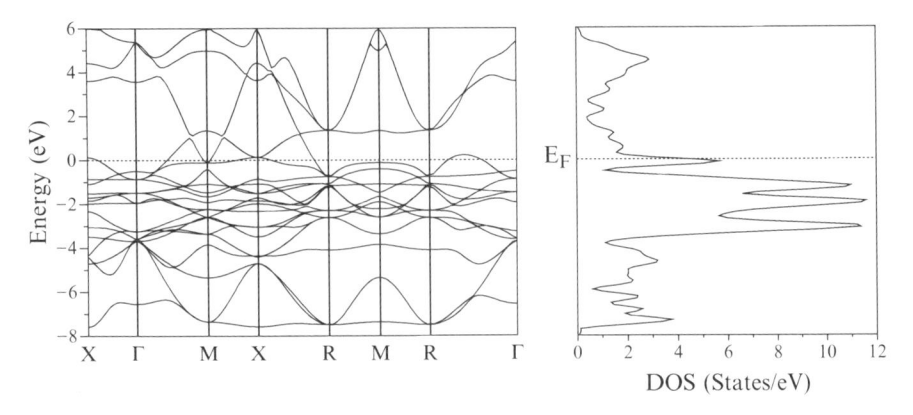

Fig. 1.7 The electronic band structure and density of states for MgCNi$_3$ obtained within the local density approximation (LDA). The fermi energy is set at 0 eV

by a strong-coupling, rather than the BCS weak-coupling. This scenario requires more than one electronic crossing of the Fermi surface. If there is a sharp peak in the electronic density of states close to the Fermi energy E_F, strong electron–phonon coupling can be expected. This suggests that a careful description of the band structure of MgCNi$_3$ is important. Photoemission and X-ray absorption measurements [35] have been made to measure the electronic structure of MgCNi$_3$. These experimental studies clearly show that there is a peak in the electronic density of states of MgCNi$_3$ which lies 0.1 eV below the Fermi level. Several theoretical methods have been used to investigate the electronic properties of MgCNi$_3$, including the self-consistent tight-binding linear muffin-tin orbital (TB-LMTO) [36], the linear muffin-tin orbital (LMTO) [37, 38], and density-functional theory within the local-density approximation [39–42]. Results from these works have indicated that the presence of a sharp peak in the electronic density of states just below the Fermi level.

The electronic band structure and density of states for MgCNi$_3$ are displayed in Fig. 1.7. The electronic structure shows the metallic nature with two bands crossing the Fermi level along the Γ-M, Γ-R, and X-R symmetry directions. However, there is a clear separation between occupied and unoccupied energy bands along the R-M symmetry direction. The total density of electronic states shows a peak at around 0.1 eV below the E_F, which is due to an extremely flat band around M along the Γ-M, X-M, and R-M symmetry directions. In agreement with experimental results [35], two more peaks in the density of states are found at energies -1.2 and -2.0 eV below the Fermi level. The former peak can be related to a flat band along the X-R and R-M symmetry directions, while the latter peak is dominated by a flat band along the [100] symmetry direction.

The LDA results [42] of the phonon spectrum for MgCNi$_3$ are displayed in Fig. 1.8. The top three optical branches modes are separated from the rest of the branches. From the density of states, this gap is calculated to be 26 meV, which

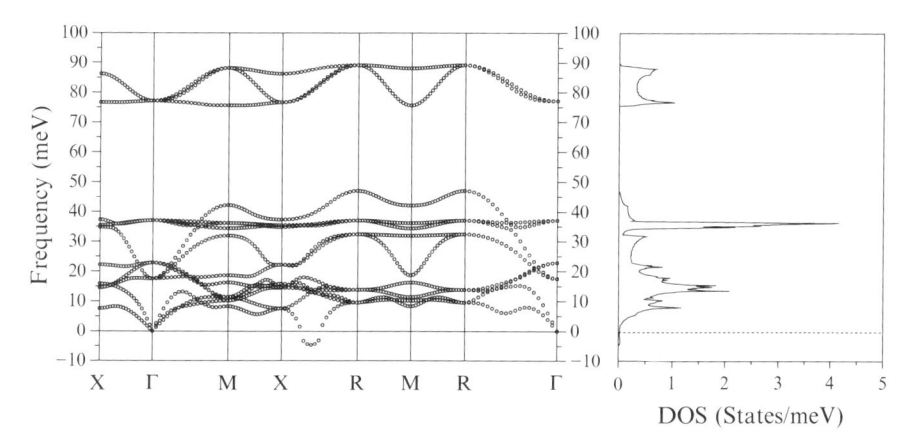

1 Lattice Dynamics of Solids, Surfaces, and Nanostructures

Fig. 1.8 *Ab initio* phonon dispersions and density of states for MgCNi$_3$

compares well with an experimental measurement of 23 meV [43]. Modes in all of the top three branches result almost exclusively involving vibrations of the lighter C atom. The lowest of this group of branches is nearly flat along the Γ-M, Γ-X, and M-X directions. The highest optical phonon branch is dispersive along the main symmetry directions [100], [110], and [111]. Below the gap region, some of the optical branches are quite dispersive, but one particular branch is almost dispersionless at 36 meV. The three acoustic phonon branches originate from the vibrations of the heavier Ni and Mg atoms. The peaks in the theoretical phonon density of states are at 8, 10, 15, 22, 33, 36, 78, and 89 meV. These results are in good agreement with the clearly resolved peaks at around 12, 16, 35, 80 meV in the neutron scattering measurements [43].

The anomaly in the lowest acoustic branch can be established as a dip at (0.275,0.275,0,275) in the [111] direction [42]. At this point in the Brillouin zone, the electron–phonon coupling parameter is calculated to be λ=1.54. The frequency of this phonon mode is 5.90 meV, involving vibrations of Mg and Ni atoms. At the zone edge R, rotational and bond-stretching phonon modes have been identified [42] at energies 13.82 meV and 47.03 meV, respectively. These modes were predicted at energies 13.00 meV and 43.30 meV in a previous theoretical work [39]. The electron–phonon coupling parameter, together with schematic representations of eleven vibrational phonon motions, at the M point is presented in Fig. 1.9. The largest value of the electron–phonon coupling parameter $\lambda = 1.173$ has been obtained for the lowest acoustic phonon mode at 8.27 meV. In agreement with previous *ab initio* calculations [43], this phonon mode includes a rotational character due to opposing motion of Ni atoms. However, it appears to be a stable phonon mode in the more recent theoretical work [42]. The large values of λ discussed above are supported by the experimental investigations [44].

- Mg
○ C
○ Ni

$\nu = 8.27$ meV
$\lambda = 1.173$

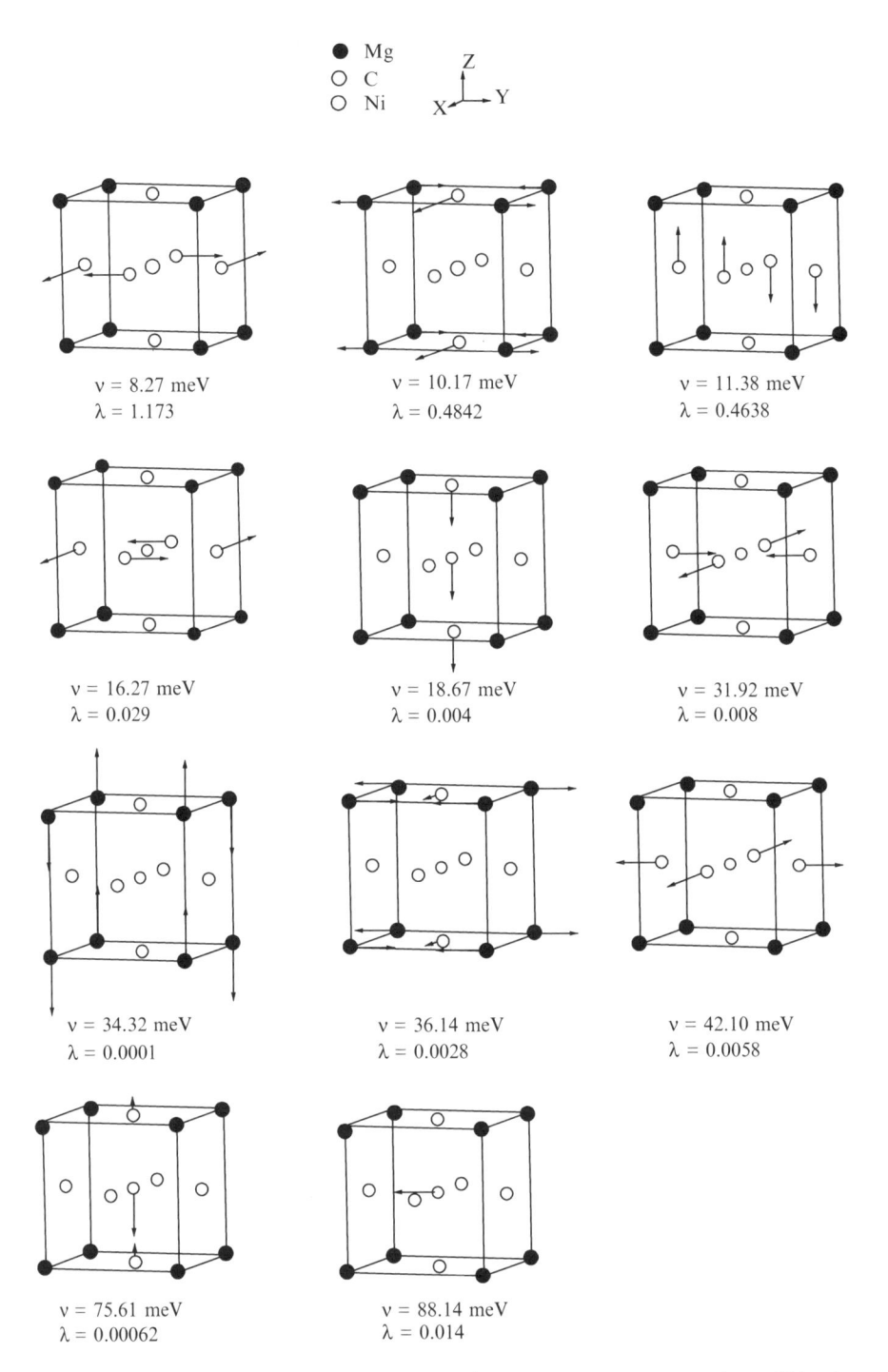

$\nu = 10.17$ meV
$\lambda = 0.4842$

$\nu = 11.38$ meV
$\lambda = 0.4638$

$\nu = 16.27$ meV
$\lambda = 0.029$

$\nu = 18.67$ meV
$\lambda = 0.004$

$\nu = 31.92$ meV
$\lambda = 0.008$

$\nu = 34.32$ meV
$\lambda = 0.0001$

$\nu = 36.14$ meV
$\lambda = 0.0028$

$\nu = 42.10$ meV
$\lambda = 0.0058$

$\nu = 75.61$ meV
$\lambda = 0.00062$

$\nu = 88.14$ meV
$\lambda = 0.014$

Fig. 1.9 The electron–phonon coupling parameter λ and eigenvector representations of M-point phonon modes in MgCNi$_3$. Taken from [42]

1.4 Surface Phonons

In the previous section we have discussed phonon eigensolutions (dispersion curves and characteristic atomic displacement patterns) in three-dimensional bulk materials and changes in eigensolutions due to change in crystal symmetry. In this section we present phonon results for solid surfaces, which can be considered as examples of two-dimensional systems. In particular, we will restrict ourselves to the (2 × 1) reconstruction of group-IV(001) surfaces and discuss results using both *ab initio* and BCM theories.

Group-IV(001) semiconductor surfaces have attracted a great deal of attention in the past twenty years, mainly due to their scientific and technological importance. The structural properties of these surfaces have been the focus of attention in various experimental investigations such as low energy electron diffraction [45–48], scanning tunnelling microscopy [49–51], X-ray diffraction [52], X-ray standing waves [53], in situ X-ray scattering [54], and He-diffraction [55]. On the theoretical side, structural properties of these surfaces have been investigated using the *ab initio* pseudopotential scheme [56–66]. Experimental as well as theoretical works have indicated that the basic reconstruction mechanism for stable structures of Si(001), Ge(001), and Sn(001) surfaces involves formation of buckled dimers. Statistical arrangement of asymmetric dimers is found to result in the c(4×2) and (2×1) reconstructions at low and high temperatures, respectively. Surface phonons have been measured using electron energy loss spectroscopy (EELS) for the Si(001)(2×1) surface [67] and C(001)(2×1) surface [68]. A variety of theoretical techniques have employed to study surface phonon modes on the (001)(2×1) surface of Si, Ge, C, and α-Sn. These include a tight binding [69–73], the BCM [61, 63, 65, 66], and *ab initio* linear-response [59, 62, 65, 66] methods. The *ab initio* and BCM phonon dispersion curves of Si(001)(2 × 1) [59,61] and C(001)(2×1) [65] surfaces compare very well with those of recent EELS results [67, 68].

1.4.1 Si(001)(2×1)

The surface phonon calculations were made by considering a repeated slab geometry, with the (2×1) periodicity on the (001) plane and an artificial periodicity along the [001] direction. The supercell along [001] contained a few Si(001) layers and an appropriate vacuum region. The BCM surface phonon spectrum of the Si(001)(2×1) surface is displayed in Fig. 1.10. The points $\overline{\Gamma}$ and $\overline{J}$ label the surface zone centre and zone edge points, respectively. The results for the asymmetric dimer geometry are shown by full lines, while the results for the symmetric dimer geometry are shown by dashed lines. For surface phonons with wave vector along $\overline{\Gamma}$-$\overline{J}$, no appreciable change is found in the dispersion between the two geometrical models, while for those with a finite wave vector component along $\overline{\Gamma}$-$\overline{J'}$ (the dimer row direction) there is noticeable change in the dispersion. In general the tilt of the dimer results in the energy of zone-edge surface phonons dropping by up

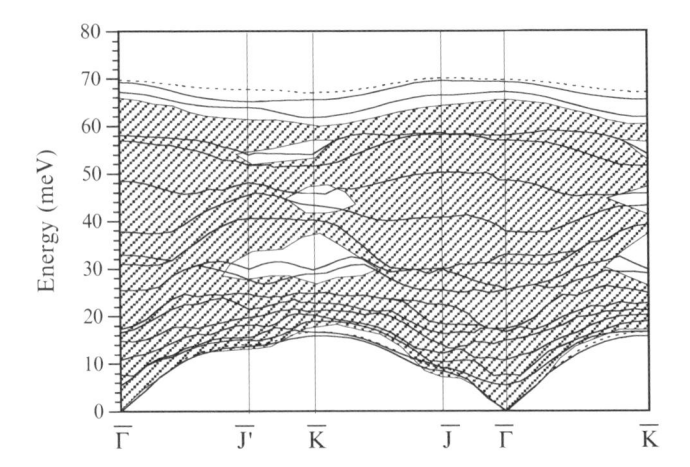

Fig. 1.10 Phonon dispersion curves for the Si(001)(2×1) surface calculated from the BCM [61]. The results for the asymmetric surface geometry are shown by *thick solid curves* while the results for the symmetric surface geometry are shown by *dashed curves*. The bulk results are plotted by *hatched regions*

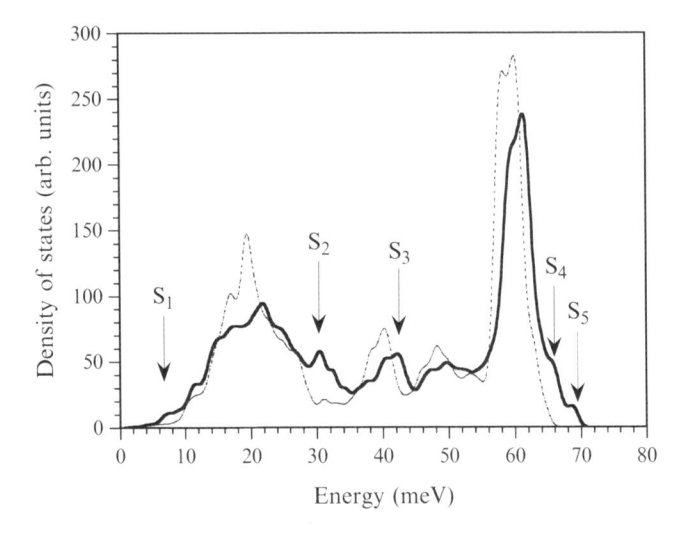

Fig. 1.11 The BCM density of phonon states for the slab supercell with the asymmetric surface geometry (*solid curve*) is compared with the BCM bulk density of states (*dotted curve*). S_1-S_5 represent surface peaks. Taken from [61]

to 2 meV. In particular, at $\overline{K}$ point the lowest surface phonon mode (known as the Rayleigh phonon mode) is slightly more separated from the bulk continuum when the asymmetric dimer model is employed.

The phonon density of states of the repeated slab with the asymmetric relaxed surface geometry is shown in Fig. 1.11. For comparison, the phonon density of states

1 Lattice Dynamics of Solids, Surfaces, and Nanostructures

Table 1.3 BCM surface-phonon frequencies of Si(001)(2×1) at the $\overline{\Gamma}$ point and their comparison with *ab initio* and experimental calculations. The symmetry is defined with respect to the mirror plane perpendicular to the dimer rows (+ is even, − is odd). Frequencies are given in meV

Symmetry	+	−	+	+	+	+	−	−	+
BCM [61]	14.86	17.28	25.49	32.89	37.74	48.50	58.05	61.32	69.25
Ab initio [59]			21.2						63.6
EELS [67]	12		20	30	33	49			64

for bulk Si is plotted by the dashed curve. Some peaks characterised by atomic vibrations on the surface are labelled S_1 to S_5 in the figure. The peak S_1 at around 7 meV is due to the Rayleigh waves mainly along the $\overline{\Gamma} - \overline{J}$ direction, while S_2 and S_3 are peaks due to localised stomach gap phonon states. The peaks labelled S_4 and S_5 are the two surface states lying above the bulk continuum. (see also Fig. 1.10).

Table 1.3 presents theoretical and experimental surface phonon modes on the Si(001)(2×1) surface. In general, the agreement between BCM and EELS results is good. The BCM phonon mode at 14.86 has a complex character with vibrations from the top two layer atoms. The BCM phonon mode at 25.49 is a dimer rocking mode with opposing motion of the surface layer atoms. This phonon mode has been measured at 20 meV by the EELS technique [67]. The phonon mode at 32.89 meV is characterised by parallel motion of the first layer atoms, while the second layer atoms vibrate perpendicular to each other. From EELS this mode is measured at 30 meV [67]. The BCM phonon mode at 48.50 meV is in excellent agreement with the experimental value of 49 meV [67]. The phonon mode at 58.05 meV comes from the vibrations of the second layer atoms perpendicular to each other while the first layer atoms vibrate in the dimer row direction. The phonon mode at 61.32 meV is characterised by opposing vibrations of the top two layer atoms in the dimer row direction. The BCM result for the highest surface optical phonon mode at the $\overline{\Gamma}$ is 69.25 meV. This phonon mode has a dimer stretching character due to opposing motion of surface layer atoms in the dimer bond direction.

Away from the $\overline{\Gamma}$ point, it is convenient to describe surface phonon modes as having either sagittal plane (SP) or shear horizontal (SH) polarisation. The sagittal plane is defined as the plane containing both the two-dimensional wave vector and surface normal direction, while shear horizontal vibrations are those which occur perpendicular to the sagittal plane. Table 1.4 presents the BCM surface phonon frequencies and polarisation characters (calculated including contributions from all 16 layers in the slab) for some modes on Si(001)(2×1) at the $\overline{J}'$ point. The BCM results are also compared with some other theoretical results [59, 70, 71, 73]. The BCM Rayleigh wave (RW) mode at 13.60 compares very well with the experimental value [67] of 13 meV. This mode is totally polarised as SP with parallel vibrations of the first layer atoms in the surface normal direction while the second layer atoms move in the dimer row direction. A similar displacement pattern for this phonon mode has also been found by Fritsch and Pavone [59] with a much lower energy. The phonon mode at 14.98 meV has a displacement pattern similar to the RW mode. The phonon mode at 15.55 meV is mainly due to the second layer atoms

Table 1.4 Comparison of BCM surface phonons on Si(001)(2×1) at $\overline{J'}$ point with results of other theoretical and experimental calculations. Frequencies are given in meV

Source	Modes at $\overline{J'}$								
BCM [61]	13.60	14.98	15.55	21.56	29.83	38.10	45.94	51.66	53.12 65.12
$[\sum U_{SP}^2]$	1.00	0.91	0.88	0.37	0.80	0.92	0.56	0.77	0.80 0.28
$[\sum U_{SH}^2]$	0.00	0.09	0.12	0.63	0.20	0.08	0.44	0.23	0.20 0.72
Ab initio [59]	9.24		12.53	20.63					
Tight-binding [71]	13.6								61.1
Tight-binding [70]									~68.7
Tight-binding [73]	~13.7								~64.0
EELS [67]	13								

vibrating in phase along the surface normal direction while the first layer atoms vibrate in phase in the dimer row direction. Fritsch and Pavone also obtained a phonon mode at 12.53 meV with a similar displacement pattern. The energy location and polarisation behavior of phonon mode at 21.56 meV agrees with the phonon mode at 20.63 meV in the theoretical work of Fritsch and Pavone. For this phonon mode, the first layer atoms vibrate against each other in the surface normal direction while they move parallel to each other in the dimer bond direction. The BCM phonon mode at 29.83 meV is a true localised surface phonon mode with 80 % SP polarisation. The BCM phonon mode at 38.10 meV corresponds to the motion of the first layer atoms in the dimer row direction while the second layer atoms vibrate against each other in the dimer bond direction. The BCM phonon mode at 45.94 meV is a dimer stretching phonon mode with opposing motion of first layer atoms in the dimer bond direction. The BCM phonon mode at 51.66 meV has 77 % SP character with vibrations of the down dimer atoms and its second-layer neighbour. The BCM phonon mode at 63.87 meV is a true localised phonon mode above the bulk continuum. The highest BCM surface optical mode at 65.12 meV has 72 % SH character and mainly corresponds to the vibrations of second layer atoms with components in both the surface normal and dimer bond directions.

1.4.2 Ge(001)(2×1)

The phonon spectrum of the Ge(001)(2×1) surface, obtained from BCM calculations [63], is displayed in Fig. 1.12. The results for the relaxed geometry are shown by full lines while the surface acoustic branch for the unrelaxed surface geometry is shown by dashed lines. The energy location of the lowest surface phonon mode (Rayleigh wave) remains nearly the same for both the relaxed and unrelaxed surface geometries. At high frequencies, in contrast, the geometry relaxation is critical, creating two surface phonon modes above the bulk continuum, where there were none for the unrelaxed surface.

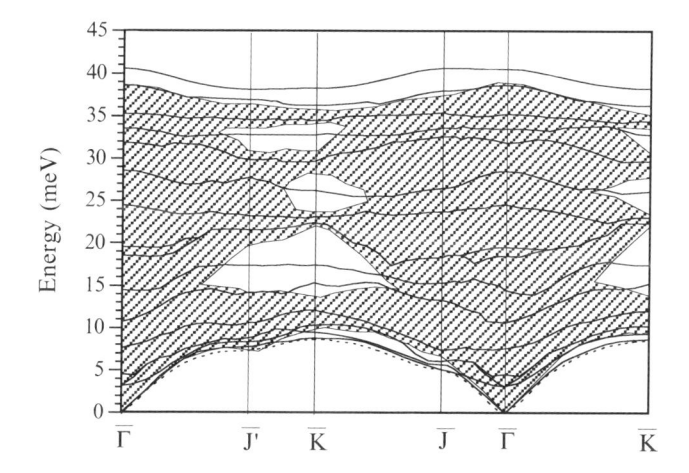

1 Lattice Dynamics of Solids, Surfaces, and Nanostructures 21

Fig. 1.12 Dispersion of BCM surface phonons on the Ge(001)(2×1) surface. The calculated results are shown by *full lines* while the lowest surface phonon mode for the unrelaxed surface geometry are shown by *dashed lines*. *Shaded areas* present the surface projection of bulk states. Taken from [63]

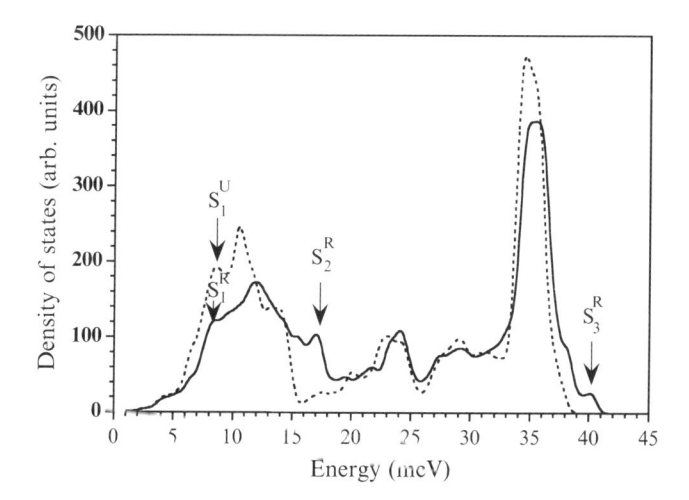

Fig. 1.13 The BCM phonon density of states of Ge(001)(2×1) surface for the relaxed surface geometry (*solid curve*) and the unrelaxed surface geometry (*dashed curve*). Taken from [63]

The phonon density of states for the unrelaxed (dashed curve) and relaxed (solid curve) surface geometries is shown in Fig. 1.13. The peaks labelled S_1^U and S_1^R are due to the RW phonons mainly along the $\overline{\Gamma} - \overline{K}$ direction. The peak S_2^R is the result of the formation of surface dimers and cannot be observed for the unrelaxed surface geometry. The peak S_3^R at 40 meV corresponds to the highest energy surface phonons and, again, is only seen in the relaxed surface geometry. Clearly the dimerized geometry produces noticeable changes for phonon modes in the stomach gap region and above the bulk continuum.

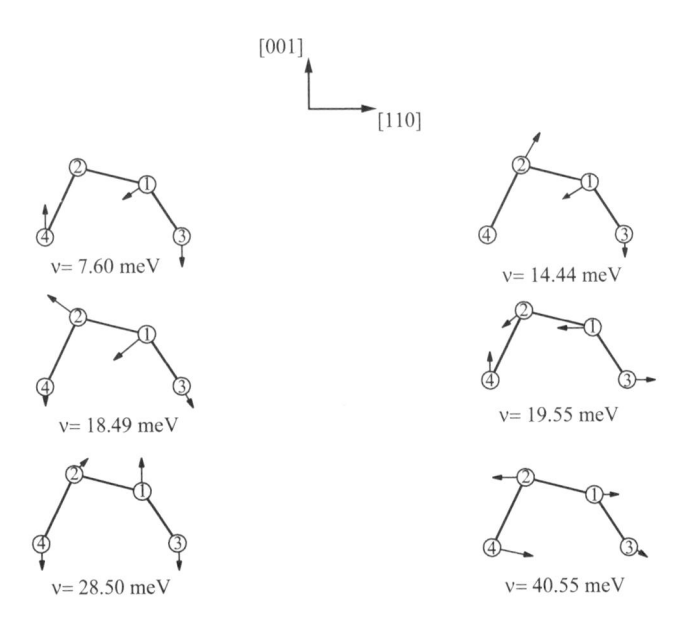

Fig. 1.14 Atomic displacement patterns of selected BCM surface phonon modes on Ge(001)(2×1) at the $\overline{\Gamma}$ point. Taken from [63]

Atomic displacement patterns of six surface phonon modes at the edcomm zone centre $\overline{\Gamma}$ are summarised in Fig. 1.14. The lowest surface optical frequency is at 7.60 meV and features opposing vibrations of the first layer and second layer atoms. The phonon mode at 14.44 meV is a dimer rocking mode and corresponds to opposing motion of the first layer atoms. As noted in Sect. 1.4.1, a corresponding phonon mode appears at 25.49 meV for the Si(001)(2×1) surface. There are also surface optical phonon modes at 18.49 and 19.55 meV. The phonon mode at 28.50 meV comes primarily from opposing vibrations of atom 1 (the buckled down dimer atom) and second layer atoms in the surface normal direction. Finally, the highest surface optical frequency is identified at 40.55 meV which mainly results from opposing motion of atom 2 and atom 4 (the buckled up dimer atom and its neighbour in the sub-surface layer) in the dimer bond direction. This mode also exhibits the stretch behaviour of the surface dimer bond.

1.4.3 C(001)(2×1)

Figure 1.15 presents the phonon dispersion curves of the C(001)(2×1) surface. The results calculated from the application of the *ab initio* linear response density functional perturbation theory (LR-DFPT) scheme [65] are shown by thick lines while the results from EELS measurements [68] along the symmetry directions

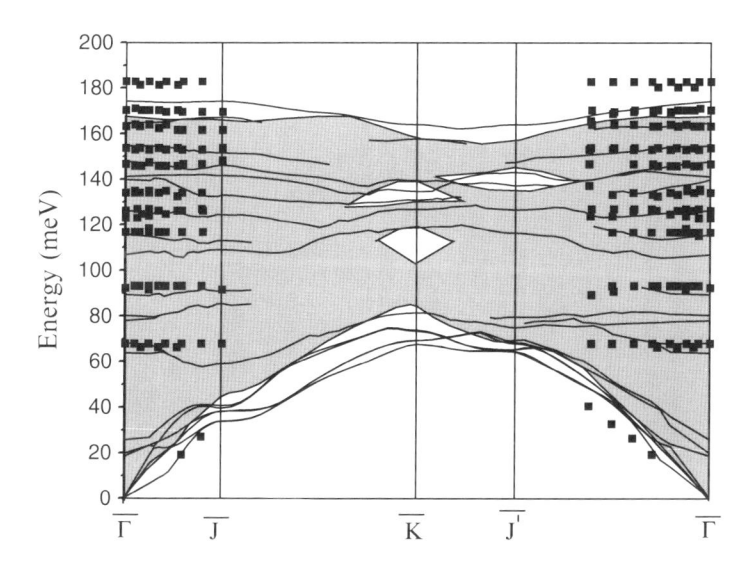

1 Lattice Dynamics of Solids, Surfaces, and Nanostructures

Fig. 1.15 Dispersion of phonon modes on the C(001)(2×1) surface obtained from the LR-DFPT method. The results for surface phonon modes are shown by *thick lines* while the EELS results [68] are shown by *filled squares*. The projected bulk phonon energies are shown *hatched regions*. Taken from [65]

$\overline{\Gamma}$-$\overline{J}$ and $\overline{\Gamma}$-$\overline{J'}$ are shown by filled squares. The highest surface optical phonon mode and the lowest surface acoustic phonon mode are observed as localised for all parts of the surface Brillouin zone. In general, the *ab initio* results are in good agreement with the experimental results.

In contrast to Si(001)-(2×1) and Ge(001)-(2×1) surfaces, the C(001)-(2×1) surface assumes the symmetric dimer formation and thus exhibits the C_{2v} (or, 2mm) symmetry [74]. One of the mirror planes, perpendicular to the dimer rows, is also present for other IV(001)-(2×1) surfaces which are characterised by tilted dimers. In order to discuss comparison with Si(001)-(2×1) and Ge(001)-(2×1) surfaces, we will exclusively consider only the mirror plane perpendicular to the dimer rows, and where possible refer to phonon modes of even and odd symmetries with respect to this plane. The polarisation characteristics of a selected number of zone-centre phonon modes are displayed in Fig. 1.16, and their even and odd symmetries are indicated. We have identified four even modes at 78, 107, 148, 174 meV and four odd modes at 80, 141, 154, and 164 meV. One of the interesting phonon modes in the lattice dynamics of the group-IV(001)(2×1) surfaces [59, 61–63] is the dimer rocking phonon mode which results from the opposing motion of top-layer atoms leading to a *rocking* motion of dimers. *Ab initio* calculations have observed several phonon modes which involve opposing motion of dimer atoms. The energies of these phonon modes are 64, 68, 78, 89, and 148 meV. The phonon modes at 68 and 78 meV are mainly localised on the dimer atoms while other phonon modes include atomic vibrations up to the third layer. The higher-energy rocking phonon mode at

Even Phonon Modes

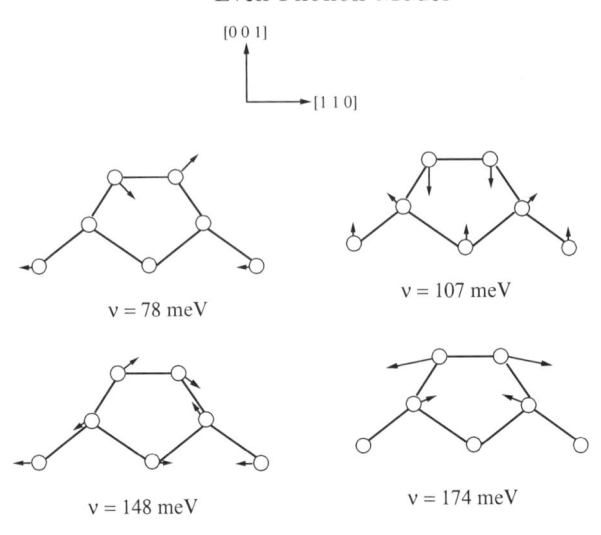

$\nu = 78$ meV

$\nu = 107$ meV

$\nu = 148$ meV

$\nu = 174$ meV

Odd Phonon Modes

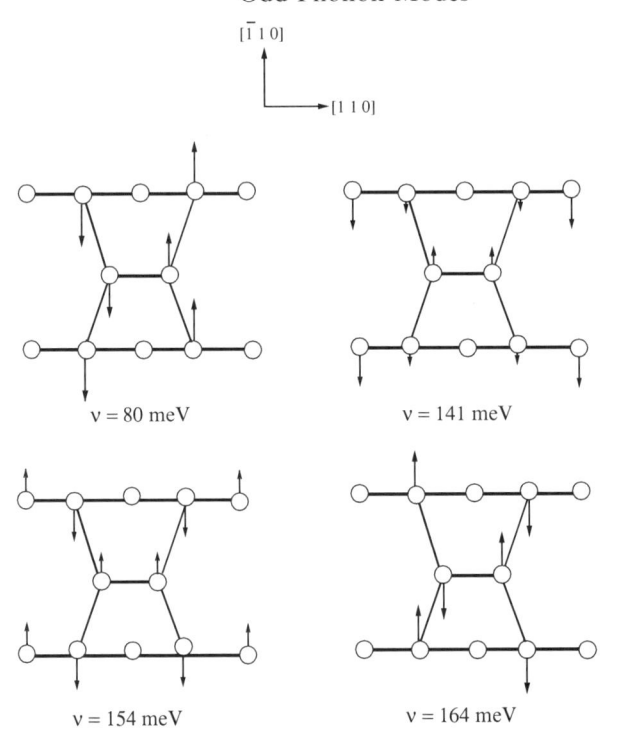

$\nu = 80$ meV

$\nu = 141$ meV

$\nu = 154$ meV

$\nu = 164$ meV

Fig. 1.16 The atomic displacement patterns of selected zone-centre phonon modes on the C(001)-(2×1) obtained from the LR-DFPT method. The modes have been classified according to the symmetry with respect to the mirror plane normal to the dimer row. Taken from [65]

1 Lattice Dynamics of Solids, Surfaces, and Nanostructures

Odd Phonon Modes   Even Phonon Modes

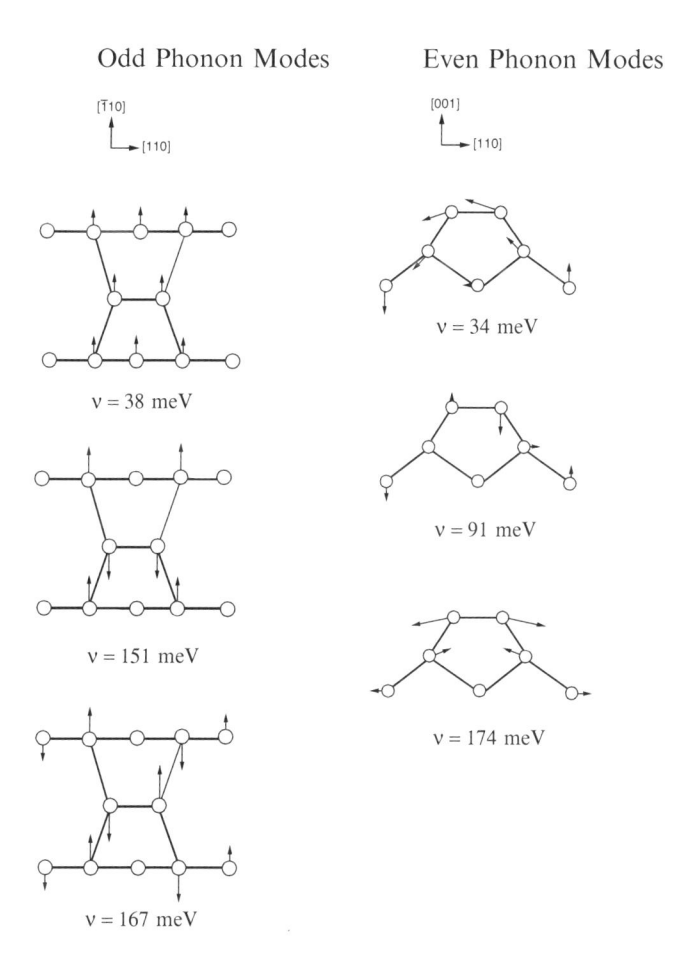

$\nu = 38$ meV

$\nu = 151$ meV

$\nu = 167$ meV

$\nu = 34$ meV

$\nu = 91$ meV

$\nu = 174$ meV

Fig. 1.17 Schematic representation of the atomic displacement patterns of selected *ab initio* even and odd phonon modes on C(001)-(2×1) at the $\overline{J}$. The modes have been classified according to the symmetry with respect to the mirror plane normal to the dimer row. Taken from [65]

148 meV can be compared with the rocking phonon mode at 151 meV in the work of Alfonso et al. [75,76]. We have observed a phonon mode at 107 meV which involves dimer *bouncing* motion and is strongly localised on the dimer atoms. The surface vibrations at 141 and 154 meV show a *swinging* character of dimer atoms. The higher-lying swinging mode compares very well with the phonon mode at 153 meV in the work of Alfonso et al. The highest lying surface state is a *dimer stretch* mode at 174 meV, characterised by the even symmetry. The energy of this phonon mode is in very good agreement with the experimental result [68] at 172 meV.

Some of the *ab initio* vibrational phonon modes at $\overline{J}$ are shown in Fig. 1.17. At this **q** point, the calculated energies of surface acoustic phonon modes are 34 and 38 meV. The lower one is the RW and results from in-phase motion of dimer

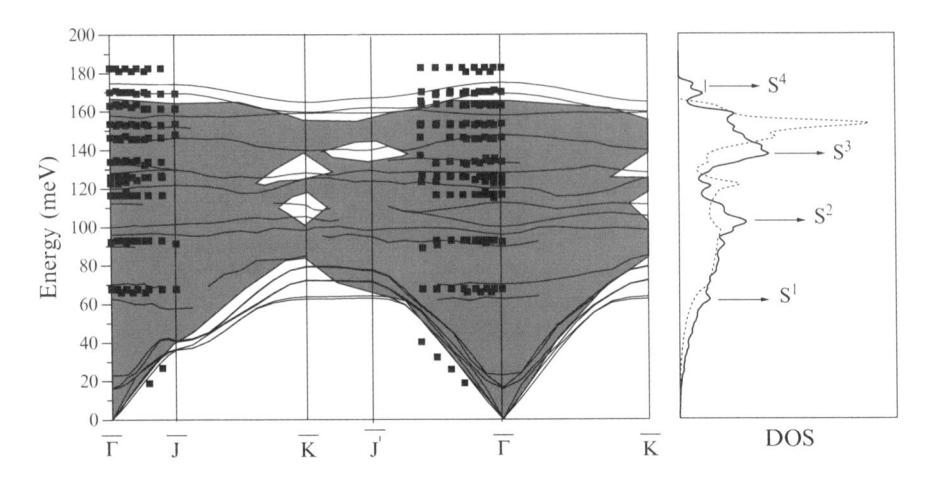

Fig. 1.18 Dispersion of phonon modes and density of sates (DOS) on the C(001)-(2×1) surface calculated from the adiabatic bond-charge model. The calculated surface results are shown by *solid lines* while EELS results are shown by *closed squares*. The bulk results are shown by *hatched regions and dashed line* (in the DOS). Peaks in the DOS related to the formation of the surface are indicated as S^1, S^2, S^3, and S^4. Taken from [65]

atoms in the dimer bond direction. For this mode there are large contributions from second and third layer atoms. The phonon mode at 91 meV agrees very well with an experimental data (see Fig. 1.15). The phonon mode at 151 meV with the odd symmetry involves the swinging motion of the dimer atoms. There is a dimer twist phonon mode at 167 meV. For this phonon mode, the dimer atoms move against each other in the dimer row direction. This is an odd symmetry surface mode, which also includes opposing motion of dimer atoms and their neighbouring substrate atoms. The highest phonon mode at 174 meV has a dimer stretch character. This is also an even symmetry surface backbond mode due to opposing motion of the dimer atoms and their neighbouring subsurface atoms.

While *ab initio* methods (such as LR-DFPT) do not contain any adjustable parameters, their application to calculate phonon modes at a general wave vector, and in particular for a complex surface reconstruction, can turn out to be computationally very demanding. The complete surface phonon spectrum of the C(001)(2×1) surface resulting from BCM calculations is plotted in Fig. 1.18. Similar to the LR-DFPT results, the BCM results are also in good agreement with the experimental results. In agreement with LR-DFPT calculations, BCM calculations show that surface acoustic waves lie below the bulk continuum along $\overline{J}$-$\overline{K}$-$\overline{J'}$. BCM calculations also find localised phonon modes in the stomach gap region. The two localised modes above the bulk continuum from the BCM are quite close to the two highest modes obtained from the LR-DFPT method. Small differences in the energy locations of these modes obtained from the two models can be rectified by simply modifying some of the surface force constant parameters for the BCM calculations (in particular, the second derivative of the central ion–ion

1 Lattice Dynamics of Solids, Surfaces, and Nanostructures

Table 1.5 Calculated (LDA) zone-centre surface phonon frequencies (meV) for the C(001)(2×1) surface from the LR-DFPT and BCM calculations. The modes have been classified according to the symmetry with respect to the mirror plane normal to the dimer row. The results are also compared with recent experimental data. Taken from [65]

Method	Odd modes in meV				Even modes in meV								
LR-DFPT	80		141	154	164	64	68	78	89	126	140	148		174
BCM	96	121	142	157		63	70	75	90	128	133	148	169	176
EELS			146	154						126	135	146		172

potential). The peaks related to atomic vibrations on the C(001)-(2×1) surface are labelled S^1 through to S^4. The surface acoustic waves along the $\overline{K}$-$\overline{J'}$ direction characterise the peak S^1. The peak S^2 at 100 meV corresponds to a rather dispersion-less phonon branch along the $\overline{\Gamma}$-$\overline{J}$ and $\overline{J}$-$\overline{K}$ directions. The localised gap phonon modes along the $\overline{K}$-$\overline{J'}$ direction dominate the peak S^3 while the peak S^4 results from the localised phonon modes above the bulk continuum.

Some calculated zone-centre phonon frequencies from the BCM calculations are compared with the LR-DFPT calculations in Table 1.5. The comparison with experimental data in this table has been made according to the C_{2v} point group symmetry [68]. The symmetry characteristics of most of the modes from the BCM calculations are in general agreement with the corresponding modes obtained from the LR-DFPT calculations. However, there are a few exceptions. For example, the second highest surface optical phonon mode from the BCM calculations has an even character while this phonon mode has an odd character for the LR-DFPT calculations. It is possible that we have not been able to identify all the surface modes at the top end of the spectrum, due to all but the highest mode being resonant within the LR-DFPT. With this in mind, and remembering that both theoretical models have an uncertainty of 1–2 meV, we believe that the results from the two theories would lead to similar polarisation characteristics of similar modes. The odd phonon modes at 142 and 157 meV compare well with our LR-DFPT results 141 meV and 154 meV, respectively. Using the BCM we have identified a phonon mode at 96 meV which has a displacement pattern similar to the LR-DFPT phonon mode at 80 meV. It may be that these are two separate odd-symmetry modes, and only one of these has been identified from each of the methods.

Some BCM and *ab initio* phonon frequencies and their polarisation characters at $\overline{K}$ are given in Table 1.6. While in general the LR-DFPT and DCM results are in good agreement with each other, a few differences are noted. First, BCM calculations predict the RW phonon mode to be totally polarised as SP while the LR-DFPT work shows only 35% SP polarisation for this phonon mode. The BCM phonon mode at 139 meV can be compared with the *ab initio* phonon mode at 136 meV. Both methods indicate localisation in the second and third layer atoms. However, the BCM mode mainly includes atomic vibrations in the dimer row direction while the LR-DFPT mode is dominated by atomic vibrations of the second and third layer atoms perpendicular to this direction. The highest surface optical phonon mode from the BCM model is mainly localised on the second and

Table 1.6 Comparison of the LR-DFPT and BCM results for the C(001)-(2×1) surface at the $\overline{K}$ point. The sums of the sagittal (SP) and shear horizontal (SH) components of the modes are also given. Taken from [65]

Method	Phonon frequency in meV						
ab initio	67	81	126	130	136	158	162
$[\sum U_{SP}^2]$	1.00	0.85	0.75	0.60	0.50	0.43	0.20
$[\sum U_{SH}^2]$	0.00	0.15	0.25	0.40	0.50	0.57	0.80
BCM	64	80	125	128	139	160	164
$[\sum U_{SP}^2]$	0.35	0.50	0.80	0.74	0.85	0.28	0.55
$[\sum U_{SH}^2]$	0.65	0.50	0.20	0.26	0.15	0.72	0.45

third layer atoms while this phonon mode is mainly due to opposing motion of dimer atoms in the dimer bond direction from the LR-DFPT. A comparison of the highest surface optical phonon mode indicates that it always includes large atomic vibrations from the second and third layer atoms within the BCM model while it always shows a dimer stretch character within the LR-DFPT. As a result, one can say that the two methods predict different amounts of mixing of SP and SH polarisations for some phonon modes at the symmetry points. In general, along $\overline{\Gamma}$-$\overline{J'}$ and $\overline{\Gamma}$-$\overline{K}$ the mixing of the SP and SH polarisations of modes is very sensitive to the electronic charge distribution and also to the level of accuracy in setting up the dynamical matrix. While the two methods inherit levels of inaccuracy in setting up the dynamical matrix, it is well recognised that the electronic charge distribution is not as accurately accounted for within BCM as in the LR-DFPT.

1.4.4 α-Sn(001)(2×1)

The *ab initio* vibrational spectrum of the α-Sn(001)(2×1) surface [66] is displayed in Fig. 1.19. The hatched region is the projection of the bulk phonon bands onto the folded 2×1 surface Brillouin zone while the solid lines are the results for surface phonons along the principal symmetry directions. It can be noticed that internal (or, stomach) gaps arise in the bulk bands along the $\overline{J}$-$\overline{K}$ and $\overline{K}$-$\overline{J'}$ directions. Several true surface modes are found in these gap regions. In contrast to the (001)(2×1) surface of Si and Ge, there are up to three localised gap phonon modes for α-Sn above the bulk phonon spectrum. While the highest one is rather dispersionless along these symmetry directions, the other two show appreciable amount of dispersion. The Rayleigh wave lies below the bulk bands along all symmetry directions.

Figure 1.20 presents the polarisation characteristics of five zone-centre modes, together with even and odd symmetries with respect to the mirror plane perpendicular to the dimer rows. In increasing order of frequency, there is a dimer stretch (S) mode at 18.60 meV, a dimer rocking (R) mode at 21.56 meV, a surface back

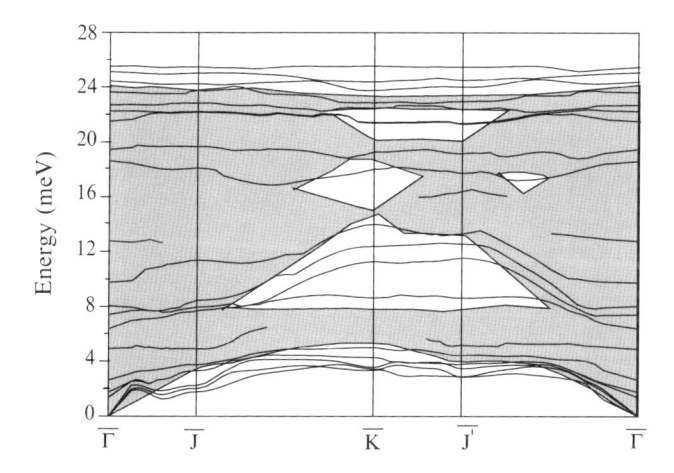

1 Lattice Dynamics of Solids, Surfaces, and Nanostructures 29

Fig. 1.19 Phonon dispersion curves for the α-Sn(001)(2×1) surface, obtained from *ab inito* calculations. The projected bulk phonon spectrum is represented by *hatched regions*. Taken from [66]

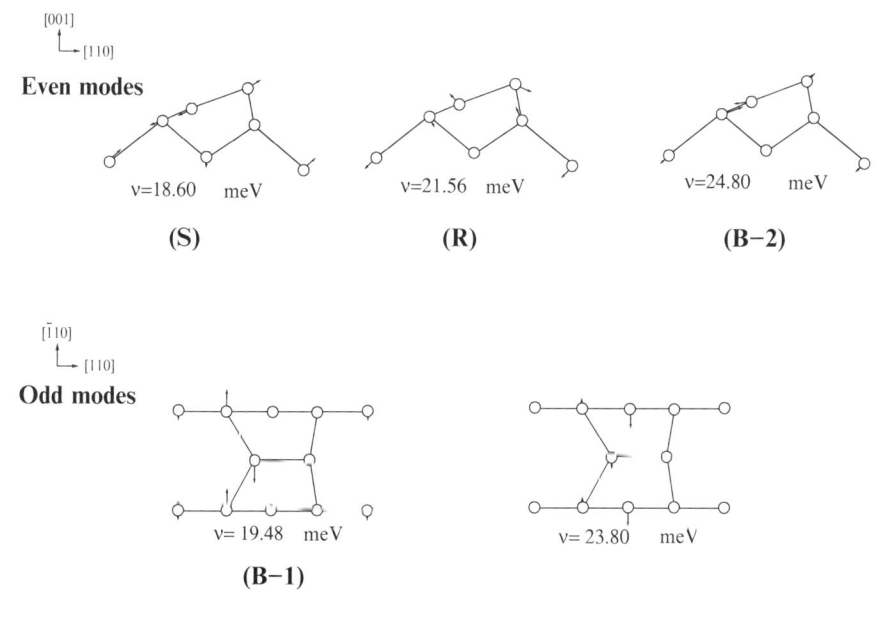

Even modes

ν=18.60 meV
(S)

ν=21.56 meV
(R)

ν=24.80 meV
(B−2)

Odd modes

ν=19.48 meV
(B−1)

ν=23.80 meV

Fig. 1.20 Eigenvector representations of five selected zone-centre phonon modes on the α-Sn (001)(2×1) surface. The even and odd symmetries of the modes with respect to the mirror plane perpendicular to the dimer rows are also indicated. The surface stretch, rocking, back bond 1, and back bond 2 modes are labelled as (S), (R), (B-1), (B-2), respectively. Taken from [66]

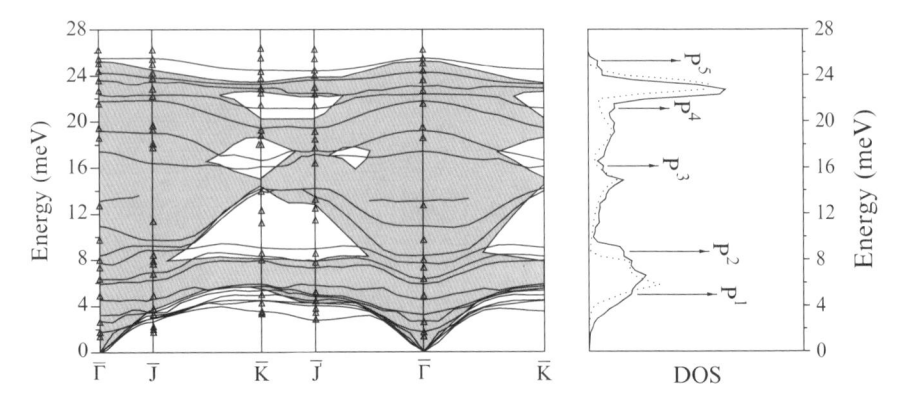

Fig. 1.21 Phonon dispersion curves and the density of phonon states for α-Sn(001)(2×1) surface calculated from the BCM. The bulk results are plotted by *hatched regions and dashed lines* (in the DOS). The *ab initio* results are also shown by *open triangles* at the symmetry points. Taken from [66]

bond (B-1) mode at 19.48 meV, and another surface back bond (B-2) mode at 24.80 meV. While the back bond mode B-1 involving the upper dimer atom has an odd symmetry, the B-2 mode involving the lower dimer atom has the even symmetry and is the highest surface phonon frequency. The second highest surface mode at 23.80 meV lies above the bulk continuum and is localised on the third layer atoms. Similar phonon modes have also been obtained in previous theoretical works on C(001)(2×1) [65], Si(001)(2×1)[59], and Ge(001)(2×1) [62].

The Rayleigh wave (RW) phonon branch turns into a flat mode away from the zone centre along the $\overline{\Gamma}$-$\overline{J'}$ direction. The energy of this phonon mode is found to be 2.85 meV at the zone-edge of this symmetry direction. Three localised stomach gap phonon modes are found along the $\overline{K}$-$\overline{J'}$ direction. All of these phonon modes are rather dispersionless along this symmetry direction. The lowest one is a dimer rocking phonon mode with opposing motion of first layer atoms in the surface normal direction. At this symmetry point, another rocking phonon mode has been found with an energy of 21.24 meV. This phonon mode also includes a dimer-stretch character due to opposing motion of surface layer atoms in the dimer bond direction. The phonon mode at 25.50 meV belongs to the highest surface optical phonon branch lying above the projected bulk phonon spectrum.

The phonon dispersion curves for the α-Sn(001)(2×1) surface obtained from the adiabatic bond charge model (BCM) are presented in Fig. 1.21. As this theoretical scheme is computationally less demanding, we have also plotted and compared in the same figure the density of phonon states for the slab supercell and the bulk. For comparison, the *ab initio* results for important surface modes at symmetry points are shown by open triangles. It is found that in general, the BCM results are in good agreement with the *ab initio* results. The density of states in this figure clearly shows that five peaks can be related to atomic vibrations on this surface. The peak marked P^1 comes from surface acoustic phonons which lie below the bulk projected phonon

1 Lattice Dynamics of Solids, Surfaces, and Nanostructures   31

Table 1.7 Calculated BCM and *ab initio* phonon results (in meV) for the α-Sn(001) (2×1) surface at the zone-centre. Results from these models are compared according to the energy location and polarisation characters of selected surface phonon modes. Taken from [66]

	Odd modes			Even modes					
BCM	4.52	22.30	24.25	7.74	10.93	13.11	19.21	21.76	25.52
ab initio	4.92	19.48	23.80	7.35	9.75	12.76	18.60	21.56	25.48

spectrum. True localised phonon mode in the stomach gap region creates the peak marked P^2 with an energy of 9 meV. The peaks P^3 and P^4 are also due to localised gap phonon modes while the highest surface optical phonon mode generates the peak P^5 at about 25 meV.

A comparison of the BCM and the *ab initio* results for the zone-centre surface phonons is presented in Table 1.7. This comparison is made according to the energy location and displacement characters of the surface phonon modes. In general, the BCM results are in good agreement with the *ab initio* findings. In particular, the dimer stretch (S) and the rocking (R) phonon modes predicted in the *ab initio* calculations are also identified in BCM calculations with energies of 19.21 and 21.76 meV, respectively.

1.4.5 Comparison of Results for IV(001)-(2×1) Surfaces

Now, we can compare the *ab initio* results for the (001)(2×1) surface of α-Sn, Ge, Si, and C between each other. First, in general, due to bigger mass and lattice constant, phonon modes on the α-Sn(001)(2×1) surface lie at lower energies than their counterparts observed for C(001)(2×1) [65], Si(001)(2×1) [59] and Ge(001)(2×1) [62] surfaces. Secondly, the Rayleigh-wave branch lies below the acoustic continuum towards the surface Brillouin zone edges for all the considered surfaces. Thirdly, in agreement with Si(001)(2×1) [59] and Ge(001)(2×1) [62], the dimer rocking phonon branch on α-Sn(001)(2×1) shows a flat dispersion along $\overline{\Gamma} - \overline{J'}$ (the dimer row direction). Three phonon branches lie in the stomach gap region for the α-Sn(001)(2×1) surface, while only one phonon branch has been reported for the Si(001)(2×1) [59] and Ge(001)(2×1) [62] surfaces. As the (001)(2×1) surface has similar atomic geometry for Si, Ge, and α-Sn, it is believed that this difference can be related to very heavy mass of the Sn atom. The highest surface optical branch is nearly flat for the α-Sn(001)(2×1) surface along the surface Brillouin zone, while this phonon branch shows reasonable dispersion along the $\overline{\Gamma}$-$\overline{J}$ and $\overline{\Gamma}$-$\overline{J'}$ symmetry directions for the C, Si, and Ge surfaces. As a general remark we note that the dimer rocking (R) and bond-stretching (S) phonon modes predicted for the C, Si, and Ge surfaces have also been identified for the Sn surface. There is a difference in the localisation behaviour of the highest surface optical phonon mode: while for the Si, Ge, and Sn surfaces it is mainly localised in the

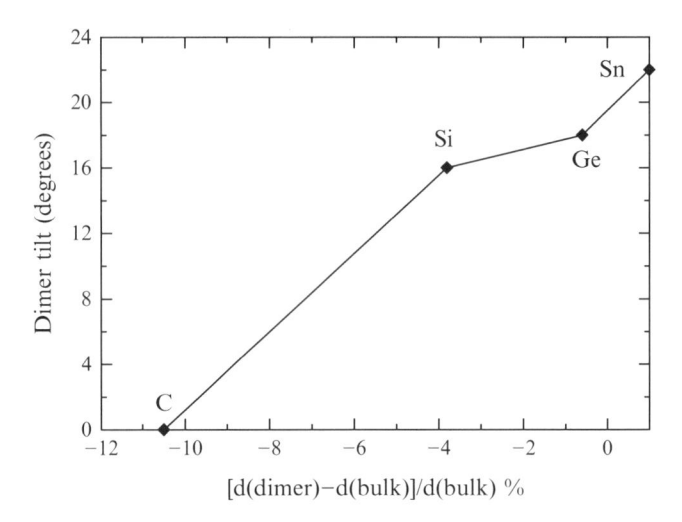

Fig. 1.22 Correlation between surface dimer tilt and percentage difference between the dimer and bulk bond lengths on IV(001)(2×1) surfaces. Taken from [66]

Table 1.8 Dimer stretch mode energy for group-IV(001)(2×1) surfaces. Also presented is the variation of the dimer effective force constant $\mu\omega^2$ for these surfaces. Taken from [66]

Surface	Dimer stretch energy [meV]	Dimer effective force constant $[10^5(\text{meV})^2$ amu]
C(001) [65]	174	1.80
Si(001) [59]	46	0.30
Ge(001) [62]	26	0.25
α-Sn(001) [66]	19	0.20

substrate, with very little motion of dimer atoms, this phonon mode includes large atomic vibrations from dimer atoms on the C(001)(2×1) surface. This difference can be related to a symmetric dimer of the C(001)(2×1) surface.

Following the trend presented in Fig. 1.22 in the dimer tilt across the IV(001) surfaces, it is particularly interesting to examine the variation of the *dimer stretch* (S) mode across the group IV column from C towards α-Sn. From the *ab inito* work on α-Sn, and previous works on C [65], Si [59] and Ge [62], we note that the energy of the *dimer stretch* (S) mode decreases along the sequence C-Si-Ge-Sn. In Table 1.8 we have presented the energy $\hbar\omega$ of this mode, as well as the quantity $\mu\omega^2$, with μ being the reduced mass. Regarding $\mu\omega^2$ as a measure of dimer force constant, we reach the conclusion that the dimer bond weakens across the surface sequence C(001)-Si(001)-Ge(001)-α-Sn(001). Clearly, therefore, the dimer bond on α-Sn(001) is nearly an order of magnitude weaker than on C(001).

Using the BCM method, the dimer twist and the dimer stretch phonon modes have been identified for all the group IV(001) surfaces. These phonon modes are dominated by opposing motion of dimer atoms in the dimer row and dimer bond

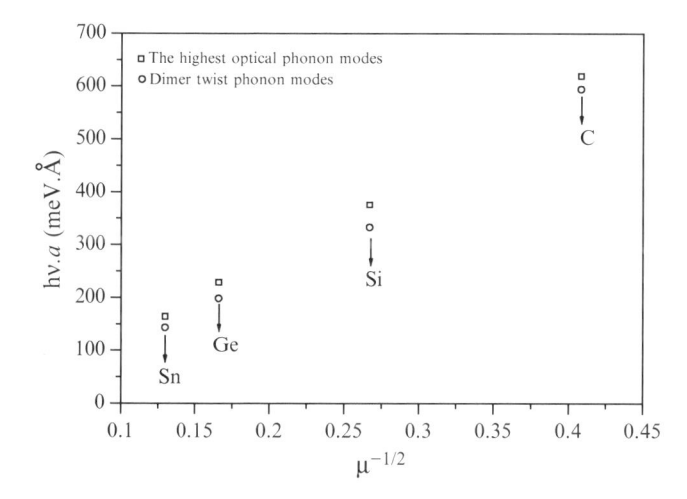

1 Lattice Dynamics of Solids, Surfaces, and Nanostructures

Fig. 1.23 Variation of the zone-centre highest surface optical (*open squares*) and dimer twist (*open circles*) phonon modes on IV(001)(2×1) surfaces with the square root of the reduced mass. Taken from [66]

directions, respectively. Figure 1.23 clearly shows that at the zone centre the energy of both the dimer twist (R) mode and the dimer stretch (S) mode scale linearly with $1/a\sqrt{\mu}$, where a is the cubic lattice constant. The similarity of the mass variation of these two modes indicates that the effective force constants of dimer twist and bond-stretching phonon modes are very similar to each other for all the surfaces considered.

1.5 Phonons in Nanostructures

So far in this chapter we have discussed phonon eigensolutions for three-dimensional (3D) bulk and two-dimensional (2D) surface structures. Usually such structures exist with dimensions exceeding several microns if not millimetres. With recent advances in growth technology it is possible to grow/fabricate low-dimensional systems, such as thin films/slabs, thin superlattices, wires and dots, all with size in the nanometre range. Phonon characteristics of such materials are expected to be vastly different from those of corresponding 3D bulk systems. In this section we discuss some main features of phonon spectra in nanostructured materials. In particular, we present results of BCM calculations for phonons in nanostructures made from semiconducting materials, such as Si/Ge superlattices, Si nanoslabs, Si nanowires, and Si nanodots. We will explain the concepts of "folded branches", "confined modes", and generation of "gaps" in phonon spectrum.

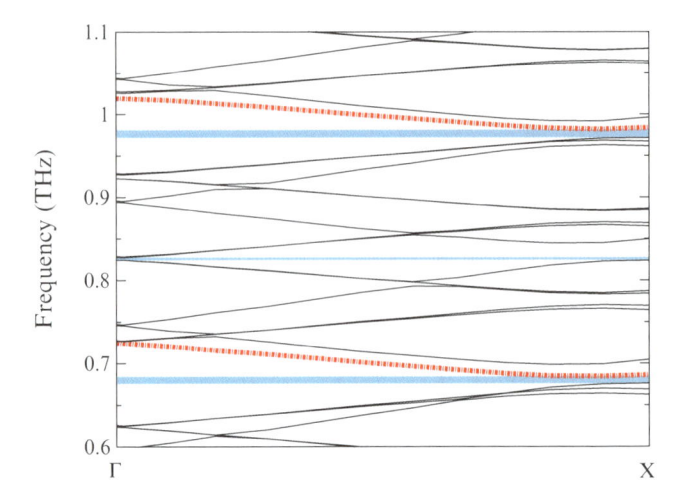

Fig. 1.24 Low-frequency phonon dispersion curves for the Si(27 nm)/Ge(17 nm)[100] superlattice. Taken from [77]

(i) Si/Ge[001] superlattice:

Figure 1.24 shows the phonon dispersion curves, in the low frequency range 0.6–1.1 THz, for the Si(27 nm)/Ge(17 nm) superlattice along the [001] growth direction. From a comparison of these with the bulk dispersion curves in Fig. 1.2, we can immediately appreciate the concept of "branch folding" upon superlattice formation. The number of times a bulk phonon branch folds depends upon the ratio of the length of the superlattice and bulk Brillouin zone along the superlattice growth direction. There are three "gaps" (shown in blue colour) in the phonon spectrum for the superlatice, at 680, 828, and 974 GHz, of widths 10, 3, and 10 GHz, respectively. Propagating atomic vibrational waves cannot be found in these frequency regions. Presence of such phonon gaps is the central concept of "phononics".

(ii) Si nanoslabs, nanowires, and nanodots:

For the calculation of phonon spectrum of Si nanostructures, we chose to construct artificially periodic supercell geometries. Specifically we constructed a quasi-two-dimensional nanoslabs (2D) with thickness d; quasi-one-dimensional nanowires (1D) with dimensions $d \times d$; quasi-zero-dimensional nanodots (0D) with square cross-sections of dimensions $d \times d \times d$. Within a supercell, there was a vacuum region large enough to prevent interaction between neighbouring nanostructures. The vacuum region was considered to consist of an embedding material with the same atomic network as the nanostructure, but with atomic mass less than 10^{-6} of a single silicon atom. With this simple consideration the nanostructure retains its unreconstructed geometry at its interface with the vacuum, and the dangling bonds at the surface(s) remain in their bulk positions. Thus the silicon bulk parameters for

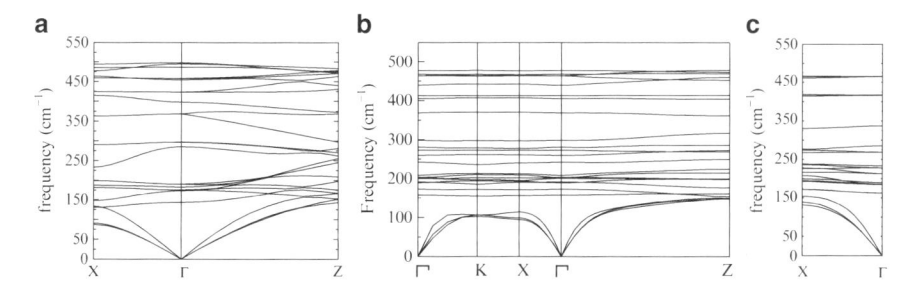

1 Lattice Dynamics of Solids, Surfaces, and Nanostructures

Fig. 1.25 Phonon dispersion curves for ultrathin Si nanostructures: (**a**) nanoslab of thickness 0.543 nm; (**b**) nanowire of cross section 0.543×0.543 nm; (**c**) nanodot of size 0.543×0.543× 0.543 nm. Taken from [78]

the adiabatic bond charge model are employed without any alteration. The phonon frequencies due to the embedding material lie far above the frequency regions for the bulk and nanostructure, and are easily discarded from the analysis of results.

The phonon dispersion curves for ultrathin Si nanostructures in the form of (a) nanoslab of thickness 0.543 nm, (b) nanowire of cross-section 0.543×0.543 nm, and (c) nanodot of size 0.543×0.543×0.543 nm are presented in Fig. 1.25. For the silicon nanoslab the phonon branches above about 300 cm^{-1} are flat and dispersionless along ΓX, the direction of confinement. For the nanowire, all the non-acoustic modes (above about 150 cm^{-1}) are almost flat (i.e. dispersionless), even along Γ-Z, the direction of propagation. The size confinement effect has led to the development of several clear band gaps in the spectrum. The largest gap occurs between 300 and 350 cm^{-1}. For the nanodot the phonon dispersion curves for non-acoustic branches (above about 150 cm^{-1}) are dispersionless in all directions. The three-dimensional geometrical confinement has led to atomic-like discrete dispersion curves in this range. The band gaps for the nanodot are much wider than those for the nanowire.

The number and width of gaps are a unique feature of ultrasmall nanostructures. Both the number and width of the gaps reduce rapidly as the nanostructure size increases. We can appreciate this by examining the DOS for nanoslabs and nanodots of two different sizes, as shown in Fig. 1.26. For the ultrathin slab of thickness 0.543 nm there is a single gap in the DOS near 400 cm^{-1} [panel (a)] and there are very low values at around 300 cm^{-1} and 250 cm^{-1} [panel (c)]. For the slab thickness of 32.520 nm, there are no clear gaps in the DOS, and low values develop at different frequencies. Similarly, the atomically sharp features in the DOS for the ultrathin nanodot [panel (b)] have broadened, with no clear gap, for the nanodot of size 1.629×1.629×1.629 nm [panel (d)]. It is also interesting to note that it is possible to describe the DOS of these nanostructures in the acoustic region following the prediction of the Debye continuum model (i.e. $g(\omega) \propto \omega$, $g(\omega) \propto \omega^0$, and $g(\omega) \propto \omega^{-1}$, for nanoslabs, nanowires, and nanodots as text book examples of 2D, 1D, and 0D systems, respectively.

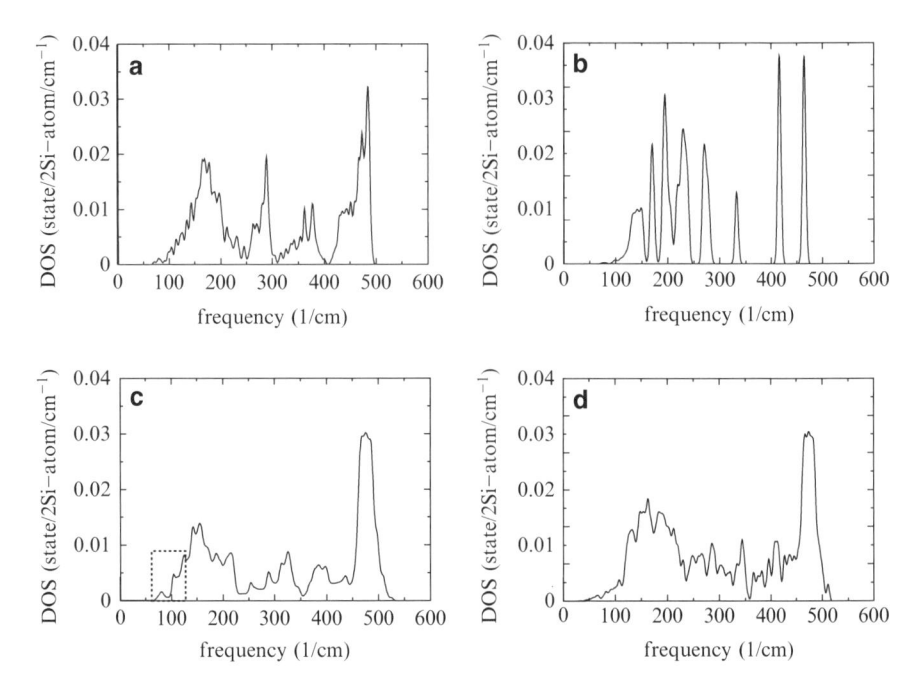

Fig. 1.26 Phonon density of states for ultrathin Si nanostructures: (**a**) nanoslab of thickness 0.543 nm; (**b**) nanodot of size 0.543×0.543×0.543 nm; (**c**) nanoslab of thickness 32.520 nm; (**d**) nanodot of size 1.629×1.629×1.629 nm. Taken from [78]

It is of interest to discuss the variation of the highest and lowest optical modes with the structure size. The highest optical mode in nanostructures shows a marked departure from the highest bulk frequency. Using a diatomic linear chain model, the downward shift $\Delta\omega$ of the highest mode frequency in a 2D nanostructure (such as the nanoslab) can be expressed as [79]

$$\Delta\omega^2 = \frac{\alpha}{d^2}, \qquad (1.6)$$

where α is a constant, and $\Delta\omega^2 = \omega_{bulk}^2 - \omega_{optical}^2$ with ω_{bulk} being the frequency of the highest optical mode in bulk. By extending this approach to two dimensions, the downward shift $\Delta\omega$ of the highest mode frequency in a 1D nanostructure (such as the nanowire) can be expressed as [78]

$$\Delta\omega^2 = \frac{\beta_1}{d^2} + \frac{\beta_2}{d^3}, \qquad (1.7)$$

1 Lattice Dynamics of Solids, Surfaces, and Nanostructures

where β_i are constants. With a further extension of the approach to three dimensions, we can express, for a 0D structure (such as a nanodot) [78]

$$\Delta\omega^2 = \frac{\eta_1}{d^2} + \frac{\eta_2}{d^3} + \frac{\eta_3}{d^4}, \tag{1.8}$$

where η_i are constants.

Numerical results [80], based on the application of the BCM, suggest that the lowest nonzero zone-centre frequency varies with the size d of Si nanostructures as

$$\omega_\sigma = A \frac{1}{d^\alpha}, \tag{1.9}$$

where $\alpha = \frac{3}{4} \pm 0.04, \frac{2}{3} \pm 0.06$, and $\frac{1}{2} \pm 0.08$, and the proportionality constant A takes values $3.57\,\text{nm}^{3/4}$ THz, $3.41\,\text{nm}^{2/3}$ THz, and $3.60\,\text{nm}^{1/2}$ THz, for slabs (2D), wires (1D) and dots (0D), respectively.

In another chapter Wang et al. have discussed the continuum (elastic and dielectric) theories of phonon in low-dimensional systems. The atomistic results discussed earlier in this section, when extrapolated from thin structures or applied to appropriately thick nanostructures, should prove useful in providing parameters for establishing accurate continuum models.

1.6 Summary

In this chapter we have presented results of accurate atomic-scale theoretical calculations for phonon eigensolutions, i.e. dispersion relations and atomic displacement patterns, for crystalline solids, semiconductor surfaces, and semiconductor nanostructures.

By considering the example of lanthanum metal in fcc and dhcp structures it has been shown that bulk phonon characteristics can be controlled by changing crystal symmetry. By considering the example of $MgCNi_3$ we have shown that role of phonon anomaly can be important in establishing large electron–phonon coupling parameter, which in turn can be useful in studying BCS superconductivity.

It has been shown that formation of surfaces and nanostructures lead to the existence of new phonon modes. The concepts of confined, rotational, and localised phonons on surfaces, and of folded and confined phonons in nanostructures have been presented with several examples. With the help of a few results we have attempted to provide a discussion of dimensionality and symmetry dependence of phonons.

It is hoped that accurate *ab initio* calculations can be carried out routinely and symmetry as well as length-dependent phonon properties of solid materials can be firmly established.

Acknowledgements We acknowledge many helpful discussions from our collaborators (students and postdocs) on the works presented here.

References

1. Born M, Huang K (1954) Dynamical theory of crystal lattices. Oxford University Press, Oxford
2. Lord Rayleigh (1887) Proc Lond Math Soc 17:4
3. Lifshitz IM, Rosenzweig LM (1948) Zh Eksp Teor Fiz 18:1012
4. Kress W, de Wette F (eds) (1991) Surface phonons. Springer, Berlin
5. Hulpke E (ed) (1992) Helium atom scattering from surfaces. Springer, Berlin
6. Yu PY, Cardona M (2010) Fundamentals of semiconductors, 4th edn. Springer, Heidelberg
7. Weber W (1974) Phys Rev Lett 33:371
8. Rustagi KC, Weber W (1976) Solid State Commun 18:673
9. Srivastava GP (1990) The physics of phonons. Adam Hilger, Bristol
10. Srivastava GP (1999) Theoretical modelling of semiconductor surfaces. World Scientific, Singapore
11. Martin RM (2004) Electronic structure: basic theory and practical methods. Cambridge University Press, Cambridge
12. Yin MT, Cohen ML (1982) Phys Rev B 26:3259
13. Kunc K, Martin RM (1982) Phys Rev Lett 48:406
14. Giannozzi P, de Gironcoli S, Pavone P, Baroni S (1991) Phys Rev B 43:7231
15. Price DL, Rowe JM, Nicklow RM (1971) Phys Rev B 3:1268
16. Nilsson G, Nelin G (1971) Phys Rev B 3:364
17. Nilsson G, Nelin G (1972) Phys Rev B 6:3777
18. Warren JL, Yarnell JL, Dolling G, Cowley RA (1967) Phys Rev 158:805
19. Bağcı S, Tütüncü HM, Duman S, Srivastava GP (2010) Phys Rev B 81:144507
20. Pickett CW, Freeman AJ, Koelling DD (1980) Phys Rev 22:2695
21. Lu ZW, Singh DJ, Krakauer H (1989) Phys Rev B 39:4921
22. Takeda T, Kübler J (1979) J Phys F Met Phys 9:661
23. Gao GY, Niu YL, Cui T, Zhang J, Li Y, Xie Y, He Z, Ma YM, zou GT (2007) J Phys Condens Matter 19:425234
24. Syassen K, Holzapfel WB (1975) Solid State Commun 16:533
25. Singh N, Singh SP (1990) Phys Rev B 42:1652
26. Nixon LW, Papaconstantopoulos DA, Mehl MJ (2008) Phys Rev B 78:214510
27. Delin A, Fast L, Johansson B, Eriksson O, Wills JM (1998) Phys Rev B 58:4345
28. Kittel C (1996) Introduction to solid state physics, 7th edn. Wiley, New York
29. Stassis C, Loong C-K, Zarestky J (1982) Phys Rev B 26:5426
30. McMillian WL (1968) Phys Rev 167:331
31. Rapp Ö, Sundqvist B (1981) Phys Rev B 24:144
32. He T, Huang Q, Ramirez AP, Wang Y, Regan KA, Rogado N, Hayward MA, Haas MK, Slusky JJ, Inumara K, Zandbergen HW, Ong NP, Cava RJ (2001) Nature (London) 411:54
33. Li SY, Fan R, Chen XH, Wang CH, Mo WQ, Ruan KQ, Xiong YM, Luo XG, Zhang HT, Li L, Sun Z, Cao LZ (2001) Phys Rev B 64:132505
34. Mao ZQ, Rosario MM, Nelson KD, Wu K, Deac IG, Schiffer P, Liu Y, He T, Regan KA, Cava RJ (2003) Phys Rev B 67:094502
35. Kim JH, Ahn JS, Kim J, Park M-S, Lee SI, Choi EJ, Oh S-J (2002) Phys Rev B 66:172507
36. Szajek A (2001) J Phys Condens Matter 13:L595
37. Dugdale SB, Jarlborg T (2001) Phys Rev B 64:100508
38. Shim JH, Kwon SK, Min BI (2001) Phys Rev B 64:180510
39. Singh DJ, Mazin II (2001) Phys Rev B 64:140507
40. Hase I (2004) Phys Rev B 70:033105
41. Johannes MD, Pickett WE (2004) Phys Rev B 70:060507(R)
42. Tütüncü HM, Srivastava GP (2006) J Phys Condens Matter 18:11089
43. Heid R, Renker B, Schober H, Adelmann P, Ernst D, Bohnen K-P (2004) Phys Rev B 69:092511

44. Wälte A, Fuchs G, Müller KH, Handstein A, Nenkov K, Narozhnyi VN, Drechsler SL, Shulga S, Schultz L, Rosner H (2004) Phys Rev B 70:174503
45. Schier RE, Farnsworth HE (1959) J Chem Phys 30:917
46. Fernandez C, Ying WS, Shih HD, Jona F, Jepsen D, Narcus PM (1981) J Phys C 14:L55
47. Kubby JA, Griffith JE, Becker RS, Vickers JS (1987) Phys Rev B 36:6079
48. Yuen WT, Liu WK, Joyce BA, Stradling RA (1990) Semicond Sci Technol 5:373
49. Wolkow RA (1992) Phys Rev Lett 68:2636
50. Hamers RJ, Tromp RM, Demuth JE (1986) Phys Rev B 34:5343
51. Munz AW, Ziegler Ch, Göpel W (1995) Phys Rev Lett 74:2244
52. Rossmann R, Meyerheim HL, Jahns V, Wever J, Moritz W, Wolf D, Dornich D, Schulz H (1992) Surf Sci 279:199
53. Fontes E, Patel JR, Comin F (1993) Phys Rev Lett 70:2790
54. Lucas CA, Dower CS, McMorrow DF, Wong GCL, Lamelas FJ, Fuoss PH (1993) Phys Rev B 47:10375
55. Lambert WR, Trevor PL, Cardillo MJ, Sakai A, Hamann DR (1987) Phys Rev B 35:8055
56. Dabrowski J, Scheffler M (1992) Appl Surf Sci 56–58:15
57. Krüger P, Pollmann J (1993) Phys Rev B 47:1898
58. Krüger P, Pollmann J (1995) Phys Rev Lett 74:1155
59. Fritsch J, Pavone P (1995) Surf Sci 344:159
60. Jenkins SJ, Srivastava GP (1996) J Phys Condens Matter 8:6641
61. Tütüncü HM, Jenkins SJ, Srivastava GP (1997) Phys Rev B 56:4656
62. Stigler W, Pavone P, Fritsch J (1998) Phys Rev B 58:13686
63. Tütüncü HM, Jenkins SJ, Srivastava GP (1998) Phys Rev B 57:4649
64. Lu Z-Y, Chiarotti GL, Scandolo S, Tosatti E (1998) Phys Rev B 58:13698
65. Tütüncü HM, Bağcı S, Srivastava GP (2004) Phys Rev B 70:195401
66. Tütüncü HM, Duman S, Bağcı S, Srivastava GP (2004) Phys Rev B 72:085327
67. Takagi N, Shimonaka S, Aruga T, Nishijima M (1999) Phys Rev B 60:10919
68. Thachepan S, Okuyama H, Aruga T, Nishijima M, Ando T, Bağcı S, Tütüncü HM, Srivastava GP (2003) Phys Rev B 68:033310
69. Mazur A, Pollmann J (1986) Phys Rev Lett 57:1811
70. Pollman J, Kalla R, Krüger P, Mazur A, Wolfgarten G (1986) Appl Phys A Solids Surf 41:21
71. Alerhand OL, Mele EJ (1987) Phys Rev B 59:657
72. Alerhan OL, Mele EJ (1987) Phys Rev B 35:5533
73. Mazur A, Pollmann J (1990) Surf Sci 255:72
74. Boardman AD, O'Connor DE, Young PA (1973) Symmetry and its applications in science. McGraw Hill, Maidenhead
75. Alfonso DR, Drabold DA, Ulloa SE (1995) Phys Rev B 51:1989
76. Alfonso DR, Drabold DA, Ulloa SE (1995) Phys Rev B 51:14669
77. Hepplestone SP, Srivastava GP (2008) Phys Rev Lett 101:105502
78. Hepplestone SP, Srivastava GP (2006) Nanotechnol 17:3288
79. Jusserand B, Paquet D, Mollot F, Alexandre F, Le Roux G (1987) Phys Rev B 35:2808
80. Hepplestone SP, Srivastava GP (2005) Appl Phys Lett 87.231905

Chapter 2
Phonons in Bulk and Low-Dimensional Systems

Zhiping Wang, Kitt Reinhardt, Mitra Dutta, and Michael A. Stroscio

Abstract This review highlights selected advances of the last decade in the theory of acoustic and optical phonons in dimensionally confined structures. The basic concepts of the elastic continuum and dielectric continuum models are reviewed. Following this review, specific examples of phonon confinement in dimensionally confined structures are highlighted. These examples include: phonons in single wall carbon nanotubes (CNTs), phonons in multi wall nanotubes, graphene sheets, graphene nanoribbons, graphene quantum dots, graphite confined along the c-axis, and wurtzite structures including quantum wells and quantum dots. The review also covers a number of mechanisms underlying carrier–phonon scattering processes. Finally, this review summarizes the mode amplitudes for a variety of nanostructures.

Z. Wang
Department of Electrical and Computer Engineering, University of Illinois at Chicago, Chicago, IL 60607, USA

Department of Physics, Inner Mongolia University, Hohhot 010021, China
e-mail: wangzhiping1974@gmail.com

K. Reinhardt
Air Force Research Laboratory, Dayton, OH 45433, USA
e-mail: kitt.Reinhardt@wpafb.af.mil

M. Dutta
Department of Electrical and Computer Engineering, University of Illinois at Chicago, Chicago, IL 60607, USA

Department of Physics, University of Illinois at Chicago, Chicago, IL 60607, USA
e-mail: dutta@uic.edu

M.A. Stroscio (✉)
Department of Electrical and Computer Engineering, University of Illinois at Chicago, Chicago, IL 60607, USA

Department of Physics, University of Illinois at Chicago, Chicago, IL 60607, USA

Department of Bioengineering, University of Illinois at Chicago, Chicago, IL 60607, USA
e-mail: stroscio@uic.edu

2.1 Introduction

The application of continuum models in describing acoustic and optical phonon phenomena has been surveyed by Stroscio and Dutta [1]. Over the last decade there have been additional applications of these continuum models to describe phonon phenomena both in conventional and in novel materials systems. Such applications have been extended to carbon nanotubes (CNTs), graphene, and graphite. In addition, over the last decade there have been many applications to conventional materials including those with cubic and wurtzite structure. This review highlights selected applications of continuum models to describe phonon effects in bulk and low-dimensional structures. In this section, a very brief review of basic material for elastic continuum model and dielectric continuum models will be summarized.

2.1.1 Elastic Continuum Model of Phonons

The elastic continuum model of acoustic phonons provides an adequate description of acoustic phonons for nanostructures having confined dimensions of about two atomic monolayers or greater [1]. The case of a longitudinal acoustic mode propagating in a quasi-one-dimensional structure provides an illuminating application of the elastic continuum model.

Consider an element dx on a quasi-one-dimensional structure located between x and $x + dx$. Let us take a longitudinal displacement as an example. The stress $T(x)$, which is the force per unit area in this structure of area A, is given following from Hooke's law:

$$T = Ye, \qquad (2.1)$$

where $e = du/dx$ is the strain, $u(x,t)$ describes the uniform longitudinal displacement of the element dx, and Y is a proportionality constant known as Young's modulus. In such a case, Newton's second law,

$$\rho(x) A dx \frac{\partial^2 u(x,t)}{\partial t^2} = [T(x+dx) - T(x)] A, \qquad (2.2)$$

describes the dynamics of the element dx of density $\rho(x)$. Moreover $\rho A dx$ is the mass associated with the element dx and $\partial^2 u / \partial t^2$. Using Hooke's law

$$T(x+dx) - T(x) = \left(\frac{\partial t}{\partial x}\right) dx = \left(Y\frac{\partial e}{\partial x}\right) dx = \left(Y\frac{\partial^2 u}{\partial x^2}\right) dx, \qquad (2.3)$$

and from Eq. (2.2) it follows that

$$\frac{\partial^2 u}{\partial x^2} = \left(\frac{\rho(x)}{Y}\right) \frac{\partial^2 u}{\partial t^2}, \qquad (2.4)$$

which is recognized as the one-dimensional wave equation, also known as the classical Helmholtz equation.

To generalize these results to three dimensions, we make the replacements $u(x) \to \mathbf{u}(x,y,z) = (u,\upsilon,w)$ and $T = Ye \to T = c : S$ with $T_i = c_{ij}S_j$. In these three-dimensional expressions, Y is replaced by a 6×6 matrix of elastic constants c_{ij}; e is replaced by a six-component object S_j; T is replaced by a six-component object T_i. For the commonly encountered and practical cases of cubic, zincblende, and wurtzite crystals the most general form of the stress–strain relation, $T_{ij} = c_{ijkl}S_{kl}$, where i, j, k, l run over x, y, z, may be represented by $T_i = c_{ij}S_j$. In this last result, i and j run over the integers from one to six. The resulting forms for the stress, T_i, are [2]

$$T_1 = T_{xx}, \quad T_2 = T_{yy}, \quad T_3 = T_{zz},$$
$$T_4 = T_{yz} = T_{zy}, \quad T_5 = T_{xz} = T_{zx}, \quad T_6 = T_{xy} = T_{yx}. \qquad (2.5)$$

For the strain, S_j, the forms are

$$S_1 = S_{xx} = \frac{\partial u}{\partial x}, \quad S_2 = S_{yy} = \frac{\partial u}{\partial y}, \quad S_3 = S_{zz} = \frac{\partial u}{\partial z},$$

$$S_4 = S_{yz} = S_{zy} = \frac{1}{2}\left(\frac{\partial w}{\partial y} + \frac{\partial \upsilon}{\partial z}\right),$$

$$S_5 = S_{xz} = S_{zx} = \frac{1}{2}\left(\frac{\partial u}{\partial z} + \frac{\partial w}{\partial x}\right),$$

$$S_6 = S_{xy} = S_{yx} = \frac{1}{2}\left(\frac{\partial u}{\partial y} + \frac{\partial \upsilon}{\partial x}\right). \qquad (2.6)$$

In many cases of practical and widespread interest in electronics and optoelectronics, nanostructures are generally fabricated from zincblende and wurtzite crystals. Taking into account that the elastic energy to be single valued $c_{ij} = c_{ji}$, only 21 distinct elements are necessary to define the 6×6 matrix c_{ij}. For cubic crystals, the matrix, c_{ij}, takes the form [2]

$$\begin{pmatrix} c_{11} & c_{12} & c_{12} & 0 & 0 & 0 \\ c_{12} & c_{11} & c_{12} & 0 & 0 & 0 \\ c_{12} & c_{12} & c_{11} & 0 & 0 & 0 \\ 0 & 0 & 0 & c_{44} & 0 & 0 \\ 0 & 0 & 0 & 0 & c_{44} & 0 \\ 0 & 0 & 0 & 0 & 0 & c_{44} \end{pmatrix} \qquad (2.7)$$

and for wurtzite crystals, c_{ij}, may be expressed in the form [2]

$$\begin{pmatrix} c_{11} & c_{12} & c_{13} & 0 & 0 & 0 \\ c_{12} & c_{11} & c_{13} & 0 & 0 & 0 \\ c_{13} & c_{13} & c_{33} & 0 & 0 & 0 \\ 0 & 0 & 0 & c_{44} & 0 & 0 \\ 0 & 0 & 0 & 0 & c_{44} & 0 \\ 0 & 0 & 0 & 0 & 0 & (c_{11}-c_{12})/2 \end{pmatrix}. \quad (2.8)$$

As is well known, cubic crystal, including zincblende crystals, may be described in terms of only three independent elastic constants, c_{11}, c_{12}, and c_{44}. Indeed, these independent constants replace Y; c_{11}, relates the compressive stress to the strain along the same direction, c_{44}, relates the shear stress and the strain in the same direction, and c_{12}, relates the compressive stress in one direction and the strain in another direction.

Furthermore, for an isotropic cubic medium, $c_{12} = c_{11} - 2c_{44}$. Thus, only two constants are necessary to specify the c_{ij}:

$$c_{12} = c_{13} = c_{21} = c_{23} = c_{31} = c_{32} = \lambda,$$

$$c_{44} = c_{55} = c_{66} = \frac{1}{2}(c_{11} - c_{22}) = \mu,$$

$$c_{11} = c_{22} = c_{33} = \lambda + 2\mu. \quad (2.9)$$

In the previous results, the constants λ and μ are recognized as the well-known Lamé's constants. In the isotropic case, the previous results imply that

$$T_{xx} = \lambda(S_{xx} + S_{yy} + S_{zz}) + 2\mu S_{xx} = \lambda \Delta + 2\mu S_{xx},$$
$$T_{yy} = \lambda(S_{xx} + S_{yy} + S_{zz}) + 2\mu S_{yy} = \lambda \Delta + 2\mu S_{yy},$$
$$T_{zz} = \lambda(S_{xx} + S_{yy} + S_{zz}) + 2\mu S_{zz} = \lambda \Delta + 2\mu S_{zz},$$
$$T_{yz} = T_{zy} = \mu S_{yz}, \quad T_{zx} = T_{xz} = \mu S_{zx}, \quad T_{xy} = T_{yx} = \mu S_{xy}, \quad (2.10)$$

where $\Delta = \partial u/\partial x + \partial v/\partial y + \partial w/\partial z$ is known as the dilatation of the medium. Then, the three-dimensional generalization of Eq. (2.2) is given by

$$\rho \frac{\partial^2 u}{\partial t^2} = \frac{\partial T_{xx}}{\partial x} + \frac{\partial T_{yx}}{\partial y} + \frac{\partial T_{zx}}{\partial z} = (\lambda + \mu)\frac{\partial \Delta}{\partial x} + \mu \nabla^2 u,$$

$$\rho \frac{\partial^2 v}{\partial t^2} = \frac{\partial T_{xy}}{\partial x} + \frac{\partial T_{yy}}{\partial y} + \frac{\partial T_{zy}}{\partial z} = (\lambda + \mu)\frac{\partial \Delta}{\partial y} + \mu \nabla^2 v,$$

$$\rho \frac{\partial^2 w}{\partial t^2} = \frac{\partial T_{xz}}{\partial x} + \frac{\partial T_{yz}}{\partial y} + \frac{\partial T_{zz}}{\partial z} = (\lambda + \mu)\frac{\partial \Delta}{\partial z} + \mu \nabla^2 w, \quad (2.11)$$

where $\nabla^2 = \partial^2/\partial x^2 + \partial^2/\partial y^2 + \partial^2/\partial z^2$ is Laplace operator.

There are two highly useful alternative forms of the three-dimensional force equations that are encountered frequently in the literature. The first of these forms may be obtained by writing the components of $\mathbf{u}(x,y,z) = (u_x, u_y, u_z)$ as u_α. Accordingly, the three force equations take the form

$$\rho \frac{\partial^2 u_\alpha}{\partial t^2} = \frac{\partial T_{\alpha\beta}}{\partial r_\beta}, \tag{2.12}$$

with

$$T_{\alpha\beta} = \lambda S_{\alpha\alpha} \delta_{\alpha\beta} + 2\mu S_{\alpha\beta}, \tag{2.13}$$

where α, $\beta = x, y, z$, and $\delta_{\alpha\beta}$ is the Kronecker delta function. Following the Einstein summation convention, a repeated index in a term implies summation.

In a second highly useful alternative form, the three force equations may be summarized in terms of a single vector equation

$$\frac{\partial^2 \mathbf{u}}{\partial t^2} = c_t^2 \nabla^2 \mathbf{u} + \left(c_l^2 - c_t^2\right) \operatorname{grad}(\nabla \cdot \mathbf{u}), \tag{2.14}$$

where c_t and c_l are the transverse and longitudinal sound speeds, respectively, and where

$$c_t^2 = \frac{\lambda}{\rho} \quad \text{and} \quad c_l^2 = \frac{\lambda + 2\mu}{\rho}. \tag{2.15}$$

In the classical theory of acoustics, the solutions for the displacement fields may be specified in terms of two potential functions: a vector potential, $\boldsymbol{\Psi} = (\Psi_x, \Psi_y, \Psi_z)$ as Ψ_i ($i = x, y, z$), and a scalar potential ϕ, through

$$u = \frac{\partial \phi}{\partial x} + \frac{\partial \Psi_x}{\partial y} - \frac{\partial \Psi_y}{\partial z},$$

$$v = \frac{\partial \phi}{\partial y} + \frac{\partial \Psi_x}{\partial z} - \frac{\partial \Psi_z}{\partial x},$$

$$w = \frac{\partial \phi}{\partial z} + \frac{\partial \Psi_y}{\partial x} - \frac{\partial \Psi_x}{\partial y}, \tag{2.16}$$

where Ψ_i and ϕ obey simple wave equations of the form

$$\nabla^2 \Psi_i = \frac{1}{c_t^2} \frac{\partial^2 \Psi_i}{\partial t^2} \quad \text{and} \quad \nabla^2 \phi = \frac{1}{c_l^2} \frac{\partial^2 \phi}{\partial t^2}. \tag{2.17}$$

The vector potential represents the "rotational" fields, and the scalar potential represents the "irrotational" part of the solution. The irrotational part is referred to in a number of different ways in the literature; specifically, it is referred to as the longitudinal, compressional, or dilatational solutions. Likewise, the rotational solutions are referred to as transverse, shear, equivoluminal, and distortional solutions.

2.1.2 Dielectric Continuum Model of Phonons

The continuum theory of optical phonons in polar materials is described by the dielectric continuum model, which is based on the fact that the associated lattice vibrations produce an electric polarization that is given in terms of the equations of electrostatics for a dielectric medium [3–11].

Assuming the volume of the structure is to be L^3 ($-L/2 \leq x, y, z \leq L/2$) and under periodic boundary conditions, the optical phonon potential, $\Phi(\mathbf{r})$, and the associated polarization field, $\mathbf{P}(\mathbf{r})$, are related by Stroscio and Dutta [1]

$$\nabla^2 \Phi(\mathbf{r}) = 4\pi \nabla \cdot \mathbf{P}(\mathbf{r}). \tag{2.18}$$

In addition, the polarization field, $\mathbf{P}(\mathbf{r})$, and the electric field, $\mathbf{E}(\mathbf{r})$, are related through the dielectric susceptibility, $\chi(\omega)$, with phonon frequency, ω, in medium g; that is,

$$\mathbf{P}(\mathbf{r}) = \chi(\omega) \mathbf{E}(\mathbf{r}), \tag{2.19}$$

where as usual,

$$\mathbf{E}(\mathbf{r}) = -\nabla \Phi(\mathbf{r}), \tag{2.20}$$

and

$$\chi(\omega) = [\varepsilon(\omega) - 1]/4\pi. \tag{2.21}$$

The dielectric function $\varepsilon(\omega)$ of medium g is given by

$$\varepsilon(\omega) = \varepsilon^\infty \frac{\omega^2 - \omega_{LO}^2}{\omega^2 - \omega_{TO}^2}, \tag{2.22}$$

in the case of a binary polar semiconductor, AB, and is expressible as

$$\varepsilon(\omega) = \varepsilon^\infty \frac{\omega^2 - \omega_{LO,a}^2}{\omega^2 - \omega_{TO,a}^2} \frac{\omega^2 - \omega_{LO,b}^2}{\omega^2 - \omega_{TO,b}^2}, \tag{2.23}$$

2 Phonons in Bulk and Low-Dimensional Systems

for the case of a ternary polar material, $A_xB_{1-x}C$, where as usual ε^∞ is the high-frequency dielectric constant. In addition, ω_{LO} and ω_{TO} are the frequency of the longitudinal optical (LO) phonons and transverse optical (TO) phonons, respectively, the subscript a denotes frequency associated with the dipole pairs AC and the subscript b denotes frequency associated with the dipole pairs BC.

The displacement field $\mathbf{u}(\mathbf{r})$ is related to the fields $\mathbf{E}(\mathbf{r})$ and $\mathbf{P}(\mathbf{r})$ through the so-called driven oscillator equation. In the case of a binary medium g,

$$-\mu\omega^2\mathbf{u}(\mathbf{r}) = -\mu\omega_0^2\mathbf{u}(\mathbf{r}) + e\mathbf{E}_{local}(\mathbf{r}),$$
$$\mathbf{P}(\mathbf{r}) = ne\mathbf{u}(\mathbf{r}) + n\alpha\mathbf{E}_{local}(\mathbf{r}), \qquad (2.24)$$

where e is the effective charge, $\mu = mM/(m+M)$ is the reduced mass, $\omega_0^2 = 2\alpha(1/m + 1/M)$ is the resonant frequency squared, and α is the electronic polarizability per unit cell, n is the number of unit cells in region g, and $\mathbf{E}_{local}(\mathbf{r})$ is the local field given by the Lorentz relation

$$\mathbf{E}_{local}(\mathbf{r}) = \mathbf{E}(\mathbf{r}) + \frac{4\pi}{3}\mathbf{P}(\mathbf{r}). \qquad (2.25)$$

Using the virtual-crystal approximation, for the dipole pairs AC (BC) in a ternary medium m, it follows that

$$-\mu_{m,a(b)}\omega^2\mathbf{u}_{m,a(b)}(\mathbf{r}) = -\mu_m\omega_{0m,a(b)}^2\mathbf{u}_{m,a(b)}(\mathbf{r}) + e_{m,a(b)}\mathbf{E}_{local}(\mathbf{r}),$$
$$\mathbf{P}(\mathbf{r}) = n_m\left[xe_{m,a(b)}\mathbf{u}_{m,a}(\mathbf{r}) + (1-x)e_{m,b}\mathbf{u}_{m,b}(\mathbf{r})\right] + n_m\alpha_m\mathbf{E}_{local}(\mathbf{r}). \qquad (2.26)$$

From Huang–Born theory [12], an alternative and useful form of these equations for the case of a diatomic polar material may be written as

$$\ddot{\mathbf{u}} = -\omega_{TO}^2\mathbf{u} + \sqrt{\frac{V}{4\pi\mu N}}\sqrt{\varepsilon^0 - \varepsilon^\infty}\omega_{TO}\mathbf{E},$$
$$\mathbf{P} = \sqrt{\frac{\mu N}{4\pi V}}\sqrt{\varepsilon^0 - \varepsilon^\infty}\omega_{TO}\mathbf{u} + \frac{\varepsilon^\infty - 1}{4\pi}\mathbf{E}, \qquad (2.27)$$

where ε^0 is static dielectric constant, V is the volume of the material, and it has been assumed that $\ddot{\mathbf{u}}$ has a general form for the time dependence and may not necessarily be simply sinusoidal in ω.

Equation (2.27) provides the basis for the derivation of the macroscopic equations describing optical phonons in polar uniaxial materials such as the hexagonal wurtzite structures of GaN, AlN, and $Ga_xAl_{1-x}N$ [13]. By introducing one dielectric constant associated with the direction parallel to the c-axis, ε_z, and another dielectric constant associated with the direction perpendicular to the c-axis, $\varepsilon_\perp$, the macroscopic equation for a uniaxial polar crystal may be derived [13]. In describing the phonon mode displacements, it is convenient to separate the displacements parallel to the *c*-axis, denoted by $\mathbf{u}_z$, and those perpendicular to *c*-axis, denoted by $\mathbf{u}_\perp$. For medium *g*, we then have

$$\ddot{\mathbf{u}}_\perp = -\omega_{TO,\perp}^2 \mathbf{u}_\perp + \sqrt{\frac{V}{4\pi\mu N}} \sqrt{\varepsilon_\perp^0 - \varepsilon_\perp^\infty} \omega_{TO,\perp} \mathbf{E}_\perp,$$

$$\mathbf{P}_\perp = \sqrt{\frac{\mu N}{4\pi V}} \sqrt{\varepsilon_\perp^0 - \varepsilon_\perp^\infty} \omega_{TO,\perp} \mathbf{u}_\perp + \frac{\varepsilon_\perp^\infty - 1}{4\pi} \mathbf{E}_\perp,$$

$$\varepsilon_\perp(\omega) = \varepsilon_\perp^\infty \frac{\omega^2 - \omega_{LO,\perp}^2}{\omega^2 - \omega_{TO,\perp}^2}, \tag{2.28}$$

and

$$\ddot{\mathbf{u}}_z = -\omega_{TO,z}^2 \mathbf{u}_z + \sqrt{\frac{V}{4\pi\mu N}} \sqrt{\varepsilon_z^0 - \varepsilon_z^\infty} \omega_{TO,z} \mathbf{E}_z,$$

$$\mathbf{P}_z = \sqrt{\frac{\mu N}{4\pi V}} \sqrt{\varepsilon_z^0 - \varepsilon_z^\infty} \omega_{TO,z} \mathbf{u}_z + \frac{\varepsilon_z^\infty - 1}{4\pi} \mathbf{E}_z,$$

$$\varepsilon_z(\omega) = \varepsilon_z^\infty \frac{\omega^2 - \omega_{LO,z}^2}{\omega^2 - \omega_{TO,z}^2}. \tag{2.29}$$

It has been assumed that $\ddot{\mathbf{u}}_\perp$ and $\ddot{\mathbf{u}}_z$ have a general form of the dependence and may not necessarily be simply sinusoidal in ω. In Loudon's model, above six equations combined with the following three equations of electrostatics for the case where there is no free charge

$$\mathbf{E}(\mathbf{r}) = -\nabla\phi(\mathbf{r}),$$
$$\mathbf{D}(\mathbf{r}) = \varepsilon \mathbf{E}(\mathbf{r}) = \mathbf{E}(\mathbf{r}) + 4\pi \mathbf{P}(\mathbf{r})$$
$$= \varepsilon_\perp(\omega) E_\perp(\mathbf{r}) \boldsymbol{\rho} + \varepsilon_z(\omega) E_z(\mathbf{r}) \mathbf{z} \nabla \cdot \mathbf{D}(\mathbf{r}) = 0, \tag{2.30}$$

where $\boldsymbol{\rho}$ and $\mathbf{z}$ are the unit vectors in the perpendicular and parallel directions, respectively.

The nine equations outlined previously provide the equations for describing the fields associated with an optical phonon, and therefore, the carrier–optical–phonon scattering in wurtzite crystals. Moreover, these equations provide a description of the case of cubic, including zincblende crystals, in the limit of $\varepsilon_\perp(\omega) = \varepsilon_z(\omega)$ as discussed in Stroscio and Dutta [1].

These phonon modes must be normalized such that the energy in each mode is equal to $\hbar\omega$. For the case of the bulk uniaxial material such as hexagonal wurtzite crystals, this normalization condition may be expressed as [14]

$$\left[\sqrt{n\mu}\mathbf{u}^*(\mathbf{q})\right] \cdot \left[\sqrt{n\mu}\mathbf{u}(\mathbf{q})\right] = \frac{\hbar}{2\omega V}. \tag{2.31}$$

here, $\mathbf{u}(\mathbf{q})$ is the Fourier transform of $\mathbf{u}(\mathbf{r})$. For the case of CNTs, this normalization may be expressed as [15,16]

$$\frac{1}{V}\int \mathbf{u}^*(\mathbf{r})\mathbf{u}(\mathbf{r})\,dV = \frac{\hbar}{2M\omega}, \qquad (2.32)$$

where M is the mass of the atoms which constitute CNTs. For the case of a two-dimensional graphene sheet, this normalization condition may be expressed as [15]

$$\frac{1}{S}\int \left(\mathbf{u}^*\cdot\mathbf{u} + \upsilon^*\cdot\upsilon\right)dx\,dy = \frac{\hbar}{M\omega}, \qquad (2.33)$$

where $\mathbf{u}$ and υ are the two displacements in the plane of the sheet. As the reader can surmise from examples and from the many references in this review dealing with phonon mode normalization, it is generally advisable to formulate a convenient mode normalization for each problem at hand.

The carrier–phonon interaction plays an important role in the properties of nanostructures. At room temperature, in low-defect polar semiconductors such as GaAs, InP, and GaN, carrier scattering is in many cases dominated by the polar–optical–phonon scattering mechanism and the polar–optical–phonon–carrier interaction is referred to as the Fröhlich interaction, which can be written as

$$H_{\text{Fr}} = -i\sqrt{\frac{2\pi e^2 \hbar \omega_{\text{LO}}}{V}}\sqrt{\frac{1}{\varepsilon^{\infty}} - \frac{1}{\varepsilon^0}}\sum_q \frac{1}{q}\left(a_\mathbf{q} e^{i\mathbf{q}\cdot\mathbf{r}} + a_\mathbf{q}^\dagger e^{-i\mathbf{q}\cdot\mathbf{r}}\right), \qquad (2.34)$$

where $a_\mathbf{q}^\dagger$ and $a_\mathbf{q}$ are the phonon creation and annihilation operators, respectively. In many cases, the LO-phonon frequency, ω_{LO}, can be taken as the zone-center value to an excellent approximation. The deformation–potential interaction is one of the most important interactions in modern semiconductor devices and it has its origin in the displacements caused by phonons. The interaction between acoustic phonon and the carrier is known as the acoustic deformation-potential, $H_{\text{aco-def}}$, and is given by [16]

$$H_{\text{aco-def}} = D_{\text{ac}}\nabla\cdot\mathbf{u}, \qquad (2.35)$$

where D_{ac} is the acoustic deformation-potential constant whose value may be estimated in a nearly free electron model [17]. For defect-free nanotubes, the carrier interactions are determined by the optical deformation–potential interaction. The optical deformation potential Hamiltonian is given by [18]

$$H_{\text{opt-def}} = \mathbf{D}_{\text{op}}\cdot\mathbf{u}, \qquad (2.36)$$

where $\mathbf{D}_{\text{op}}$ is the optical deformation potential constant. Since the axial phonons along the length of the CNT are of primary importance for transport along the nanotube axis, the deformation potential mediating transport in the axial direction can be written as

$$H_{\text{opt-def}} = |D_{\text{op}}| u_z. \tag{2.37}$$

In general, the application of an external strain to a piezoelectric crystal will produce a macroscopic polarization as a result of the displacements of ions. Thus an acoustic phonon mode will drive a macroscopic polarization in a piezoelectric crystal. The piezoelectric interaction may be modeled in terms of the interaction of carriers with the macroscopic electric potential produced by the piezoelectric field. In Cartesian coordinates, the polarization created by the piezoelectric interaction in cubic crystals, including zincblende crystals, may be written as

$$\mathbf{P}^{\text{piezo}} = \left\{ e_{x4} \frac{\partial w/\partial y + \partial v/\partial z}{2}, e_{x4} \frac{\partial u/\partial z + \partial w/\partial x}{2}, e_{x4} \frac{\partial u/\partial y + \partial v/\partial x}{2} \right\}, \tag{2.38}$$

here e_{x4} is the piezoelectric coupling constant and has been described previously, the factors multiplying e_{x4} are the components of the strain tensor that contribute to the piezoelectric polarization in a zincblende crystal. This polarization may be expressed in a more general tensor form as

$$\mathbf{P}^{\text{piezo}} = e_{\lambda,\mu\nu} S_{\mu\nu}, \tag{2.39}$$

where $S_{\mu\nu}$ is the strain tensor discussed previously and $e_{\lambda,\mu\nu}$ is known as the piezoelectric tensor. The components of this strain tensor have been summarized by Auld [2] for a variety of crystals. For a phonon of wave vector q_μ, the corresponding piezoelectric potential may be expressed as [1, 19]

$$V^{\text{piezo}}(q) = 4\pi \frac{q_\lambda e_{\lambda,\mu\nu} q_\mu}{q_\lambda e^0_{\lambda\mu} q_\mu} u_\nu, \tag{2.40}$$

where $e_{\lambda\mu}{}^0$ satisfies $D_\lambda = 4\pi e_{\lambda,\mu\nu} S_{\mu\nu} + e_{\lambda\mu}{}^0 E_\mu.$

2.2 Phonons in CNTs

CNTs discovered by Iijima [20] are single graphene sheets rolled up into cylinders of nanometer diameters along a chiral vector of the tube surface, which exhibit novel electronic and optical properties. CNTs have been studied extensively for possible uses as nanodevices such as field-effect transistors [21], flat-panel display [22], MIS capacitor [23], and infrared detectors [24].

2 Phonons in Bulk and Low-Dimensional Systems

In this section, the elastic continuum model is applied to determine the phonon modes, both optical and acoustic, in the finite length CNTs. The dispersion relations and deformation potential of the CNTs are obtained. Phonon bottleneck effects in short CNTs and thermal conductivity of CNTs are investigated in detail.

2.2.1 Phonon Modes in Single Wall Nanotubes

Herein we consider a short single wall nanotubes clamped in the ends of the tube. The approach taken here is to treat the CNT as a thin cylindrical membrane with elastic properties specified in terms of Poisson's ratio, υ, density, ρ, and Young's modulus, Y. The case has treated by Donnell [25] for an arbitrary cylindrical elastic sheet. The Donnell's equations of equilibrium for a thin wall cylindrical membrane are given by [16, 26, 27]

$$\frac{\partial^2 u_z}{\partial z^2} + \frac{1-\upsilon}{2a^2}\frac{\partial^2 u_z}{\partial \theta^2} + \frac{1+\upsilon}{2a}\frac{\partial^2 u_\theta}{\partial z \partial \theta} + \frac{\upsilon}{a}\frac{\partial u_r}{\partial z} - \frac{1-\upsilon^2}{Y}\rho\frac{\partial^2 u_z}{\partial t^2} = 0,$$

$$\frac{1+\upsilon}{2a}\frac{\partial^2 u_z}{\partial z \partial \theta} + \frac{1-\upsilon}{2}\frac{\partial^2 u_\theta}{\partial z^2} + \frac{1}{a^2}\frac{\partial^2 u_\theta}{\partial \theta^2} + \frac{1}{a^2}\frac{\partial u_r}{\partial \theta} - \frac{1-\upsilon^2}{Y}\rho\frac{\partial^2 u_\theta}{\partial t^2} = 0,$$

$$\frac{\upsilon}{a}\frac{\partial u_z}{\partial z} + \frac{1}{a^2}\frac{\partial^2 u_\theta}{\partial \theta^2} + \frac{u_r}{a^2} + \frac{d^2}{12}\nabla^4 u_r + \frac{1-\upsilon^2}{Y}\rho\frac{\partial^2 u_r}{\partial t^2} = 0, \quad (2.41)$$

where u_z is the displacement in the axial direction, u_θ the displacement in the circumferential direction, u_r the displacement in the radial direction, a the radius and d the thickness of the nanotubes, respectively.

The solution is divided into two parts—odd and even. For the odd part, the phonon displacements can be assumed as

$$u_z = A\left[-\sin\frac{\lambda}{a}\left(\frac{l}{2}-z\right) + k\sinh\frac{\lambda}{a}\left(\frac{l}{2}-z\right)\right]\cos(n\theta)\cos(\omega t),$$

$$u_\theta = B\left[\cos\frac{\lambda}{a}\left(\frac{l}{2}-z\right) + k\cosh\frac{\lambda}{a}\left(\frac{l}{2}-z\right)\right]\sin(n\theta)\cos(\omega t),$$

$$u_r = C\left[\cos\frac{\lambda}{a}\left(\frac{l}{2}-z\right) + k\cosh\frac{\lambda}{a}\left(\frac{l}{2}-z\right)\right]\cos(n\theta)\cos(\omega t), \quad (2.42)$$

where A, B, and C are arbitrary constants, λ is a variable and will be determined later, and l is the length of the tube segment.

For the case of CNT with the so-called clamped ends, the boundary conditions are taken as

$$u_z = u_\theta = u_r = \partial u_r/\partial z = 0 \quad \text{at } z = 0 \quad \text{and} \quad z = 1. \quad (2.43)$$

Applying the above condition to Eq. (2.42) leads to the following results:

$$k = \sin\left(\frac{\lambda l}{2a}\right) / \sinh\left(\frac{\lambda l}{2a}\right) \quad \text{and} \quad \tan\left(\frac{\lambda l}{2a}\right) + \tanh\left(\frac{\lambda l}{2a}\right) = 0, \quad (2.44)$$

together with the roots of the above transcendental equation, $\lambda l/a == 1.5\pi, 3.5\pi, 5.5\pi, \ldots$ corresponding to 1, 3, 5, ... axial half waves.

Substituting the displacement of Eq. (2.42) into the Donnell's equation, and solving for A/C and B/C yields

$$\frac{A}{C} = \frac{v\lambda\left(n^2 + \frac{1-v}{2}\lambda^2\frac{\theta_2}{\theta_1} - \Delta\right) + \frac{1+v}{2}\lambda n^2}{\left(\lambda^2\frac{\theta_1}{\theta_2} + \frac{1-v^2}{2}n^2 - \Delta\right)\left(n^2 + \frac{1-v}{2}\lambda^2\frac{\theta_2}{\theta_1} - \Delta\right) - \left(\frac{1+v}{2}\right)^2\lambda^2 n^2\frac{\theta_2}{\theta_1}}, \quad (2.45)$$

$$\frac{B}{C} = \frac{v\lambda^2 n\frac{1+v}{2}\frac{\theta_2}{\theta_1} - n\left(\lambda^2\frac{\theta_1}{\theta_2} + \frac{1-v^2}{2}n^2 - \Delta\right)}{\left(\lambda^2\frac{\theta_1}{\theta_2} + \frac{1-v^2}{2}n^2 - \Delta\right)\left(n^2 + \frac{1-v}{2}\lambda^2\frac{\theta_2}{\theta_1} - \Delta\right) - \left(\frac{1+v}{2}\right)^2\lambda^2 n^2\frac{\theta_2}{\theta_1}}, \quad (2.46)$$

where $\Delta = \rho a^2(1-v^2)\omega^2/Y$, $\theta_1 = 1 + k^2$, and $\theta_2 = 1 - k^2 + (2a/\lambda l)\sin(\lambda l/a)$.

The phonon displacement modes can be quantized by using Eq. (2.32)

$$\frac{1}{V}\int \left(u_z u_z^* + u_\theta u_\theta^* + u_r u_r^*\right) dV = \frac{\hbar}{2M\omega}. \quad (2.47)$$

Imposing this normalization condition leads to the results

$$C = \frac{\pi l d}{2\pi M\omega}\left\{\left(\frac{A}{C}\right)^2 I_+ + \left[\left(\frac{B}{C}\right)^2 + 1\right]I_-\right\}^{-1}, \quad (2.48)$$

with

$$I_\pm = \mp\frac{\pi}{4\lambda}e^{-\lambda l/a}\Big[4\cos\left(\tfrac{\lambda l}{2a}\right)\sin\left(\tfrac{\lambda l}{2a}\right)ae^{\lambda l/a} + 4ka\cos\left(\tfrac{\lambda l}{2a}\right)e^{\lambda l/2a}$$
$$\pm 4ka\sin\left(\tfrac{\lambda l}{2a}\right)e^{\lambda l/2a} \pm 4ka\sin\left(\tfrac{\lambda l}{2a}\right)e^{3\lambda l/2a} \mp 4ka\cos\left(\tfrac{\lambda l}{2a}\right)e^{3\lambda l/2a} \quad (2.49)$$
$$\pm k^2 a \mp k^2 a e^{2\lambda l/a} \mp 2\lambda l e^{\lambda l/a} + 2k^2\lambda l e^{\lambda l/a}\Big]$$

Carrying out similar analyses for the even modes, the even phonon mode displacements in the three directions are found easily by a similar analysis, and it is written as

$$\begin{aligned} u_z &= A\left[-\cos\tfrac{\lambda}{a}\left(\tfrac{l}{2}-z\right) + k\cosh\tfrac{\lambda}{a}\left(\tfrac{l}{2}-z\right)\right]\cos(n\theta)\cos(\omega t), \\ u_\theta &= B\left[\sin\tfrac{\lambda}{a}\left(\tfrac{l}{2}-z\right) + k\sinh\tfrac{\lambda}{a}\left(\tfrac{l}{2}-z\right)\right]\sin(n\theta)\cos(\omega t), \\ u_r &= C\left[\sin\tfrac{\lambda}{a}\left(\tfrac{l}{2}-z\right) + k\sinh\tfrac{\lambda}{a}\left(\tfrac{l}{2}-z\right)\right]\cos(n\theta)\cos(\omega t), \end{aligned} \quad (2.50)$$

2 Phonons in Bulk and Low-Dimensional Systems

where

$$k = \cos\left(\frac{\lambda l}{2a}\right) / \cosh\left(\frac{\lambda l}{2a}\right) \quad \text{and} \quad \tan\left(\frac{\lambda l}{2a}\right) - \tanh\left(\frac{\lambda l}{2a}\right) = 0, \quad (2.51)$$

together with the roots, of the above transcendental equation, $\lambda l/a = 2.5\pi, 4.5\pi, 6.5\pi, \ldots$ corresponds to $2, 4, 6, \ldots$ axial half waves, and A, B, C also satisfy Eqs. (2.45)–(2.47) with

$$I_\pm = \mp \tfrac{\pi}{4\lambda} e^{-\lambda l/a} \Big[4\cos\left(\tfrac{\lambda l}{2a}\right) \sin\left(\tfrac{\lambda l}{2a}\right) a e^{\lambda l/a} \mp 4ka\cos\left(\tfrac{\lambda l}{2a}\right) e^{\lambda l/2a}$$
$$\mp 4ka\sin\left(\tfrac{\lambda l}{2a}\right) e^{\lambda l/2a} - 4ka\cos\left(\tfrac{\lambda l}{2a}\right) e^{3\lambda l/2a} \mp 4ka\sin\left(\tfrac{\lambda l}{2a}\right) e^{3\lambda l/2a} \quad (2.52)$$
$$\mp k^2 a \pm k^2 a e^{2\lambda l/a} \pm 2\mu l e^{\lambda l/a} + 2k^2 \lambda l e^{\lambda l/a} \Big].$$

The dispersion relations for a number of different CNTs have been obtained by Raichura et al. [16, 18, 26]. The relations reveal that wave number versus the wave vector is composed of a series of points rather than a continuous function, as is to be expected since a CNT of finite length is basically a quantum dot with confinement in all three dimensions having no free, continuous wave vector. From these dispersion relations, it is found that the frequency of the well-known breathing mode for the (10, 10) nanotube is found to be about 30 THz which agrees well with that given by Suzurra and Ando [17].

From the analytic solutions given by Eqs. (2.42)–(2.51), the mode amplitudes are determined in terms of simple trigonometric and hyperbolic functions; due to the clamped boundary conditions many of the modes resemble half-wavelength confined modes and the corresponding deformation potentials have a similar character as expected from Eq. (2.35).

2.2.2 Optical Phonon Modes in Multi Wall Nanotubes

As discussed in the last section, the acoustic modes in a CNT may be determined in terms of the elastic continuum model. In this section, optical phonon modes of multi wall CNTs are derived using the method of Constantinou and Ridley [28, 29] whereby differences of acoustic mode displacements are need to construct optical mode displacements, as fires applied to CNTs by Stroscio et al. [30].

Consider a nanotube of outer radius, b, inner radius, a, and length, l, where the length is taken to be relatively large compared to any of the relevant physical scales in the system. As motivated by Born and Huang [12], herein we use the modified ionic displacement, w, rather than the actual ionic displacement, u. As discussed by Born and Huang [12], two displacements are related by

$$w = \sqrt{\mu/V}\, u, \quad (2.53)$$

where μ is the reduced mass, V is the volume of the unit cell. Moreover, the displacement, w, satisfies the classical Helmholtz equation,

$$\left(\nabla^2 + k_i^2\right) w^{(i)} = 0, \qquad (2.54)$$

where

$$k_i^2 = \left(\omega_i^2 - \omega^2\right) \beta_i^{-2}. \qquad (2.55)$$

In this last expression, ω_i is the zone-center LO-phonon frequency for material region i, and β_i is the acoustic phonon velocity.

The solutions for the three components of the optical mode displacements may be determined by Eq. (2.53) and can be written as [18]

$$\begin{aligned} w_z &= C_{mn} e^{im\theta} \left[N_m' (q_{mn}a) J_m (q_{mn}r) - J_m' (q_{mn}a) N_m (q_{mn}r) \right] \sin(q_z z), \\ w_\theta &= -\frac{C_{mn} i m e^{im\theta}}{r q_z} \left[N_m' (q_{mn}a) J_m (q_{mn}r) - J_m' (q_{mn}a) N_m (q_{mn}r) \right] \cos(q_z z), \\ w_r &= -\frac{C_{mn} q_{mn} e^{im\theta}}{q_z} \left[N_m' (q_{mn}a) J_m' (q_{mn}r) - J_m' (q_{mn}a) N_m' (q_{mn}r) \right] \cos(q_z z). \end{aligned} \qquad (2.56)$$

Furthermore, as a result of the conditions

$$\left[\frac{1}{r} \frac{\partial}{\partial r} (r w_r) + \frac{1}{r} \frac{\partial}{\partial \theta} (w_\theta) + \frac{\partial}{\partial z} (w_z) \right]_{a,b} = 0, \qquad (2.57)$$

it follows that we have the "double infinite" set of eigenvalues

$$q = q_{mn}, \quad m = 0, 1, 2, \ldots \quad n = 0, 1, 2 \ldots, \qquad (2.58)$$

obtained from the positive roots of

$$N_m' (q_{mn}a) J_m' (q_{mn}b) - J_m' (q_{mn}a) N_m' (q_{mn}b) = 0. \qquad (2.59)$$

Using the clamped boundary conditions

$$w_z = w_r = dw_r/dz = 0 \quad \text{at } z = 0, l \quad \text{and} \quad z = a, b, \qquad (2.60)$$

it is found that $q_z = k\pi/l$ with k being an integer.

Thus, the frequencies are

$$\omega_{mn}^2 = \omega_{LO}^2 - \beta^2 \left(q_{mn}^2 + q_z^2\right), \qquad (2.61)$$

where the frequency ω_{LO} is the zone-center LO-phonon frequency of the material composing the nanotube.

As discussed previously, the phonon displacement modes can be quantized by using Eq. (2.32)

$$\frac{1}{V}\int (w_z w_z^* + w_\theta w_\theta^* + w_r w_r^*)\, dV = \frac{\hbar}{\mu \omega_{mn}}, \qquad (2.62)$$

Applying this normalization condition, we find

$$C_{mn} = \frac{\sqrt{a^2 - b^2}\sqrt{l}}{\sqrt{2}} \sqrt{\frac{\hbar}{\mu \omega_{mn}}} \sqrt{\frac{1}{I_r^{(m)}}}, \qquad (2.63)$$

with

$$\begin{aligned} I_r^{(m)} &= \tfrac{l}{2}\left[I_{r1}^{(m)} + (q_{mn}/q_z)^2 I_{r2}^{(m)} + (m/q_z)^2 I_{r3}^{(m)} \right] \\ &+ \tfrac{\sin(2q_z l)}{4 q_z}\left[-I_{r1}^{(m)} + (q_{mn}/q_z)^2 I_{r2}^{(m)} + (m/q_z)^2 I_{r3}^{(m)} \right], \end{aligned} \qquad (2.64)$$

and

$$\begin{aligned} I_{r1}^{(m)} &= N'^2_m(q_{mn}a)\, I_1^{(m)} + J'^2_m(q_{mn}a)\, I_2^{(m)} - 2 N'_m(q_{mn}a)\, J'_m(q_{mn}a)\, I_3^{(m)}, \\ I_{r2}^{(m)} &= N'^2_m(q_{mn}a)\, I_4^{(m)} + J'^2_m(q_{mn}a)\, I_5^{(m)} - 2 N'_m(q_{mn}a)\, J'_m(q_{mn}a)\, I_6^{(m)}, \\ I_{r3}^{(m)} &= N'^2_m(q_{mn}a)\, I_7^{(m)} + J'^2_m(q_{mn}a)\, I_8^{(m)} - 2 N'_m(q_{mn}a)\, J'_m(q_{mn}a)\, I_9^{(m)}, \end{aligned} \qquad (2.65)$$

where

$$\begin{aligned} I_1^{(m)} &= \frac{b^2}{2}\left[J_m^2(q_{mn}b) - J_{m-1}(q_{mn}b)\, J_{m+1}(q_{mn}b) \right] \\ &\quad - \frac{a^2}{2}\left[J_m^2(q_{mn}a) - J_{m-1}(q_{mn}a)\, J_{m+1}(q_{mn}a) \right], \end{aligned}$$

$$\begin{aligned} I_2^{(m)} &= \frac{b^2}{2}\left[N_m^2(q_{mn}b) - N_{m-1}(q_{mn}b)\, N_{m+1}(q_{mn}b) \right] \\ &\quad - \frac{a^2}{2}\left[N_m^2(q_{mn}a) - N_{m-1}(q_{mn}a)\, N_{m+1}(q_{mn}a) \right], \end{aligned}$$

$$\begin{aligned} I_3^{(m)} &= \frac{b^2}{4}\big[2 J_m(q_{mn}b)\, N_m(q_{mn}b) - J_{m+1}(q_{mn}b)\, N_{m-1}(q_{mn}b) \\ &\quad - J_{m-1}(q_{mn}b)\, N_{m+1}(q_{mn}b) \big] - \frac{a^2}{4}\big[2 J_m(q_{mn}a)\, N_m(q_{mn}a) \\ &\quad - J_{m+1}(q_{mn}a)\, N_{m-1}(q_{mn}a) - J_{m-1}(q_{mn}a)\, N_{m+1}(q_{mn}a) \big], \end{aligned}$$

$$I_4^{(m)} = \frac{b^2}{2} \left[J_{m+1}^2(q_{mn}b) - J_m(q_{mn}b) N_{m+1}(q_{mn}b) \right]$$
$$- \frac{a^2}{2} \left[J_{m+1}^2(q_{mn}a) - J_m(q_{mn}a) N_{m+1}(q_{mn}a) \right]$$
$$+ \frac{m}{2q_{mn}^2} \left[-J_0^2(q_{mn}b) - J_m^2(q_{mn}b) - 2\sum_{k=1}^{m-1} J_k^2(q_{mn}b) \right.$$
$$\left. + J_0^2(q_{mn}a) - J_m^2(q_{mn}a) + 2\sum_{k=1}^{m-1} J_k^2(q_{mn}a) \right]$$
$$- \frac{2m}{q_{mn}^2} \left[-J_0^2(q_{mn}b) - \sum_{k=1}^{m} J_k^2(q_{mn}b) + J_0^2(q_{mn}a) + \sum_{k=1}^{m} J_k^2(q_{mn}a) \right],$$

$$I_5^{(m)} = \frac{b^2}{2} \left[N_{m+1}^2(q_{mn}b) - N_m(q_{mn}b) N_{m+2}(q_{mn}b) \right]$$
$$- \frac{a^2}{2} \left[N_{m+1}^2(q_{mn}a) - N_m(q_{mn}a) N_{m+2}(q_{mn}a) \right]$$
$$+ \frac{m}{2q_{mn}^2} \left[-N_0^2(q_{mn}b) - N_m^2(q_{mn}b) \right.$$
$$\left. -2\sum_{k=1}^{m-1} N_k^2(q_{mn}b) + N_0^2(q_{mn}a) - N_m^2(q_{mn}a) - 2\sum_{k=1}^{m-1} N_k^2(q_{mn}a) \right] \quad (2.66)$$
$$- \frac{2m}{q_{mn}} \int_a^b N_{m+1}(q_{mn}r) N_m(q_{mn}r) \, dr,$$

$$I_6^{(m)} = \frac{b^2}{4} \left[2J_{m+1}(q_{mn}b) N_{m+1}(q_{mn}b) - J_{m+1}(q_{mn}b) N_{m-1}(q_{mn}b) \right.$$
$$\left. - J_{m-1}(q_{mn}b) N_{m+1}(q_{mn}b) \right] - \frac{a^2}{4} \left[2J_{m+1}(q_{mn}a) N_{m+1}(q_{mn}a) \right.$$
$$\left. - J_{m+1}(q_{mn}a) N_{m-1}(q_{mn}a) - J_{m-1}(q_{mn}a) N_{m+1}(q_{mn}a) \right]$$

$$- \frac{m}{2q_{mn}^2} \left[J_0(q_{mn}b) N_0(q_{mn}b) + J_m(q_{mn}b) N_m(q_{mn}b) \right.$$
$$\left. + 2\sum_{k=1}^{m-1} J_k(q_{mn}b) J_k(q_{mn}b) \right] + \frac{m}{2q_{mn}^2} \left[J_0(q_{mn}a) N_0(q_{mn}a) \right.$$
$$\left. + J_m(q_{mn}a) N_m(q_{mn}a) + 2\sum_{k=1}^{m-1} J_k(q_{mn}a) J_k(q_{mn}a) \right]$$
$$- \frac{m}{q_{mn}} \int_a^b N_{m+1}(q_{mn}r) J_m(q_{mn}r) \, dr - \frac{m}{q_{mn}} \int_a^b J_{m+1}(q_{mn}r) N_m(q_{mn}r) \, dr,$$

2 Phonons in Bulk and Low-Dimensional Systems

$$I_7^{(m)} = -\frac{1}{2m}\left[J_0^2(q_{mn}b) + J_m^2(q_{mn}b) + 2\sum_{k=1}^{m-1} J_k^2(q_{mn}a)\right.$$
$$\left. - J_0^2(q_{mn}a) - J_m^2(q_{mn}a) - 2\sum_{k=1}^{m-1} J_k^2(q_{mn}a)\right],$$

$$I_8^{(m)} = -\frac{1}{2m}\left[N_0^2(q_{mn}b) + N_m^2(q_{mn}b) + 2\sum_{k=1}^{m-1} N_k^2(q_{mn}a)\right.$$
$$\left. - N_0^2(q_{mn}a) - N_m^2(q_{mn}a) - 2\sum_{k=1}^{m-1} N_k^2(q_{mn}a)\right],$$

$$I_9^{(m)} = -\frac{1}{2m}\left[J_0(q_{mn}b)N_0(q_{mn}b) + J_m(q_{mn}b)N_m(q_{mn}b)\right.$$
$$\left. + 2\sum_{k=1}^{m-1} J_k(q_{mn}b)J_k(q_{mn}b)\right] + \frac{1}{2m}\left[J_0(q_{mn}a)N_0(q_{mn}a)\right.$$
$$\left. + J_m(q_{mn}a)N_m(q_{mn}a) + 2\sum_{k=1}^{m-1} J_k(q_{mn}a)J_k(q_{mn}a)\right].$$

For the lowest azimuthal mode, $m = 0$, and

$$C_{0n} = \frac{\sqrt{a^2-b^2}\sqrt{l}}{\sqrt{2}}\sqrt{\frac{\hbar}{\mu\omega_{0n}}}\sqrt{\frac{1}{I_r^{(0)}}}, \qquad (2.67)$$

where

$$I_r^{(0)} = \frac{l}{2}\left[I_{r1}^{(0)} + \left(\frac{q_{0n}}{q_z}\right)^2 I_{r2}^{(0)}\right] + \frac{\sin(2q_z l)}{4q_z}\left[-I_{r1}^{(0)} + \left(\frac{q_{0n}}{q_z}\right)^2 I_{r2}^{(0)}\right], \quad (2.68)$$

and

$$\begin{aligned}I_{r1}^{(0)} &= N_0'^2(q_{0n}a)I_1^{(m)} + J_0'^2(q_{0n}a)I_2^{(0)} - 2N_0'(q_{0n}a)J_0'(q_{0n}a)I_3^{(0)},\\ I_{r2}^{(0)} &= N_0'^2(q_{0n}a)I_4^{(0)} + J_0'^2(q_{0n}a)I_5^{(0)} - 2N_0'(q_{0n}a)J_0'(q_{0n}a)I_6^{(0)},\end{aligned} \quad (2.69)$$

with

$$I_1^{(0)} = \frac{b^2}{2}\left[J_0^2(q_{0n}b) + J_1^2(q_{0n}b)\right] - \frac{a^2}{2}\cdot\left[J_0^2(q_{0n}a) + J_1^2(q_{0n}a)\right],$$

Table 2.1 The roots, q_{mn}, for $m = 0,1,2$ and $n = 0,1,2$ for radii ratios $b/a = 1.5$, 2.0 and 2.5

		$n = 0$	$n = 1$	$n = 2$
$b/a = 1.5$	$m = 0$	0.39953479	6.32429094	2.58589799
	$m = 1$	0.80513571	6.37648698	12.61258158
	$m = 2$	1.60796911	6.53788074	12.69223312
$b/a = 2.0$	$m = 0$	0.3239608	3.19842986	6.31220220
	$m = 1$	0.67724429	3.28240592	6.35291530
	$m = 2$	1.34058549	3.53129775	6.47312708
$b/a = 2.5$	$m = 0$	0.27107423	2.15646816	4.22295780
	$m = 1$	0.58499003	2.26364240	4.27287650
	$m = 2$	1.13689860	2.56697694	4.42339550

$$I_2^{(0)} = \frac{b^2}{2}\left[N_0^2(q_{0n}b) + N_1^2(q_{0n}b)\right] - \frac{a^2}{2}\left[N_0^2(q_{0n}a) + N_1^2(q_{0n}a)\right],$$

$$I_3^{(0)} = \frac{b^2}{2}\left[J_0(q_{0n}b)N_0(q_{0n}b) + J_1(q_{0n}b)N_1(q_{0n}b)\right] - \frac{a^2}{2}\left[J_0(q_{0n}a)N_0(q_{0n}a) + J_1(q_{0n}a)N_1(q_{0n}a)\right],$$

$$I_4^{(0)} = \frac{b^2}{2}\left[J_1^2(q_{0n}b) - J_0(q_{0n}b)J_2(q_{0n}b)\right] - \frac{a^2}{2}\left[J_1^2(q_{0n}a) - J_0(q_{0n}a)J_2(q_{0n}a)\right], \quad (2.70)$$

$$I_5^{(0)} = \frac{b^2}{2}\left[N_1^2(q_{0n}b) - N_0(q_{0n}b)N_2(q_{0n}b)\right] - \frac{a^2}{2}\left[N_1^2(q_{0n}a) - N_0(q_{0n}a)N_2(q_{0n}a)\right],$$

$$I_6^{(0)} = \frac{b^2}{4}\left[2J_1(q_{0n}b)N_1(q_{0n}b) - J_2(q_{0n}b)N_0(q_{0n}b) - J_0(q_{0n}b)N_2(q_{0n}b)\right] - \frac{a^2}{4}\left[2J_1(q_{0n}a)N_1(q_{0n}a) - J_2(q_{0n}a)N_0(q_{0n}a) - J_0(q_{0n}a)N_2(q_{0n}a)\right].$$

The dispersion relation is found from Eq. (2.61) by substituting the values of q_{mn} for values of $m = 0,1,2,\ldots$ and $n = 0,1,2,\ldots$. Table 2.1 summarizes selected values of q_{mn} for the lowest modes in n and m.

2.2.3 Phonon Bottlenecks in Short CNTs

The quantization of the phonons in CNTs of finite length results in a phonon bottlenecks of importance in the use of finite length CNTs used as infrared detectors [24] and as transistors [31]. As described previously, short CNTs have discrete dispersion relations analogous to those of quantum dots. Moreover, the electronic bands in these CNTs are also discrete due to the dimensional confinement in three dimensions. The discreteness of both the phonon and carrier dispersion relations implies that it is difficult to satisfy both momentum and energy conservation in phonon emission and absorption processes. To analyze the severity of this expected phonon bottleneck, Raichura et al. [16, 27] have calculated the discrete phonon dispersion curves for a number of different CNTs and they have quantized the tight-binding electron energy dispersion curves of Pennington and Goldsman [32] along the CNT axis to take into account dimensional confinement for short CNTs. In particular, Raichura et al. [16, 27] have noted that the wavevector for quantized electronic states in the axial direction since the nanotube is short and is given by $k_z = n' \pi/l$ where n' labels the multiples of half-wavelength standing carrier modes in the axial direction [33]. For the given case of a (10, 0) nanotube, the circumferential unit cell width is $T_c = \pi(2a)/10 = 0.251$ nm and the unit cell in the axial direction is given by $T = 0.43$ nm. In particular for a (10, 0) nanotube that is ten unit cells in circumference and nine unit cells in length, the allowed values of k_z are $k_z = n' \pi/9T$ and the maximum value of k_z is π/T corresponding to $n' = 9$.

In this previous work, Raichura et al. [16, 27] examined different possible electronic transitions for intravalley intrasubband (1↔1, 2↔2 and 3↔3) and intravalley intersubband (1↔2, 2↔3 and 1↔3) transitions. In order to avoid a phonon bottleneck and for an electron to make any one of the possible transitions, there should be a phonon available that allows momentum conservation and whose energy is equal to the difference in the energy of the two transition bands. As described by Pennington and Goldsman [32], the subbranch quantum number for the phonons is $n = 0$ for intrasubband transition, $n = 1$ for intersubband (1↔2, 1↔3) and $n = 2$ for 2↔3 intersubband transition. The phonon energies for $n = 0, 1, 2$ are shown for the breathing modes, torsional modes, and the axial modes have been calculated by Raichura et al. [16, 27] in order to assess the degree of the phonon bottleneck; these authors considered a (10, 0) nanotube of length 4 nm. Raichura et al. [16, 27] found that for the intravalley intrasubband transition, there are no phonons available to assist the electron transition. Furthermore they found that for the intravalley intersubband 1↔2 case, allowing for a 10 % broadening of the transition energies, there is only one allowed transition involving an axial phonon with energy of 55 meV. Finally they found that for the 1↔3 intravalley intersubband case, there is a phonon (torsional)-assisted transition allowed at 39 meV. For the 2↔3 intravalley intersubband case, there are only two allowed transitions. One of those involves an axial phonon at 87 meV and the other, a torsional mode phonon at 65 meV.

From these results for short CNTs, it is evident that the electron scattering due to phonons is very limited due to the confinement-related phase space restriction of phonon energies. This is, of course, different from the case of infinite length nanotubes, where most of the transitions are allowed due to the continuous nature of the phonon energies. These phonon bottleneck effects are expected to lead to reduced scattering and the suppression of phonon-assisted transitions in infrared devices using nanotubes as well as the consequent enhancement of quasi ballistic transport in nanotube-based transistors.

2.2.4 Thermal Conductivity

The thermal conductivities of both CNTs and other graphene-based structures have considerable practical interests. In this section, the thermal conductivity of a CNT with metallic and contacts is considered using the Klemens model. In a subsequent section, these Klemens model results will be combined with continuum model results that predict dimensional confinement effects in few-larger graphite structures. Herein, we discuss the case of CNT in contact with metallic regions at each end [34]. By using the Klemens's model for thermal transport in graphite [35], Sun et al. [34] investigated the thermal transport properties of CNTs at high temperatures based on a Debye-type continuum model that assumes equipartition. From the graphite dispersion relations [36], it is observed that above 4 THz, lattice waves propagate in the basal plane and a two-dimensional (2D) phonon gas model is appropriate. As discussed by Klemens [35], the phonon spectral specific heat in two dimensions is

$$C_2(f) = \frac{4k_B f}{a^3 f_d^2}, \tag{2.71}$$

where a^3 is the volume of one molecular group of the solid, k_B is Boltzmann's constant, f is the phonon frequency, and $f_d = 4.6 \times 10^{13}$ Hz is the Debye frequency.

In CNTs, there are three dominant phonon scattering mechanisms: intrinsic, point-defect, and grain boundary scattering. As discussed by Klemens, the expression for the two-dimensional intrinsic mean free path is

$$l_i(f, T) = \frac{M v^3 f_d}{4\pi \gamma^2 k_B T f^2} \tag{2.72}$$

where v is the phonon velocity and γ is the Grüneisen parameter. In this discussion, $\gamma^2 = 4$ is adopted, based on Klemens' treatment [35]. Accordingly, to the thermal conductivity in the basal plane of graphite is thus given by

$$\lambda_i(T) = \frac{1}{2} \int_{f_c}^{f_d} C_2(f) v l_i(f, T) \, df, \tag{2.73}$$

where $f_c = 4$ THz, the low-frequency limit below which lattice waves propagate along the c-axis in three dimensions. As discussed by Klemens, the expression for the 2D thermal conductivity is given by

$$\lambda_i(T) = \frac{M v^4}{2\pi a^3 \gamma^2 T f_d} (\ln f_d - \ln f)_c. \tag{2.74}$$

Hence, as the temperature increases the intrinsic thermal conductivity decreases. This result also predicts thermal conductivities in the range of approximately

2,000 W/mK at 300 K and 1,000 W/mK at 600 K. In making these estimates, the group velocity is taken as $\upsilon = 1.86 \times 10^4$ m/s, which is obtained by averaging the longitudinal (2.36×10^4 m/s) and fast transverse (1.59×10^4 m/s) velocities from the phonon dispersion relation for the basal plane as calculated by Klemens (1994)

$$\frac{2}{\langle\upsilon\rangle^2} = \frac{1}{\langle\upsilon_{LA}\rangle^2} + \frac{1}{\langle\upsilon_{TA}\rangle^2}. \tag{2.75}$$

As discussed by Sun et al. [34], for the case of frequency below 4 THz, significant interplanar vibrations make a three-dimensional model necessary, which takes into account the LA and TA branches with velocities of 1,960 and 700 m/s, respectively, as calculated from the phonon dispersion relation. Accordingly, when this approach is applied to CNTs, it is necessary to make certain modifications. In the limit of a long CNT, the wave vectors along the tube axis are continuous and the dispersion relation is the same as that of graphene. However, as expected, in the perpendicular or the circumferential direction, the wave vectors are discrete. Moreover, at low temperatures it is important to consider the effects of the discreteness on the specific heat and the three-phonon interactions. However, at high temperatures where equipartition is assumed, these effects are unimportant. Thus, it is not essential to consider the effects of graphene sheet orientation with respect to the tube axis in a Debye continuum model. Indeed, the effects of graphene sheet orientation on the acoustic modes at high temperatures are small. For such a scenario, the lower cutoff frequency for CNTs is determined by the breathing modes that can decay anharmonically into two acoustic modes. As argued clearly by Klemens, the equivalent cutoff frequency is about $f_c = 3$ THz, below which the two-dimensional mean free path is greatly reduced.

As discussed by Klemens, in the 2D case, point defects scatter as the third power of frequency. For substitutional atoms which differ only in their masses from the carbon atoms by ΔM, the point defect mean free path is given as

$$\frac{1}{l_p(f)} = Af^3 = c\left(\frac{\Delta M}{M}\right)^2 4\pi^2 \frac{f^3}{\upsilon f_d^2}, \tag{2.76}$$

where A depends on properties and the number of defects relative to the number of carbon atoms, c, of the point defects. As is well known, point defects reduce the thermal conductivity relative to the intrinsic thermal conductivity. Indeed, it has been established that CNTs have low densities of point defects ($1/10^{12}$ atoms, [37]). Because the point defect concentration is very low along the length of the tubes, the point defect scattering events occur primarily as end effects for CNTs of submicron lengths and, accordingly, the point defect concentration is a function of the tube length.

To estimate the point defect density, consider unrolling the CNT. Perpendicular to the tube axis, a row is defined such that it has $2N$ atoms, where N is the number of circumferential atoms in the original CNT, assumed to be ten for specificity. Since for a tube of one row, its length is 0.5 d, where d is the length of the side of the hexagon equal to 0.142 nm and for n rows, the tube length is

$$l_n = 0.5d + (n-1) \times 1.5d, \tag{2.77}$$

the number of rows is thus

$$n = (1 + l_n/d)/1.5. \tag{2.78}$$

The total number of atoms in the CNT is $2Nn$. With one defect at each end of the tube the average effective point defect concentration for a CNT with defects at each end is

$$c = \frac{2}{2Nn} = \frac{1.5}{10(1+l_n/d)}. \tag{2.79}$$

Accordingly the CNT thermal conductivity with point defect end effects is thus given by

$$\begin{aligned}\lambda_p(T) &= \tfrac{1}{2}\int_{f_c}^{f_d} C_2(f)\upsilon l_t(f,T)\,df \\ &= \tfrac{2.34\times 10^5}{T}\{\ln(f_d/f_c) - \ln[(f_d+f_0)/(f_c+f_0)]\},\end{aligned} \tag{2.80}$$

where the total mean free path is given in terms of the intrinsic and point-defect-related mean free path by

$$\frac{1}{l_t} = \frac{1}{l_i} + \frac{1}{l_p}, \tag{2.81}$$

with l_p being point defect mean free path, and

$$f_0 = 2.556 \times 10^{-6} f_d T (1 + l_n/d)/1.35. \tag{2.82}$$

Based on this analysis, the thermal conductivity with end effects as a function of tube length may be estimated. First of all, as the tube length increases, the thermal conductivity increases and the end effects become less important as the intrinsic mean free path becomes dominant. For tube lengths of a few hundred nanometers, the thermal conductivity has values of approximately 2,000 W/mK at 300 K, 1,000 W/mK at 500 K, and 800 W/mK at 800 K; more detailed results are given by Sun et al. [34]. Sun et al. also consider the effects produced by grain boundary scattering at the ends of the CNT, the reader is referred to Sun et al. [34] for additional details.

2.3 Phonons in Graphene

The physical properties of graphene have been investigated extensively since it was first successfully made by Novoselov et al. [38]. A typical structure suitable for potential device applications is a graphene nanoribbon (GNR), which can be visualized as an unwrapped CNT. Rana et al. [39] have examined generation and recombination events intravalley and intervalley phonon scattering in graphene. Herein we review the work of Qian et al. [15] on phonon confinement effects in graphene.

This section highlights the results of Qian et al. [15] on the use of the elastic continuum model to derive analytical displacements and dispersion relations of optical phonons for a graphene sheet. In addition, these results provide expressions for the confined optical phonons for graphene are obtained as well as the optical deformation potential interactions for graphene.

2.3.1 Optical Phonons in Graphene Sheets

Consider a GNR with x and y axes in the plane of the graphene such that they are perpendicular and parallel to the "clamping" boundaries, respectively; the z-axis is out-of-plane and perpendicular to the graphene surface. The width of the GNR is L, the length is infinite, and the thickness is neglected. Using the elastic continuum model for the phonon, the GNR is treated as a 2D elastic sheet in the long wavelength limit. The unit length along the x-axis is $a\cos\theta$, where $a = \sqrt{3}a_{c-c} = 2.46$ Å is the length of the unit vector in terms of the distance between adjacent carbon atoms, a_{c-c}, and θ is the chiral angle from the unit vector a with respect to the x-axis. In a convenient formal analogy with CNTs, two kinds of specific GNRs are defined: armchair-end and zigzag-end GNRs, depending on the carbon atom arrangement on the free-standing sides, which is identical to the definition of the CNTs [40]. The armchair-end and zigzag-end GNRs correspond to $\theta = 30°$ and $0°$, respectively. Therefore, the unit lengths along the x-axis of armchair-end and zigzag-end GNRs are $\sqrt{3}a/2$ and a [15].

Goupalov (2005) has derived generalized mechanical equations describing the relative displacement of the two sublattices in graphene sheet for the long wavelength optical phonon modes. Moreover, Babiker [41] has formulated a method of treating optical modes based on the difference between the displacements of these sublattices. The equation describing this relative displacement is written as

$$\ddot{\mathbf{u}} = \widehat{\Lambda}^{\text{opt}} \mathbf{u}, \tag{2.83}$$

where $\widehat{\Lambda}^{\text{opt}}$ is the operator for optical phonons given as

$$-\widehat{\Lambda}^{\text{opt}}_{\text{2DG}} = \omega^2_{\text{TO}} + \tfrac{\beta^2_T - \beta^2_L}{2} J^2_z \nabla^2_\perp + \tfrac{\beta^2_T + \beta^2_L}{2} \left(\nabla^2_- J^2_+ + \nabla^2_+ J^2_- \right)$$
$$+ \tfrac{\lambda^2}{2} \left(\nabla^2_- J^2_+ + \nabla^2_+ J^2_- \right) \nabla^2_\perp - \tfrac{\lambda^2}{2} J^2_z \nabla^4_\perp - \beta^2_z \left(J^2_z - 1 \right) \nabla^2_\perp \quad (2.84)$$
$$+ \left(\omega^2_{\text{TO}} - \omega^2_{\text{ZO}} \right) \left(J^2_z - 1 \right)$$

where $J_\pm = \mp 1/\sqrt{2}\left(J_x \pm i J_y \right)$, $J_x = \begin{pmatrix} 0 & 0 & 0 \\ 0 & 0 & -i \\ 0 & i & 0 \end{pmatrix}$, $J_y = \begin{pmatrix} 0 & 0 & i \\ 0 & 0 & 0 \\ -i & 0 & 0 \end{pmatrix}$, $J_z = \begin{pmatrix} 0 & -i & 0 \\ i & 0 & 0 \\ 0 & 0 & 0 \end{pmatrix}$ are the matrices of projections of the spin operator $J = 1$ in the Cartesian basis. The parameters ω_{TO}, β_T, β_L, λ are approximated by fitting the dispersion curves of TO and LO phonons in 2D graphite in the Γ–M direction from Maultzsch et al. [42], while the ω_{ZO} and β_Z are by ZO results from Wirtz and Rubio [43]. For the in-plane TO and LO modes and the out-of-plane ZO mode, the phonon dispersions are fitted by

$$\omega^2(k) = \begin{cases} \omega^2_{\text{TO}} - \beta^2_T k^2 & \text{TO} \\ \omega^2_{\text{LO}} - \lambda^2 k^4 + \beta^2_L k^2 & \text{LO} \\ \omega^2_{\text{ZO}} - \beta^2_Z k^2 & \text{ZO}, \end{cases} \quad (2.85)$$

For non-polar graphene, $\omega_{\text{TO}} = \omega_{\text{LO}} = 1{,}581$ cm^{-1}, $\omega_{\text{ZO}} = 893$ cm^{-1} are the frequencies of TO, LO, and ZO modes at the center, Γ point, of reciprocal lattices of the graphene; $\beta_T = 9.6 \times 10^5$ cm/s, $\beta_L = 7.8 \times 10^5$ cm/s, $\beta_Z = 7.9 \times 10^5$ cm/s, and $\lambda = 9.3 \times 10^5$ cm^2/s.

Explicit expressions are obtained for the two in-plane and single out-of-plane displacements upon applying the above operator to the relative displacements:

$$\begin{cases} -\ddot{u} = \left[\omega^2_{\text{TO}} - \beta^2_L \nabla^2_x + \beta^2_T \nabla^2_y - \lambda^2 \nabla^2_x \left(\nabla^2_x + \nabla^2_y \right) \right] u \\ \quad + \left[-\left(\beta^2_L + \beta^2_T \right) \nabla_x \nabla_y - \lambda^2 \nabla_x \nabla_y \left(\nabla^2_x + \nabla^2_y \right) \right] \upsilon \\ -\ddot{\upsilon} = \left[\omega^2_{\text{TO}} - \beta^2_L \nabla^2_y + \beta^2_T \nabla^2_x - \lambda^2 \nabla^2_y \left(\nabla^2_x + \nabla^2_y \right) \right] \upsilon \quad (2.86) \\ \quad + \left[-\left(\beta^2_L + \beta^2_T \right) \nabla_x \nabla_y - \lambda^2 \nabla_x \nabla_y \left(\nabla^2_x + \nabla^2_y \right) \right] u \\ -\ddot{w} = \left[\omega^2_{\text{ZO}} + \beta^2_z \left(\nabla^2_x + \nabla^2_y \right) \right] w. \end{cases}$$

Thus, it is seen that the in-plane vibrations, u and υ, and out-of-plane vibration, w, are decoupled, while the in-plane vibrations, u and υ, are coupled each other. Considering solutions in the form of traveling waves,

$$\mathbf{u}(x, y, t) = \mathbf{A} \exp\left[i\left(q_x x + q_y y - \omega t\right)\right], \tag{2.87}$$

it follows that the displacement equations can be written as

$$\begin{cases} \omega^2 u = \left[\omega_{TO}^2 + \beta_L^2 q_x^2 - \beta_T^2 q_y^2 - \lambda^2 q_x^2 \left(q_x^2 + q_y^2\right)\right] u \\ \qquad + \left[\left(\beta_L^2 + \beta_T^2\right) q_x q_y - \lambda^2 q_x q_y \left(q_x^2 + q_y^2\right)\right] v \\ \omega^2 v = \left[\left(\beta_L^2 + \beta_T^2\right) q_x q_y - \lambda^2 q_x q_y \left(q_x^2 + q_y^2\right)\right] u \\ \qquad + \left[\omega_{TO}^2 + \beta_L^2 q_y^2 - \beta_T^2 q_x^2 - \lambda^2 q_y^2 \left(q_x^2 + q_y^2\right)\right] v \\ \omega^2 w = \left[\omega_{ZO}^2 - \beta_Z^2 \left(q_x^2 + q_y^2\right)\right] w. \end{cases} \tag{2.88}$$

The dispersion relationships for these modes are found to be those of Goupalov and they have the form

$$\omega^2(q) = \begin{cases} \omega_{TO}^2 - \beta_T^2 \left(q_x^2 + q_y^2\right) \\ \omega_{TO}^2 - \lambda^2 \left(q_x^2 + q_y^2\right)^2 + \beta_L^2 \left(q_x^2 + q_y^2\right) \\ \omega_{ZO}^2 - \beta_Z^2 \left(q_x^2 + q_y^2\right). \end{cases} \tag{2.89}$$

Accordingly, for phonon propagation in these graphene sheets, the continuum model has the same dispersion curves for arbitrary propagation directions.

2.3.2 Confined Optical Phonons in Clamped GNRs

In this treatment, the displacements at boundaries are taken to be zero, in an approximation commonly described as "clamped" boundary conditions; that is,

$$u = 0 \quad \text{at } x = 0 \text{ and } L. \tag{2.90}$$

These boundary conditions result in confined modes and they imply that $q_x = q_n = n\pi/L$, and let $q_y = q$. Accordingly the displacements are given by

$$\mathbf{u}(x, y, t) = \mathbf{A} \exp\left[i\left(q_n x + q y - \omega t\right)\right]. \tag{2.91}$$

The displacement along the x-axis, u, is confined between the two boundaries, u is given by the standing wave equation

$$u = u_0 \sin(q_n x). \tag{2.92}$$

The dispersion relations with confined wavevectors follow immediately from the above dispersion relations with $q_x = q_n$:

$$\omega_n^2 = \begin{cases} \omega_{TO}^2 - \beta_T^2 \left(q_n^2 + q^2\right) \\ \omega_{TO}^2 - \lambda^2 \left(q_n^2 + q^2\right)^2 + \beta_L^2 \left(q_n^2 + q^2\right) \\ \omega_{ZO}^2 - \beta_Z^2 \left(q_n^2 + q^2\right). \end{cases} \quad (2.93)$$

As illustrative examples armchair-end and zigzag-end GNRs with N periods are analyzed herein. For the armchair-end GNR, the widths are $N\sqrt{3}a/2$ when N is even, and $(N + 1/3)\sqrt{3}a/2$ when N is odd. In the case of the zigzag-end GNR, the width is Na for all N. As expected from the continuum model these are dispersion curves for optical phonon modes with quantized number n from 0 to N for q_x, since the smallest quantized wave vector should be larger than the unit length along the x-axis.

Qian et al. [15] have considered the dispersion curves of LO, TO, and ZO vibration modes for armchair-end GNR of ten periods. It is found that the larger the quantized number n, the smaller the vibration frequencies of the phonon modes for TO and ZO modes over the whole wave vector domain. At the zone-center, when n equals four, there exists a maximum frequency 1,590.9 cm^{-1}. When n is smaller than four, the LO frequencies increase first and then decrease, while, for n is larger than four, the LO frequencies decrease directly.

Qian et al. [15] have also considered the dispersion curves for LO, TO, and ZO vibration modes for ten units zigzag-end GNR. The dispersion curves manifest the same properties as for the armchair case; there is a critical quantized number existing for LO modes, 1,590.9 cm^{-1} when n equals five.

By substituting the dispersion relation $\omega_n^2 = \omega_{TO}^2 - \beta_T^2(q_n^2 + q^2)$ of the TO mode in the first displacement equation of Eq. (2.88), it follows that there is a relationship between u and υ:

$$\upsilon/u = -q_n/q. \quad (2.94)$$

Taking the clamping boundaries along the x-axis, the displacements equations for u and υ are

$$\begin{cases} u = u_0 \sin(q_n x) \\ \upsilon = -(q/q_n) u_0 \sin(q_n x). \end{cases} \quad (2.95)$$

As discussed previously in this review, the mode amplitudes are given by the condition that the energy in each mode is $\hbar\omega_n$; that is,

$$\frac{1}{S}\int_s (u \cdot u* + \upsilon \cdot \upsilon*)\,dx\,dy = \frac{\hbar}{M\omega_n}. \quad (2.96)$$

Using this normalization condition, the quantized displacement amplitudes are

$$u_0 = \left(\frac{2\hbar}{M\omega_n} \frac{q^2}{q_n^2 + q^2}\right)^{1/2} \quad \text{and} \quad v_0 = -\left(\frac{2\hbar}{M\omega_n} \frac{q_n^2}{q_n^2 + q^2}\right)^{1/2}. \tag{2.97}$$

In a similar manner for the LO modes, by substituting $\omega^2 = \omega_{TO}^2 - \lambda^2(q_n^2 + q^2)^2 + \beta_L^2(q_n^2 + q^2)$ of the LO mode in the first displacement equation of Eq. (2.88), it follows that the relationship between u and v is

$$v/u = -q/q_n, \tag{2.98}$$

and the quantized displacement amplitudes are

$$u_0 = \left(\frac{2\hbar}{M\omega_n} \frac{q_n^2}{q_n^2 + q^2}\right)^{1/2} \quad \text{and} \quad v_0 = -\left(\frac{2\hbar}{M\omega_n} \frac{q^2}{q_n^2 + q^2}\right)^{1/2}. \tag{2.99}$$

2.3.3 Deformation Potential in GNRs

As discussed previously in this review, the optical deformation potential Hamiltonian describing carrier optical-phonon scattering can be determined from those normalized continuum modes. Replacing **u** by u and v and substituting into Eq. (2.36), one can obtain

$$H_{\text{opt-def}} = |D_{\text{op}}|(u + v), \tag{2.100}$$

where D_{op} is the optical-phonon deformation potential constant as equal to 8.89 eV for graphene [44]. Accordingly the deformation potential Hamiltonian for the TO phonon is given by

$$H_{\text{opt-def}} = |D_{\text{op}}|\left(\frac{2\hbar}{M\omega_n}\right)^{\frac{1}{2}} \frac{q - q_n}{\sqrt{q_n^2 + q^2}} \sin(q_n x), \tag{2.101}$$

while the deformation potential Hamiltonian for the LO phonon is given by

$$H_{\text{opt-def}} = |D_{\text{op}}|\left(\frac{2\hbar}{M\omega_n}\right)^{\frac{1}{2}} \frac{q_n - q}{\sqrt{q_n^2 + q^2}} \sin(q_n x). \tag{2.102}$$

As is well known, the carrier–phonon interaction described by this optical deformation potential Hamiltonian is caused by the interaction of a charge carrier in a medium where the energy band structure is modulated alternately to higher and lower energies. For a more in-depth treatment of the optical deformation potential, and the related acoustic deformation potential, see Stroscio and Dutta [1].

2.4 Phonons in Graphite

In this section, we discuss the c-axis thermal conductivity of thin graphite layers with emphasis on the fact that taking into account phonon confinement [45] results in substantial changes in the c-axis thermal conductivity. As described by Klemens [35], the phonon specific heat in one dimension is

$$C_1(f) = \frac{k_B N}{2\pi f_d} = \frac{k_B}{2\pi f_d} G_1^2 G_3, \qquad (2.103)$$

in which $G_3 = 1/a_3$ is the number of layers in a crystal of unit thickness and unit width, each layer having $G_1^2 = 1/a_2^2$ atoms per unit area, and the number of phonon states $N = G_1^2 G_3$. In this formulation, the one-dimensional (1D) specific heat is not a function of the phonon frequency. It is constant for the c-axis transport. Therefore, 1D intrinsic thermal conductivity is given by

$$\lambda_i = \int_{f_{\text{lower}}}^{f_{\text{upper}}} C_1 \upsilon l_i \, df, \qquad (2.104)$$

where f_{lower} and f_{upper} are determined by the c-axis phonon dispersion relation. In this discussion, the expression used for the mean free path is that derived by Klemens (1994) for the anharmonic relaxation process. In the work of Sun et al. [45], it has been evaluated for the velocity along the c-axis. As commented by Sun et al., the Grüneisen constant for the 1D mean free path is, to the best of our knowledge, not known with precision and, accordingly, the same Grüneisen constant as for the basal plane is assumed. There is, however, some evidence that the c-axis and basal plane Grüneisen constants are within a factor of 2 of each other as indicated by Sun et al. [45]. For film thickness decreases below 10 nm, phonon quantization is expected to occur resulting in discrete dispersion relations. Here we take $f_{\text{upper}} = 2.75$ THz and $f_{\text{lower}} = f_{\text{upper}}/L$, where L is the number of atoms in the one-dimensional chain of carbon atoms. The frequency range over which the thermal conductivity is integrated is about 20 times smaller than that used for the 2D case. Thus, the expression for 1D thermal conductivity is

$$\lambda_i(T, L) = \frac{\rho \upsilon^4}{8\pi^2 \gamma^2 T} \left(\frac{1}{f_{\text{upper}}/L} - \frac{1}{f_{\text{upper}}} \right). \qquad (2.105)$$

For this situation, instead of a logarithmic function, as for the 2D case, an inverse relation is obtained for c-axis thermal conductivity and frequency. Equation (2.105) predicts a pronounced dependence on the thickness of the graphite lager as a consequence of phonon confinement.

Using this formulation, Sun et al. [45] have calculated the ratio of the 2D thermal conductivity to the 1D thermal conductivity. These authors found that the intrinsic thermal conductivity is about four orders of magnitude smaller in one dimension than in two dimensions. This finding is consistent with the following scaling argument. Indeed, since the velocity in one dimension is about 20 times smaller than that in two dimensions and since the mean free path varies with the third power of the velocity, it is not unreasonable that the thermal conductivity is four orders of magnitude smaller in 1D than in 2D.

Based on this formulation Sun et al. [45] perform a numerical analysis of the 2D thermal conductivity to 1D thermal conductivity in c-axis graphite; over a wide range of temperatures (300–1,000 K), they find that ratio is approximately 1.4×10^5, 6×10^4, and 3×10^4 for graphite with a thickness of three, six, and ten atoms, respectively.

These results reveal a dramatic effect caused by phonon dimensional confinement that enters through the lower limit of the thermal conductivity integral.

2.5 Phonons in Wurtzite Structures

As has been discussed by Stroscio and Dutta [1], the treatment of the phonons in wurtzite structures is more complicated than for the zincblendes, but it is still possible to obtain a description based on the continuum approach. As is well known, there are four atoms per unit cell for wurtzites and they have lower symmetry than zincblende crystals. Furthermore there are nine optical and three acoustic modes associated with wurtzite crystals. Due to the uniaxial character of wurtzite crystals and the associated optical anisotropy, the long wavelength lattice vibrations may be classified in terms of orientation with respect to the c-axis, the phonon wave vector **q**, the electric field **E**, and the polarization **P** [13] Loudon's treatment divides the lattice vibrations into two groups of phonons: ordinary and extraordinary. In the case of the dispersionless ordinary phonons, **E** and **P** are both perpendicular to **q** and the c-axis. In the case of the extraordinary phonons, the phonon frequencies are dependent upon the angle between the phonon wave vector and the c-axis.

This section reviews the basic properties of phonon modes in bulk wurtzite structures, wurtzite quantum wells (QWs), and wurtzite quantum dots (QDs), respectively. Moreover, phonon frequencies and electron–phonon interaction Hamiltonian of both bulk wurtzite structures and wurtzite QWs are presented. In addition, the formal analytical solutions for interface modes in wurtzite QDs are obtained.

2.5.1 *Phonons in Bulk Wurtzite Structures*

Within the context of Loudon's model [1], after some algebraic manipulation, the phonon frequency for ordinary phonons has a trivial solution $\omega = \omega_\perp$ and $\mathbf{E}(\mathbf{r}) = 0$. The phonon frequencies for extraordinary phonons satisfy [14, 46]

$$\varepsilon_\perp(\omega)\sin^2(\theta) + \varepsilon_z(\omega)\cos^2(\theta) = 0, \qquad (2.106)$$

such that the direction-related dielectric constants $\varepsilon_\perp(\omega)$ and $\varepsilon_z(\omega)$, are given by

$$\varepsilon_\perp(\omega) = \varepsilon_\perp^\infty \frac{\omega^2 - \omega_{LO,\perp}^2}{\omega^2 - \omega_\perp^2}, \qquad (2.107)$$

$$\varepsilon_z(\omega) = \varepsilon_z^\infty \frac{\omega^2 - \omega_{LO,z}^2}{\omega^2 - \omega_z^2}, \qquad (2.108)$$

when $|\omega_{LO,\perp} - \omega_{LO,z}|$, $|\omega_\perp - \omega_z| << |\omega_{LO,\perp} - \omega_\perp|$, $|\omega_{LO,z} - \omega_z|$, which is the case for the wurtzite-based GaN, ZnO, CdS, and CdSe materials, the solutions become

$$\omega_{LO}^2 = \omega_{LO,z}^2 \cos^2(\theta) + \omega_{LO,\perp}^2 \sin^2(\theta), \qquad (2.109)$$

$$\omega_{TO,z}^2 = \omega_z^2 \sin^2(\theta) + \omega_\perp^2 \cos^2(\theta). \qquad (2.110)$$

These are predominantly longitudinal and transverse modes, respectively. Herein, the phonon frequencies of GaN will be based on the experimental results in Hellwege et al. [47].

From these results Chen et al. [14, 46] showed that the LO-like mode frequency is nearly constant as a function of θ, indicating that the mode is weakly dependent upon the direction. In contrast, the TO-like mode exhibits weak anisotropy.

As has been discussed previously in this review, the phonon modes must be normalized so that the energy in each mode is $\hbar\omega$. As indicated by Eq. (2.31), the normalized electron–optical–phonon Hamiltonian for the bulk uniaxial material is given as

$$H = \sum_q \left[\frac{4\pi e^2 \hbar V^{-1}}{\frac{\partial}{\partial \omega}\left[\varepsilon_\perp(\omega)\sin^2(\theta) + \varepsilon_z(\omega)\cos^2(\theta)\right]} \right]^{1/2} \frac{1}{q}\left(a_q e^{i\mathbf{q}\cdot\mathbf{r}} + a_q^\dagger e^{-i\mathbf{q}\cdot\mathbf{r}}\right)$$

$$= \sum_q \sqrt{\frac{2\pi e^2 \hbar}{V\omega}} \frac{1}{q}\left(a_q e^{i\mathbf{q}\cdot\mathbf{r}} + a_q^\dagger e^{-i\mathbf{q}\cdot\mathbf{r}}\right)$$

$$\times \frac{(\omega_\perp^2 - \omega^2)(\omega_z^2 - \omega^2)}{\left[(\varepsilon_\perp^0 - \varepsilon_\perp^\infty)\omega_\perp^2(\omega_z^2-\omega^2)^2\sin^2(\theta) + (\varepsilon_z^0-\varepsilon_z^\infty)\omega_z^2(\omega_\perp^2-\omega^2)^2\cos^2(\theta)\right]^{1/2}}.$$

$$(2.111)$$

2 Phonons in Bulk and Low-Dimensional Systems

Moreover, using the Fermi golden rule, the transition probability from electron state $\mathbf{k}$ to $\mathbf{k}'$ per unit time, $W(\mathbf{k}, \mathbf{k}')$, is given by

$$W(\mathbf{k}, \mathbf{k}') = \frac{2\pi}{\hbar}|M_\mathbf{q}|^2 \delta\left(E_{\mathbf{k}'} - E_\mathbf{k} \pm \hbar\omega_\mathbf{q}\right), \tag{2.112}$$

where $E_\mathbf{k}$ is the initial electron energy and $E_{\mathbf{k}'}$ is the final electron energy. $\mathbf{q} = \mathbf{k} - \mathbf{k}'$ is a transferred momentum and $\hbar\omega_\mathbf{q}$ is transition energy, with the upper sign "+" (the lower sign "−") corresponding to phonon emission (absorption). As usual the transition matrix element, $M_\mathbf{q}$, may be written as

$$|M_\mathbf{q}|^2 = \langle \mathbf{k}' | H | \mathbf{k} \rangle, \tag{2.113}$$

such that the electron states are plane-wave states normalized in volume V,

$$\begin{aligned}|\mathbf{k}\rangle &= e^{i\mathbf{k}\cdot\mathbf{r}}/\sqrt{V} \\ \langle \mathbf{k}'| &= e^{-i\mathbf{k}'\cdot\mathbf{r}}/\sqrt{V}.\end{aligned} \tag{2.114}$$

As described in Chen et al. [14, 46],

$$|M_\mathbf{q}|^2 = \frac{2\pi e^2 \hbar}{V\omega}\frac{1}{q^2}\left(n_{\mathrm{ph}} + \tfrac{1}{2} \pm \tfrac{1}{2}\right) \frac{(\omega_\perp^2 - \omega^2)^2(\omega_z^2 - \omega^2)^2}{(\varepsilon_\perp^0 - \varepsilon_\perp^\infty)\omega_\perp^2(\omega_z^2 - \omega^2)^2 \sin^2\theta + (\varepsilon_z^0 - \varepsilon_z^\infty)\omega_z^2(\omega_\perp^2 - \omega^2)^2 \cos^2\theta}, \tag{2.115}$$

where $n_{\mathrm{ph}} = [\exp(\hbar\omega/k_B T) - 1]^{-1}$ is the phonon occupation number. Based on this equation, both the LO- and TO-like modes exhibit anisotropy. Moreover, the anisotropy for the TO-like mode is very pronounced, but smaller than those for the LO-like mode.

Following the standard procedures of Chen et al. [14, 46], the scattering rate, $W(\mathbf{k})$, may be written as

$$W(\mathbf{k}) = \sum_{\mathbf{k}'} W\left(\mathbf{k}, \mathbf{k}'\right)$$

$$= \frac{e^2\sqrt{m}}{4\sqrt{2\pi^2}\hbar}\int_0^{2\pi}\int_0^\pi d\theta\, d\varphi \frac{1}{\omega}\left(n_{\mathrm{ph}} + \frac{1}{2} \pm \frac{1}{2}\right)\sin\theta \frac{\sigma}{\sqrt{E_\mathbf{k}\cos^2(\mathbf{k}\cdot\mathbf{q}) \mp \hbar!}} \tag{2.116}$$

$$\frac{(\omega_\perp^2 - \omega^2)^2(\omega_z^2 - \omega^2)^2}{(\varepsilon_\perp^0 - \varepsilon_\perp^\infty)\omega_\perp^2(\omega_z^2 - \omega^2)^2\sin^2\theta + (\varepsilon_z^0 - \varepsilon_z^\infty)\omega_z^2(\omega_\perp^2 - \omega^2)^2\cos^2\theta},$$

with $\cos(\mathbf{k}\cdot\mathbf{q}) = \sin\theta\sin\theta_k\cos\varphi + \cos\theta\cos\theta_k$, such that θ_k represents the angle between the initial electron wave vector k and the c-axis. In the case of phonon emission, σ is a step function given by

$$\sigma = \begin{cases} 0 & \text{for } \cos(\mathbf{k} \cdot \mathbf{q}) < \sqrt{\hbar! / E_\mathbf{k}} \\ 2 & \text{otherwise,} \end{cases} \quad (2.117)$$

and for the case of phonon absorption, $\sigma \equiv 1$.

The total scattering rates for LO-like and TO-like phonons for bulk GaN were evaluated numerically in by Chen et al. [14, 46]. The scattering rates as a function of the incident angle of the electron with respect to the c-axis, θ_k, with an initial electron energy of 0.3 eV for GaN, vary little with θ_k for LO-like phonon emission, LO-like phonon absorption, and TO-like phonon absorption, which the TO-like phonon emission rate exhibits variations of a factor of 2 or 3 over the full range of θ_k.

These authors found that the emission of LO-like phonons have the highest scattering rate being of the order of 10^{13}–10^{14} s^{-1}, while the absorption of TO-like phonons has the lowest scattering rate being of the order of 10^{10}–10^{11} s^{-1}. LO-like mode absorption rates were found to be in the range of 10^{12}–10^{13} s^{-1} and TO-like mode emission rates were found to be smaller by about one order of magnitude.

2.5.2 Phonons in Wurtzite Quantum Wells

As an illustrative example of phonons in wurtzite QWs, this section will review the results of Chen et al. [46] for the GaN/ZnO material system. Following these authors, we consider a double-heterointerface system where material region one occupies $|z| < \pm d/2$, and material region two occupies both $z < -d/2$ and $z > d/2$, where d is the thickness. As has been discussed previously [1], in such a double-heterointerface system, there are four distinct classes of optical-phonon modes; these are the interface (IF), confined, half-space, and propagating modes.

As a result of translational symmetry perpendicular to the z-axis, the Fröhlich potential may be written as

$$\Phi(r) = \sum_q \Phi(\mathbf{q}, z) \exp(i \mathbf{q} \cdot \boldsymbol{\rho}), \quad (2.118)$$

where $\mathbf{q} = (q_x, q_y)$, $\boldsymbol{\rho} = (x,y)$, and $\Phi(\mathbf{q},z)$ is the electron–optical–phonon interaction potential. As is standard for the dielectric continuum model, the boundary conditions at the interfaces require that the tangential components of $\mathbf{E}$, E_t, and D_z be continuous at $z = \pm d/2$. Moreover, the normalization condition is given by

$$\int \sqrt{n\mu} \mathbf{u}^*(\mathbf{q}, z) \cdot \sqrt{n\mu} \mathbf{u}(\mathbf{q}, z) dz = \frac{\hbar}{2\omega L^2}. \quad (2.119)$$

In addition, the electron–optical–phonon Hamiltonian may be written

$$H = \sum_q -e\Phi(\mathbf{q}, z)\left(a_\mathbf{q} e^{i\mathbf{q}\cdot\mathbf{r}} + a_\mathbf{q}^\dagger e^{-i\mathbf{q}\cdot\mathbf{r}}\right). \tag{2.120}$$

As for zincblende heterostructures, in these wurtzite structures, there are also symmetric and antisymmetric IF modes. The solutions can be found for the dispersion relation of the symmetric case [11]

$$\sqrt{\varepsilon_{1\perp}\varepsilon_{1z}} \tanh\left(\sqrt{\varepsilon_{1\perp}\varepsilon_{1z}}qd/2\right) - \sqrt{\varepsilon_{2\perp}\varepsilon_{2z}} = 0 \quad \text{with } \varepsilon_{1z}\varepsilon_{2z} < 0, \tag{2.121}$$

and of the antisymmetric case

$$\sqrt{\varepsilon_{1\perp}\varepsilon_{1z}} \coth\left(\sqrt{\varepsilon_{1\perp}\varepsilon_{1z}}qd/2\right) - \sqrt{\varepsilon_{2\perp}\varepsilon_{2z}} = 0 \quad \text{with } \varepsilon_{1z}\varepsilon_{2z} < 0. \tag{2.122}$$

In the case of the symmetric mode, the interaction Hamiltonian is given as

$$H_{IF}^S = \sum_q \sqrt{\frac{4\pi e^2 \hbar L^{-2}}{\frac{\partial}{\partial \omega}\left(\sqrt{\varepsilon_{1\perp}\varepsilon_{1z}} \tanh\left(\sqrt{\varepsilon_{1\perp}\varepsilon_{1z}}qd/2\right) - \sqrt{\varepsilon_{2\perp}\varepsilon_{2z}}\right)}} \frac{1}{\sqrt{2q}}$$

$$\times \left(a_\mathbf{q} e^{i\mathbf{q}\cdot\boldsymbol{\rho}} + a_\mathbf{q}^\dagger e^{-i\mathbf{q}\cdot\boldsymbol{\rho}}\right) \times \begin{cases} \cosh\left(\sqrt{\varepsilon_{1\perp}\varepsilon_{1z}}qz\right)/\cosh\left(\sqrt{\varepsilon_{1\perp}\varepsilon_{1z}}qd\right) & |z| < d/2 \\ \exp\left[-\sqrt{\varepsilon_{2\perp}\varepsilon_{2z}}q(|z|-d/2)\right] & |z| > d/2. \end{cases} \tag{2.123}$$

where the frequency ω is determined from the dispersion relationship. In the case of the antisymmetric mode, the interaction Hamiltonian is given by

$$H_{IF}^A = \sum_q \sqrt{\frac{4\pi e^2 \hbar L^{-2}}{\frac{\partial}{\partial \omega}\left(\sqrt{\varepsilon_{1\perp}\varepsilon_{1z}} \coth\left(\sqrt{\varepsilon_{1\perp}\varepsilon_{1z}}qd/2\right) - \sqrt{\varepsilon_{2\perp}\varepsilon_{2z}}\right)}} \frac{1}{\sqrt{2q}}$$

$$\times \left(a_\mathbf{q} e^{i\mathbf{q}\cdot\boldsymbol{\rho}} + a_\mathbf{q}^\dagger e^{-i\mathbf{q}\cdot\boldsymbol{\rho}}\right) \times \begin{cases} \sinh\left(\sqrt{\varepsilon_{1\perp}\varepsilon_{1z}}qz\right)/\sinh\left(\sqrt{\varepsilon_{1\perp}\varepsilon_{1z}}qd\right) & |z| < d/2 \\ \text{sgn}(z)\exp\left[-\sqrt{\varepsilon_{2\perp}\varepsilon_{2z}}q(|z|-d/2)\right] & |z| > d/2 \end{cases} \tag{2.124}$$

Based on these results, the zone center frequencies of the high-frequency symmetric mode, low-frequency symmetric mode, high-frequency antisymmetric mode, and low frequency antisymmetric mode are 705, 422, 690, and 446 cm^{-1}, respectively. In GaN/ZnO/GaN structures, Chen et al. [46] found that the zone-center frequencies of the high frequency symmetric mode, low frequency symmetric mode, high frequency antisymmetric mode, and low frequency antisymmetric mode are 692, 440, 703, and 430 cm^{-1}, respectively.

As a result of the direction-dependent dielectric constant, the confined phonon modes for wurtzite structures are more difficult to calculate than for zincblende structures. As discussed by Komirenko et al. [48], for symmetric modes (even modes), the dispersion relation becomes

$$Q_m^S = \frac{2\left\{m\pi + \xi \arctan\left[\sqrt{|\varepsilon_{2z}(\omega)\varepsilon_{2\perp}(\omega)|}/\sqrt{|\varepsilon_{1z}(\omega)\varepsilon_{1\perp}(\omega)|}\right]\right\}}{\sqrt{|\varepsilon_{1\perp}(\omega)/\varepsilon_{1z}(\omega)|}}, \quad (2.125)$$

while for antisymmetric modes (odd modes)

$$Q_m^A = \frac{2\left\{m\pi - \xi \arctan\left[\sqrt{|\varepsilon_{1z}(\omega)\varepsilon_{1\perp}(\omega)|}/\sqrt{|\varepsilon_{2z}(\omega)\varepsilon_{2\perp}(\omega)|}\right]\right\}}{\sqrt{|\varepsilon_{1\perp}(\omega)/\varepsilon_{1z}(\omega)|}}. \quad (2.126)$$

where $\xi = \text{sgn}[\varepsilon_{1z}(\omega)\varepsilon_{2z}(\omega)]$. Accordingly, for the symmetric mode, the interaction Hamiltonian is given by

$$H_C^S = \sum_q \sum_m \sqrt{\frac{4\pi e^2 \hbar L^{-2}}{\frac{\partial}{\partial \omega}\left[(\varepsilon_{1\perp}q^2 + \varepsilon_{1z}k_{1m}^2)d/2 - 2qf_S(\omega)\cos(k_{1m}d/2)\right]}} \left(a_q e^{i\mathbf{q}\cdot\boldsymbol{\rho}} + a_q^\dagger e^{-i\mathbf{q}\cdot\boldsymbol{\rho}}\right)$$
$$\times \begin{cases} \cos(k_{1m}d/z) & |z| < d/2 \\ \cos(k_{1m}d/2)\exp\left[-\sqrt{\varepsilon_{2\perp}/\varepsilon_{2z}}q(|z|-d/2)\right] & |z| > d/2, \end{cases} \quad (2.127)$$

where $f_S(\omega) = \text{sgn}(\varepsilon_{1z})\sqrt{\varepsilon_{1\perp}\varepsilon_{1z}}\sin(k_{1m}d/2) - \text{sgn}(\varepsilon_{2z})\sqrt{\varepsilon_{2\perp}\varepsilon_{2z}}\cos(k_{1m}d/2)$. k_{1m} is determined from $\varepsilon_{1z}k_{1m}\sin(k_{1m}d/2) - \varepsilon_{2z}\sqrt{\varepsilon_{2\perp}/\varepsilon_{2z}}q\cos(k_{1m}d/2) = 0$ with $2m\pi/d < k_{1m} < (2m+1)\pi/d$, and the phonon frequencies may be determined from Eqs. (2.129) and (2.130). Likewise, for the antisymmetric mode

$$H_C^A = \sum_q \sum_m \sqrt{\frac{4\pi e^2 \hbar L^{-2}}{\frac{\partial}{\partial \omega}\left[(\varepsilon_{1\perp}q^2 + \varepsilon_{1z}k_{1m}^2)d/2 - 2qf_A(\omega)\sin(k_{1m}d/2)\right]}} \left(a_q e^{i\mathbf{q}\cdot\boldsymbol{\rho}} + a_q^\dagger e^{-i\mathbf{q}\cdot\boldsymbol{\rho}}\right)$$
$$\times \begin{cases} \sin(k_{1m}d/z) & |z| < d/2 \\ \text{sgn}(z)\sin(k_{1m}d/2)\exp\left[-\sqrt{\varepsilon_{2\perp}/\varepsilon_{2z}}q(|z|-d/2)\right] & |z| > d/2, \end{cases} \quad (2.128)$$

where $f_A(\omega) = \text{sgn}(\varepsilon_{1z})\sqrt{\varepsilon_{1\perp}\varepsilon_{1z}}\cos(k_{1m}d/2) - \text{sgn}(\varepsilon_{2z})\sqrt{\varepsilon_{2\perp}\varepsilon_{2z}}\sin(k_{1m}d/2)$, k_{1m} is determined from $\varepsilon_{1z}k_{1m}\cos(k_{1m}d/2) - \varepsilon_{2z}\sqrt{\varepsilon_{2\perp}/\varepsilon_{2z}}q\sin(k_{1m}d/2) = 0$ with $(2m-1)\pi/d < k_{1m} < (2m+1)\pi/d$.

2.5.3 Phonons in Wurtzite Quantum Dots

In the case of polar wurtzite QDs, understanding the role of the confined (LO) and the surface optical (SO) phonon modes is essential [49, 50]. From Eq. (2.30), we obtain

$$\varepsilon \Delta \phi = 0, \quad (2.129)$$

where ϕ is the potential associated with the Fröhlich interaction discussed previously. The equation implies that $\varepsilon = 0$ or that $\Delta\phi = 0$. In the first case, $\varepsilon = 0$, which corresponds to the LO mode, the eigenfunctions, for a spherical QD, may be written as $B_k j_l(kr) Y_{lm}(\theta,\varphi)$, where $j_l(kr)$ is the spherical Bessel function of order l ($l = 0, 1, 2, 3 \ldots$), and $Y_{lm}(\theta, \phi)$ is the spherical harmonic function with $m = 0, \pm 1, \cdots \pm l$. B_k is the normalization constant, given by

$$B_k^{-2} = R^3 j_{l+1}^2(kR)/2. \tag{2.130}$$

Here, R is the radius of QD. Accordingly, the potential, ϕ, may be expressed as

$$\phi(r) = \sum_{l,m} \sum_k B_k j_l(kr) Y_{lm}(\theta, \varphi). \tag{2.131}$$

The continuity of ϕ and the normal component of **D** at the interface imply that ϕ vanishes outside the sphere and at its surface and can be expressed as $j_l(kR) = 0$ with $k = \alpha_{n,l}/R$, here $\alpha_{n,l}$ is the nth zero of the spherical Bessel function of order l.

In the second case, $\Delta\phi = 0$; this case corresponds to the SO modes. For this case, the solutions are

$$\begin{aligned}\phi(r) &= A_{l,m} r^l Y_{lm}(\theta,\varphi) \quad \text{for } r < R \\ \phi(r) &= B_{l,m} r^{-l-1} Y_{lm}(\theta,\varphi) \quad \text{for } r > R,\end{aligned} \tag{2.132}$$

where A_{lm} and B_{lm} are the normalization constants. The boundary conditions lead to the requirement that

$$\varepsilon_1/\varepsilon_2 = -(1+l)/l. \tag{2.133}$$

For the case of isotropic material, it follows that

$$\varepsilon_i(\omega_l) = \varepsilon_i^\infty \left(\omega_{\text{LO},i}^2 - \omega_l^2\right) / \left(\omega_{\text{TO},i}^2 - \omega_l^2\right) \quad (i = 1, 2), \tag{2.134}$$

where $i = 1$ denotes the dielectric constant of QDs with frequency ω_l, and $i = 2$ represents dielectric constant of the matrix element surrounding the dots with the same frequency. Thus, in the case where outside matrix is also a polar material, the frequency of the LO phonon follows upon solving Eq. (2.133)

$$\frac{\varepsilon_1^\infty \left(\omega_{\text{LO}1}^2 - \omega_l^2\right)\left(\omega_{\text{TO}2}^2 - \omega_l^2\right)}{\varepsilon_2^\infty \left(\omega_{\text{TO}1}^2 - \omega_l^2\right)\left(\omega_{\text{LO}2}^2 - \omega_l^2\right)} = -\frac{1+l}{l}, \tag{2.135}$$

For a non-polar surrounding material, the SO phonon frequency is

$$\omega_l = \omega_{\text{TO}} \left\{\left[\varepsilon_g + (\varepsilon_g + \varepsilon_0) l\right] / \left[\varepsilon_g + (\varepsilon_g + \varepsilon_\infty) l\right]\right\}^{1/2} \tag{2.136}$$

where the subscript g labels the surrounding matrix material. There is only one SO mode in this case.

Upon using Green's first identity

$$\int_V \nabla\varphi \cdot \nabla\psi \, d\mathbf{r} = -\int_V \varphi \nabla^2 \psi \, d\mathbf{r} + \int_S \varphi \frac{\partial \psi}{\partial n} \, dS, \quad (2.137)$$

the surface integral vanishes for the confined phonon, while the volume integral vanishes for the surface phonon [in the present case $\varphi = \psi = B_k j_l(kr) Y_{lm}(\theta,\varphi)$]. The phonon modes outside the dots vanish for the case of polar dots in a non-polar matrix; accordingly, the integral is nonzero only inside the dots.

The interaction Hamiltonian for confined phonons may be written as [50]

$$H_{I-C} = e\phi = \sum_{l,m} f^c_{lm}(k) [a_{lm}(k) j_l(kr) Y_{lm}(\theta,\varphi) + \text{H.c.}] \quad (2.138)$$

where

$$f^c_{lm}(k) = \sqrt{4\pi\hbar\omega_{LO} e^2 / [j_{l+1}^2(kR) \varepsilon^* R^3 k^2]}, \quad (2.139)$$

with $1/\varepsilon^* = 1/\varepsilon_1^\infty - 1/\varepsilon_1^0$.

According to Eqs. (2.24), (2.25), (2.30), and (2.132), we can obtain

$$\nabla\phi = (1 - 4\pi/3)\left(\omega_l^2 - \omega_{TO}^2\right) \frac{\mu}{e} \mathbf{u}. \quad (2.140)$$

By using the normalization relationship Eq. (2.31), for any given l and m, the normalized displacement is found to be

$$\mathbf{u} = \sqrt{(2l+1)\hbar / (2n\mu\omega_l l R)} \left[\nabla r^l Y_{lm}(\theta,\varphi) + \text{H.c.}\right]. \quad (2.141)$$

Thus, the interaction Hamiltonian for SO phonons may be written as

$$H_{I-SO} = e\phi = \sum_{l,m} f^s_{lm} \left[b_{lm} \left(\frac{r}{R}\right)^l Y_{lm}(\theta,\varphi) + \text{H.c.}\right], \quad (2.142)$$

where

$$f^s_{lm} = \varepsilon_1^\infty \omega_{LO} \sqrt{2\pi e^2 \hbar l (2l+1) / \omega_l \varepsilon^* R} / \left[l\varepsilon_1^\infty + (l+1)\varepsilon_1^0\right], \quad (2.143)$$

where b_{lm} is the annihilation operator, and its conjugate is the creation operator.

The scattering rate is typically formulated by using Fermi's golden rule [1]

$$W = \left(\frac{2\pi}{\hbar}\right) |\langle\Psi_m|(H_{I-C} + H_{I-SO})|\Psi_n\rangle|^2 \delta(E_m - E_n \mp \hbar\omega_l), \quad (2.144)$$

E_m, E_n are the energy eigenvalue of the final state Ψ_m and initial state Ψ_n, respectively. For phonon-assisted transitions, the energy difference between the final state and the initial state should be the energy of surface phonon energy. Clearly, in a quantum dot, this energy difference may be adjusted to be the surface phonon energy by simply selecting an appropriate radius for the dot. As discussed by Chen et al. [49], the three-dimensional confinement in the case of a QDs necessitates the estimation of a linewidth in estimating the transition rate. Numerical evaluation of the scattering rate for phonon associated transitions in QDs embedded in vacuum (Chen et al. [49]) reveals that rates as large as 10^{13}–10^{14} s^{-1} are possible for QDs with radii of 3.5 to 4.5 nm.

2.6 Summary

The review highlights selected advances of the last decade in the theory of acoustic and optical phonons in dimensionally confined structures. Among the structures considered are CNTs, graphene, graphite, and dimensionally confined wurtzite structures.

Acknowledgements The authors are grateful to AFOSR for supporting portions of the original research reported in this review. One author (Z. P. Wang) wishes to acknowledge CSC for supporting his visit to the University of Illinois at Chicago.

References

1. Stroscio MA, Dutta M (2001) Phonons in nanostructures. Cambridge University Press, Cambridge
2. Auld BA (1973) Acoustic fields and waves in solids. Wiley-Interscience, John Wiley & Sons, New York
3. Fuchs R, Kliewer KL (1965) Optical modes of vibration in an ionic crystal slab. Phys Rev 140:A2076–A2088
4. Fuchs R, Kliewer KL, Pardee WJ (1966) Optical properties of an ionic crystal slab. Phys Rev 150:589–596
5. Engleman R, Ruppin R (1968) Optical lattice vibrations in finite ionic crystals: I. J Phys C2:614–629
6. Engleman R, Ruppin R (1968) Optical lattice vibrations in finite ionic crystals: II. J Phys C2:630–643
7. Engleman R, Ruppin R (1968) Optical lattice vibrations in finite ionic crystals: III. J Phys C2:1515–1531
8. Ruppin R, Engleman R (1970) Optical phonons in small crystals. Rep Prog Phys 33:149–196
9. Licari JJ, Evrard R (1977) Electron–phonon interaction in a dielectric slab: effect of the electronic polarizability. Phys Rev B15:2254–2264
10. Wendler L (1985) Electron–phonon interaction in dielectric bilayer systems: effects of the electronic polarizability. Phys Status Solidi (b) 129:513–530

11. Mori N, Ando T (1989) Electron–optical phonon interaction in single and double heterostructures. Phys Rev B40:6175–6188
12. Born M, Huang K (1954) Dynamical theory of crystal lattices. Oxford University Press, Oxford
13. Loudon R (1964) The Raman effect in crystals. Adv Phys 13:423–482; Erratum, ibid. (1965) 14:621
14. Chen C, Dutta M, Stroscio MA (2004) Electron scattering via interaction with optical phonons in wurtzite crystals. Phys Rev B70:075316
15. Qian J, Allen MJ, Yang Y, Dutta M, Stroscio MA (2009) Quantized long-wavelength optical phonon modes in graphene nanoribbon in the elastic continuum model. Superlattice Microstruct 46:881–888
16. Raichura A, Dutta M, Stroscio MA (2004) Continuum model for acoustic phonons in nanotubes: phonon bottleneck. Phys Status Solidi (b) 241:3448–3453
17. Suzurra H, Ando T (2002) Phonons and electron–phonon scattering in carbon nanotubes. Phys Rev B65:235412
18. Raichura A, Dutta M, Stroscio MA (2004) Quantized optical vibrational modes of finite-length multi wall nanotubes: optical deformation potential. Superlattices Microstruct 35:147–153
19. Vogl P (1980) The electron–phonon interaction in semiconductors. In: Ferry DK, Barker JR, Jacoboni C (eds) Physics of nonlinear transport in semiconductors. NATO Advanced Study Institute Series. Plenum, New York
20. Iijima S (1991) Helical microtubules of graphitic carbon. Nature 354:56–58
21. Bachtold A, Hadley P, Nakanishi T, Dekker C (2001) Logic circuits with carbon nanotube transistors. Science 294:1317–1320
22. Baughman RH (2002) Carbon nanotubes—the route toward applications. Science 297:787–792
23. Guo J, Goasguen S, Lundstrom M, Datta M (2002) Metal–insulator–semiconductor electrostatics of carbon nanotubes. Appl Phys Lett 81:1486–1488
24. Xu JM (2001) Highly ordered carbon nanotube arrays and IR detection. Infrared Phys Technol 42:485–491
25. Kraus H (1967) Thin elastic shells. Wiley, New York
26. Raichura A, Dutta M, Stroscio MA (2003) Quantized acoustic vibrations of single-wall carbon nanotube. J Appl Phys 94:4060–4065
27. Raichura A, Dutta M, Stroscio MA (2005) Acoustic phonons and phonon bottleneck in single wall nanotubes. J Comput Electron 4:91–95
28. Constantinou NC, Ridley BK (1990) Interaction of electrons with the confined LO phonons of a free-standing GaAs quantum wire. Phys Rev B41:10622–10626
29. Constantinou NC, Ridley BK (1990) Guided and interface LO phonons in cylindrical GaAs/Al$_x$Ga$_{1-x}$As quantum wires. Phys Rev B41:10627–10631
30. Stroscio MA, Dutta M, Kahn D, Kim KW (2001) Continuum model of optical phonons in a nanotube. Superlattice Microstruct 29:405–409
31. Collins PG, Hersam M, Arnold M, Martel R, Avouris P (2001) Current saturation and electrical breakdown in multiwalled carbon nanotubes. Phys Rev Lett 86:3128–3131
32. Pennington G, Goldsman N (2003) Semiclassical transport and phonon scattering of electrons in semiconducting carbon nanotubes. Phys Rev B68:045426
33. Stroscio MA, Dutta M, Raichura A (2007) Conductance of nanowires: phonons effects. J Comput Electron 6:247–249
34. Sun K, Stroscio MA, Dutta M (2009) Thermal conductivity of carbon nanotubes. J Appl Phys 105:074316
35. Klemens PG (2004) Graphite, graphene and carbon Nanotubes. Thermal conductivity. In Proceeding of the 26th international thermal conductivity conference, DEStech Publications, Lancaster
36. Nicklow R, Wakabayashi N, Smith HG (1972) Lattice dynamics of pyrolytic graphite. Phys Rev B5:4951–4962
37. Fan YW, Goldsmith BR, Collins PG (2005) Identifying and counting point defects in carbon nanotubes. Nature 4:906–911

38. Novoselov KS, Geim AK, Morozov SV, Jiang D, Zhang Y, Dubonos SV, Grigorieva IV, Firsov AA (2004) Electric field effect in atomically thin carbon films. Science 306:666–669
39. Rana F, Geroge PA, Strait JH, Dawlaty J, Shivaraman S, Chandrashekhar M, Spencer MG (2009) Carrier recombination and generation rates for intravalley and intervalley phonon scattering in graphene. Phys Rev B79:115447
40. Dresselhaus MS, Dresselhaus G, Eklund P (1996) Science of fullerenes and carbon nanotubes. Academic, Salt Lake City
41. Babiker M (1986) Longitudinal polar optical modes in semiconductor quantum wells. J Phys C Solid State Phys 19:683–697
42. Maultzsch J, Reich S, Thomsen C, Requardt H, Ordejon P (2004) Phonon dispersion in graphite. Phys Rev Lett 2004:075501
43. Wirtz L, Rubio A (2004) The phonon dispersion of graphite revisited. Solid State Commun 131:141–152
44. Akturk A, Pennington G, Goldsman N, Wickenden A (2007) Electron transport and velocity oscillations in a carbon nanotube. IEEE Trans Nanotechnol 6:469–474
45. Sun K, Stroscio MA, Dutta M (2009) Graphite C-axis thermal conductivity. Superlattice Microstruct 45:60–64
46. Chen C, Dutta M, Stroscio MA (2004) Confined and interface phonon modes in GaN/ZnO heterostructures. J Appl Phys 95:2540–2546
47. Hellwege KH, Madelung O, Hellege AM (1987) Landolt–Bornstein-numerical data and functional relationships in science and technology. Springer, New York
48. Komirenko SM, Kim KW, Stroscio MA, Dutta M (1999) Dispersion of polar optical phonons in wurtzite heterostructures. Phys Rev B59:5013–5020
49. Chen C, Dutta M, Stroscio MA (2004) Surface-optical phonon assisted transitions in quantum dots. J Appl Phys 96:2049–2054
50. Klein MC, Hache F, Ricard D, Flytzanis C (1990) Size dependence of electron–phonon coupling in semiconductor nanospheres: the case of CdSe. Phys Rev B41:11123–11132

Chapter 3
Theories of Phonon Transport in Bulk and Nanostructed Solids

G.P. Srivastava

Abstract In this chapter we outline the theories that are usually employed for phonon transport in solids. In particular, we provide a detailed description of the essential steps in deriving the lattice thermal conductivity expressions within the single-mode relaxation-time approximation. Explicit expression for various phonon scattering rates, in bulk and low-dimensional solids, have been provided. Numerical evaluation of scattering rates and the conductivity expressions is detailed using both Debye's isotropic continuum scheme and a realistic Brillouin zone summation technique based upon the application of special phonon wavevectors scheme. Results of the conductivity are presented for selected bulk, superlattice, and nanostructed systems. Based on such results, we briefly discuss the concept of phonon engineering of high-efficiency thermoelectric materials.

3.1 Introduction

Techniques for experimental measurement and theoretical modelling of thermal conductivity in naturally grown bulk soilds have been established since the early 1950s. In this millenium, nanostructures, or meta materials in general, have attracted a great deal of attention. Such materials do not exist in nature but can be fabricated in laboratory using modern growth techniques. However, in general, methods of experimental measurements and theoretical modelling of some properties, including thermal conductivity, of such materials have not yet been well established.

Thermal conduction in insulating and semiconducting solid structures (bulk or low-dimensional forms) is almost exclusively contributed by phonons. A phonon, using the second-quantised notation, is a quantum of atomic vibrational waves in

---

G.P. Srivastava (✉)
School of Physics, University of Exeter, Stocker Road, Exeter EX4 4QL, UK
e-mail: G.P.Srivastava@exeter.ac.uk

crystalline solids and is represented both as a quasi-particle and an elementary excitation of energy $\hbar\omega(qs)$ corresponding to wavevector q and polarisation s. Phonon transport in a given solid is governed by details of dispersion relations (i.e. ω vs. q for all polarisation branches s) and processes that control lifetimes of these quasi-particles. An ideally perfect and purely harmonic solid would exhibit perfect heat conduction, i.e. infinite thermal conductivity, as its atoms can be viewed as executing purely harmonic vibrations, leading to the concept of independent, or non-interacting, phonons of infinitely long lifetime. Real solids are characterised by anharmonic inter-atomic forces and presence of defects. These features limit phonon lifetimes and thus thermal conductivity. In order to estimate phonon lifetimes and thermal conductivity it is important to have a detailed knowledge of phonon dispersion relations for the relevant crystal structure.

Ab initio methods for phonon dispersion calculations have been described in the chapter by Tütüncü and Srivastava, and for phonon lifetimes in bulk semiconductors in the chapters by Garg et al. and Mingo et al. In this chapter we will briefly discuss phonon dispersion relations and their usage in deriving expressions for phonon lifetime contributions from various scattering sources in bulk and nanostructured semiconductors. We will then discuss how these two ingredients can be utilised in numerical calculations of the phonon conductivity in such systems. Finally, we will briefly discuss how changes in the thermal conductivity due to nanostructuring can be engineered for increasing the efficiency (i.e. the figure-of-merit) of thermoelectric materials.

3.2 Phonon Transport Theories

The rate of heat energy flow Q per unit area normal to a finite (but small) temperature gradient ∇T across a solid is given by Fourier's law

$$Q_i = -\sum_j \kappa_{ij} \nabla T_j, \qquad (3.1)$$

with the coefficient $\{\kappa_{ij}\}$ known as the thermal conductivity tensor. Accepting this observation, conductivity expressions have been derived by following approaches at two different levels of sophistication. At one level, a statistical mechanical approach, known as the Green–Kubo linear-response approach [1], is used to express κ in terms of the time integral of the heat current autocorrelation function (i.e. canonical-ensemble average with respect to the Hamiltonian of the system) $< Q(t) \cdot Q(0) >$. This approach has been followed both at classical and quantum levels. At another level, κ is expressed by obtaining a solution of a linearised Boltzmann equation satisfied by the phonon distribution function $n_{qs}(r, t)$ in the steady state of heat flow through the solid (see, e.g., [2, 3]). As in general, the phonon Boltzmann equation cannot be solved exactly, several formulations for expressing κ have been presented [3]. In this section we will provide a brief overview of the theories for the phonon conductivity based on the Green–Kubo's linear-response approach and the Boltzmann equation formulation.

3.2.1 Theories Based on Green–Kubo's Linear-Response Approach

3.2.1.1 Classical Level

At the classical level of the Green–Kubo approach a molecular dynamical simulation is performed in real space (see, e.g., [4, 5]). In this approach, Newton's second law and the kinematic equations of motion are used, based on (semi)classically derived interatomic potential(s), to determine the classical position and momentum space trajectories of a system of particles.

The heat current is written as

$$Q = \sum_i E_i v_i + \frac{1}{2} \sum_{i,j} (F_{ij} \cdot v_i) r_{ij}, \qquad (3.2)$$

where E_i, r_i, and v_i are, respectively, the energy, position vector, and velocity of particle i, r_{ij} is the interparticle separation vector, and F_{ij} is the force between particles i and j. For a monatomic bulk crystal, the heat current autocorrelation function is fitted into a functional form

$$\langle Q(t) \cdot Q(0) \rangle = A \exp(-t/\tau), \qquad (3.3)$$

where A is a constant and τ is a time constant. The isotropic thermal conductivity expression for a bulk material is then obtained as

$$\kappa = \frac{k_B T^2 N_0 \Omega}{3} A\tau, \qquad (3.4)$$

where k_B is Boltzmann's constant, $N_0 \Omega$ represents crystal volume (with N_0 unit cells, each of volume Ω). The energy E and force F terms are usually obtained from the use of empirically derived inter-atomic potential. Details on the application of the approach can be found in McGaughey and Kaviany [4], and Huang and Kaviany [6].

3.2.1.2 Quantum Level

At quantum level the heat current is expressed as an operator in the Heisenberg representation

$$\hat{Q}(t) = \frac{1}{N_0 \Omega} \sum_{qs} \hbar\omega(qs)\hat{n}_{qs}(t)c_s(q), \quad (3.5)$$

where $c_s(q)$ is the velocity of phonon mode qs with frequency $\omega(qs)$ and $\hat{n}_{qs}$ is the phonon number operator

$$\hat{n}_{qs} = \hat{a}_{qs}^{\dagger}\hat{a}_{qs} \quad (3.6)$$

with $\hat{a}_{qs}^{\dagger}$ and $\hat{a}_{qs}$ as phonon creation and annihilation operatores, respectively. The conductivity tensor expression, therefore, reads

$$\kappa_{ij} = \frac{\hbar^2}{N_0 \Omega k_B T^2} \Re \int_0^\infty dt \sum_{qsq's'} \omega(qs)\omega(q's')c_s^i(q)c_{s'}^j(q')\mathscr{C}_{qsq's'}(t), \quad (3.7)$$

where $c_s^i(q)$ is the ith component of the velocity of a phonon mode qs and

$$\mathscr{C}_{qsq's'}(t) = \langle \hat{a}_{qs}^{\dagger}(t)\hat{a}_{q's'}(0) \rangle \quad (3.8)$$

is a *correlation function*.

The correlation function $\mathscr{C}_{qsq's'}(t)$ represents the canonical-ensemble average of the operator $\hat{a}_{qs}^{\dagger}(t)\hat{a}_{q's'}(0)$ with respect to the total phonon Hamiltonian H of the system:

$$\mathscr{C}_{qsq's'}(t) = \frac{\text{Tr}(e^{-\beta H}\hat{a}_{qs}^{\dagger}(t)\hat{a}_{q's'}(0))}{\text{Tr}(e^{-\beta H})}, \quad \beta = 1/k_B T. \quad (3.9)$$

Several techniques have been employed to evaluate the correlation function, including Zwangis–Mori's projection operator method, double-time Green's function method, and imaginary-time Green's function method. Details of the first two methods can be found in Srivastava [3] and of the third method in Ziman [7]. The final solution can be expressed as

$$\mathscr{C}_{qsq's'}(t) = \delta_{qq'}\delta ss' \bar{n}_{qs}(\bar{n}_{qs} + 1)e^{-t/\tau_{qs}}, \quad (3.10)$$

where $\bar{n}_{qs}$ is the Bose–Einstein distribution function and τ_{qs} is the relaxation time for a phonon in mode qs. The Hamiltonian required for the simulation is usually adopted from empirically chosen inter-atomic potential, but it can be made from first-principles treatments.

3.2.1.3 Extraction of Relaxation Time

An effective relaxation time τ_{qs} can be obtained for different situations of the change in the Hamiltonian from that for a perfect crystal within harmonic approximation, e.g. due to the presence of impurities and defects, and crystal anharmonicity. However, there are genuine difficulties in dealing with two situations in particular. Firstly, the contribution towards τ_{qs} from the presence of isotopic impuries requires the molecular dynamics simulation to be carried out over an excessively large unit cell containing atoms of appropriate atomic masses. Such contributions, therefore, have not yet been included satisfactorily. Secondly, it is not easy to unscramble the anharmonic contribution in the form of separate phonon contributions involved in three-phonon or four-phonon processes. Neither is it easy to establish separate roles of the anharmonic Normal (momentum conserving) and Umklapp (momentum non-conserving) processes.

3.2.2 Theories Based on Phonon Boltzmann Transport Equation

Theories of lattice thermal conductivity based on the Boltzmann transport equation make the basic assumption that the occupation number of a phonon in mode qs is governed by a distribution function $n_{qs}(r,t)$ in the neighbourhood of space position r at time t. In the general form of Boltzmann equation, for a dielectric subjected to a (small) temperature gradient ∇T, the distribution function satisfies the equation

$$-c_s(q) \cdot \nabla T \frac{\partial n_{qs}}{\partial T} + \frac{\partial n_{qs}}{\partial t}\bigg|_{\text{scatt}} = 0, \qquad (3.11)$$

with the second term on the left-hand side representing the rate of change due to phonon scattering mechanisms. The linearised Boltzmann equation

$$-c_s(q) \cdot \nabla T \frac{\partial \bar{n}_{qs}}{\partial T} + \frac{\partial n_{qs}}{\partial t}\bigg|_{\text{scatt}} = 0, \qquad (3.12)$$

represents a physically appealing simplification of Eq. (3.11), where

$$\bar{n}_{qs} = [\exp(\hbar\omega(qs)/k_BT) - 1]^{-1} \qquad (3.13)$$

is the Bose–Einstein (or equilibrium) distribution function. Expressions for the term $\partial n_{qs}/\partial t\big|_{\text{scatt}}$ corresponding to relevant phonon scattering mechanisms must be derived before Eq. (3.12) is solved for $n_{qs}(t)$ and eventually an expression for thermal conductivity is established. In general, phonon scattering mechanisms can be described as *elastic* (in which the participating phonon qs retains its identity) and *inelastic* (in which the participating phonon qs loses its identity). In general, only approximate forms of inelastic scattering rates can be derived. This is particularly the case for phonon scattering due to crystal anharmonic effects. This difficulty has led to two main routes for the derivation of thermal conductivity expression. These will be briefly described here.

3.2.2.1 Variational Principles

The scattering term in Eq. (3.12) can, in general, be expressed as

$$-\frac{\partial n_{qs}}{\partial t}\bigg|_{\text{scatt}} = \sum_{q's'} P_{qq'}^{ss'} \psi_{q'}^{s'}, \qquad (3.14)$$

where $\psi_{qs} (\equiv \psi_q^s)$ provides a measure of deviation of the phonon distribution function from its equilibrium value

$$\begin{aligned} n_{qs} &= [\exp(\hbar\omega(qs)/k_B T - \psi_{qs}) - 1]^{-1} \\ &\simeq \bar{n}_{qs} - \psi_{qs} \frac{\partial \bar{n}_{qs}}{\partial(\hbar\omega(qs))} \\ &= \bar{n}_{qs} + \psi_{qs} \bar{n}_{qs}(\bar{n}_{qs} + 1), \end{aligned} \qquad (3.15)$$

and $P_{qq'}^{ss'}$ are the elements of the phonon collision operator, providing a measure of phonon transition probabilities.

Expressions for the phonon collision operator elements $P_{qq'}^{ss'}$ can be derived by applying time-dependent perturbation theory. The deviation function ψ_{qs} can then be obtained from Eq. (3.12) provided that the inverse of the matrix operator $\{P_{qq'}^{ss'}\}$ exists. Unfortunately, only partial information is available about the nature of the anharmonic part of the collision operator (for detail, see [3, 8]). This leaves ψ_{qs} unknown in the temperature range where the role of crystal anharmonicity plays an important role. The essence of the variational method for lattice thermal conductivity is to treat ψ_{qs} as a trial function. The simplest approximation for the anharmonic contribution to ψ_{qs} is [3, 8]

$$\psi_{qs} = \boldsymbol{q} \cdot \boldsymbol{u}, \qquad (3.16)$$

where $\boldsymbol{u}$ is some constant vector parallel to the applied temperature gradient.

Using the trial function in Eq. (3.16) Ziman [2] derived a *lower bound* for the conductivity. By noting and employing the positive semi-definite property of the phonon collision operator P, Benin [9] developed a *sequence of monotonically convergent lower bounds* for the conductivity. The first term in this sequence is the Ziman limit. It was later shown by Srivastava [10] that a *sequence of monotonically convergent upper bounds* for the conductivity can also be developed. In theory, an estimate for the exact conductivity can then be confined to a small difference between an upper bound and a lower bound. An improved estimate of any conductivity bound can be made by adopting scaling and Ritz procedures [11]. The concept of obtaining both a lower bound and an upper bound for estimating a desired (but inherently unknown) quantity is called *complementary variational principles*, details of which can be found in the book by Arthurs [12].

3.2.2.2 Relaxation-Time Theories

The difficulty in expressing the scattering rate in Eq. (3.12) in terms of a possible solution for the deviation function ψ_{qs} requiring consideration of the phonon collision operator P in its full form is usually dealt with the introduction of the concept of phonon *relaxation time*. This is achieved by expressing

$$-\left.\frac{\partial n_{qs}}{\partial t}\right|_{\text{scatt}} = \frac{n_{qs} - \bar{n}_{qs}}{\tau_{qs}}, \qquad (3.17)$$

where τ_{qs} is the relaxation time for a phonon in mode qs. It should, however, be made clear that this expression is based upon a big simplifying assumption, is subject to a fundamental criticism, and is valid under certain conditions. The criticism is that the concept of relation time of a phonon mode becomes invalid in describing its participation in multi-phonon scattering events involving phonon creation and annihilation. Notwithstanding this criticism, the expression above assumes the concept of a single-mode relaxation time. To explain this point let us consider three phonons $qs, q's'$ and $q''s''$ involved in a three-phonon scattering event. In its simple form Eq. (3.17) assumes that only the phonon mode qs is described by a displaced distribution and the other two modes obey the equilibrium distribution. Such a description is referred to as the *single-mode relaxation time* (*smrt*) approximation and can be justified to some extent [13]. Essentially, τ_{smrt} can be derived from the diagonal part of the phonon collision operator P. Modifications of the *smrt* has been proposed by Callaway [14] and Srivastava [15]. The validity of the relaxation-time approach is limited by the Landau–Peirls–Ziman condition [2]

$$\omega\tau > 1; \quad \text{or} \quad \Lambda > \lambda, \qquad (3.18)$$

where Λ and λ are phonon mean-free path and wavelength, respectively. Thus the Boltzmann-equation-based relaxation time approach is unsuitable for applications to samples thinner than average phonon wavelength.

Representing τ^* as the relaxation time for a chosen model (e.g. $\tau^* = \tau_{\text{smrt}} - \tau$ within the *smrt* approximation, or $\tau^* = \tau_C$ within the Callaway model), we can express the thermal conductivity tensor as

$$\kappa_{ij} = \frac{\hbar^2}{N_0 \Omega k_B T^2} \sum_{qs} \omega^2(qs) c_s^i(q) c_s^j(q) \tau^*(qs) \bar{n}(qs)(\bar{n}(qs) + 1). \qquad (3.19)$$

This expression will be discussed later using the *smrt* and Callaway models for the relaxation time.

3.3 Ingredients for Calculation of Thermal Conductivity

We will now consider calculation of lattice thermal conductivity using the relaxation-time-based expression in Eq. (3.19). It requires knowledge of (i) lattice dynamics, i.e. phonon dispersion curves ($\omega = \omega(\boldsymbol{q}s)$) and phonon density of states ($g(\omega)$), (ii) relaxation time τ_{qs} for relevant phonon scattering processes, and (iii) an adequate procedure for carrying out summations over $\boldsymbol{q}$-vectors inside the Brillouin zone for the crystal structure under consideration.

3.3.1 Phonon Dispersion Curves and Density of States

The topic of lattice dynamics, fundamental to calculations of all properties related to atomic vibrations in solids, has been pursed for many decades. Both phenomenological and *ab initio* approaches have been employed. One of the most physically appealing phenomenological approaches for tetrahedrally bonded bulk semiconductors is the adiabatic bond charge model developed by Weber [16]. This method has been extended for application to surfaces [17] and nanostructured materials [18]. With the development of parameter-free total-energy calculations, *ab initio* methods of lattice dynamical calculations are now affordable (see, e.g., [19]). Details of such a method, based on the applications of the plane-wave pseudopotential technique, have been presented in the chapter by Tütüncü and Srivastava.

3.3.1.1 Bulk Materials

Figure 3.1 shows the phonon dispersion curves for bulk Si along the principal symmetry directions in the central primitive unit cell in momentum space (the Brillouin zone). Clearly, both the acoustic branches (characterised with $\omega = 0$ for $q = 0$) and optical branches (characterised with $\omega = $ constant for $q = 0$) are highly dispersive as well as anisotropic in the momentum space. In a vast majority of theoretical developments and numerical calculations of lattice thermal conductivity two major approximations are employed: (i) isotropic dispersion relations ($\omega(\boldsymbol{q}s) = \omega(|\boldsymbol{q}|s)$) and (ii) the continuum model ($\omega(qs) = c_s q$). Clearly, the continuum model is more suitable for low-frequency acoustic branches. In practice the isotropic continuum scheme is carried out inside a sphere (known as Debye sphere of radius q_D) as a replacement for the actual shape of the BZ. The volume of the Debye sphere should in principle be taken as the volume of the appropriate BZ. However, usually it is chosen to ensure that the integral of the density of states (DOS) equals the total number of acoustic modes. A comparison of the DOS, $g(\omega)$, results for Si obtained within the continuum approximation for an average acoustic branch (with average phase speed of 5,691 m/s) and the lattice dynamical dispersion relations is

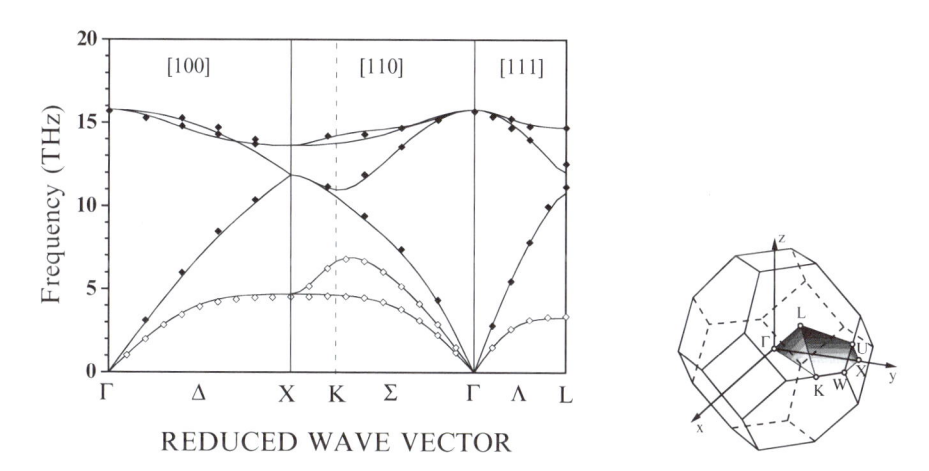

3 Theories of Phonon Transport in Bulk and Nanostructed Solids

Fig. 3.1 Phonon dispersion curves for bulk Si. Theoretical results (*solid curves*) obtained from the application of the bond charge model compare well with experimentally measured results (*filled diamond* from [20] and *open diamond* from [21]). Also shown is the Brillouin zone and an irreducible part in it

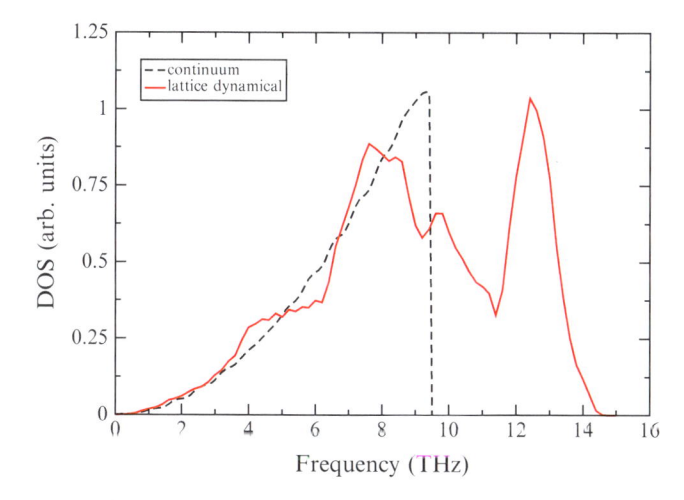

Fig. 3.2 Phonon density of states for bulk Si. The *solid curve* shows the results obtained from a full lattice dynamical calculation using the adiabatic bond charge model and the *dashed curve* is obtained from the consideration of the isotropic continuum model for an average acoustic branch

made in Fig. 3.2. The smooth quadratic rise of the continuum result, $g(\omega) \propto \omega^2$, is consistent with the realistic picture only in the low frequency range of 0–3.5 THz. Beyond this frequency range the realistic picture shows significant departure from the continuum results. In particular, the continuum model does not cover the optical frequency range and predicts a much larger DOS (a single van Hove singularity)

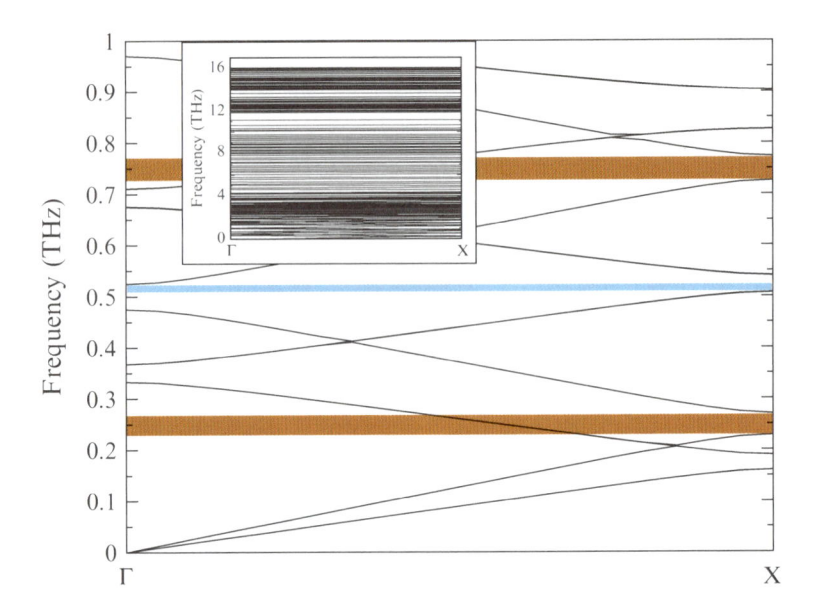

Fig. 3.3 Phonon dispersion curves for the Si(4 nm)/Si$_{0.4}$Ge$_{0.6}$(8 nm)[001] superlattice. The *inset* shows the full range of the spectrum. *Highlighted* are the lowest three LA gap regions. The *central gap region* is common to both LA and TA branches, indicating that the system is a true one-dimensional phononic along the grown direction. Taken from [18]

at the highest frequency (9.4 THz) modelled with the linear dispersion relation. In contrast, the realistic DOS shows several peaks (van Hove singularities), notably at frequencies 4, 7.6, 8.4, 9.7, and 12.5 THz.

3.3.1.2 Nanostructured Materials

Phonon dispersion curves of low-dimensional systems generally exhibit significant differences compared to bulk materials. There are at least three new features that can be expected to develop due to reduction in dimensionality: reduction in group velocity of phonons modes in all branches, creation of gaps in acoustic as well as optical branches both at zone edges and at zone centre, and confinement of higher lying acoustic and optical branches. We will explain these features by considering two examples: a thin Si/SiGe superlattice and ultrathin Si nanostructures.

Figure 3.3 shows the phonon dispersion curves, obtained from the application of an enhanced adiabatic bond charge model [18], for the Si(4 nm)/Si$_{0.4}$Ge$_{0.6}$(8 nm) superlattice along the growth direction [001]. This system exhibits several gaps in the phonon dispersion curves for the entire frequency range. Higher optical branches (with frequencies greater than 4 THz) show very flat dispersion curves (i.e. have low group velocities), indicating confinement effects within different superlattice layers. Several frequency gaps, for both longitudinal acoustic (LA) and transverse acoustic (TA) branches, are obtained. The lowest three sub-terahertz LA gaps occur as follows: zone edge gap of 48 GHz centred 252 GHz, zone centre gap of 40 GHz centred 495 GHz, and zone edge gap of 30 GHz centred 805 GHz. These results are in good agreement with the sub-picosecond spectroscopic measurements made by Ezzahri et al. [22]. The lowest three TA gaps are: zone edge gap of 30 GHz centred 175 GHz, zone centre gap of 35 GHz centred 350 GHz, and zone edge gap of 33 GHz centred 523 GHz. The phonon speeds for the lowest TA and LA branches are 4.18 km/s and 6.00 km/s, respectively. These values are smaller than the average of the corresponding speeds in bulk Si and Ge. The theoretical work also shows that there is a clear overall gap along the superlattice growth direction between 515 and 539 GHz, meaning that this superlattice is a true one-dimensional phononic and would not allow propagation of phonons in this frequency range.

Figure 3.4 shows the phonon dispersion curves, obtained from the application of the adiabatic bond charge model, for ultrathin Si nanostructures such as a nanoslab, a wire, and a dot. Looking at the results for the nanoslab we note the existence of gap opening and flatness of branches along the slab normal (i.e. along $\Gamma - X$) compared to an in-plane direction. These features become even more prominent as the dimensionality decreases from two-dimensional (slab) to one-dimensional (wire) and zero-dimensional (dot). The confinement effect produces new peaks (van Hove singularities) in the phonon density of states (DOS). A clear example of this can be seen in the right-hand panel of Fig. 3.4b, which shows a comparison of the DOS of the Si nanowire of cross-section 0.543× 0.543 nm (solid curves) with that of bulk Si (dashed curves). In general, the nanowire DOS shows several sharp delta-like peaks, particularly at approximately 220 cm^{-1} where the bulk DOS shows a dip.

3.3.2 Brillouin Zone Summation

In order to obtain numerical values of any physical property of a crystalline material it is important to develop a method of Brillouin zone summation. Several methods exist for such an exercise, at different levels of sophistication. Let us consider the summation of a general periodic function $f(\boldsymbol{q})$

$$I = \sum_{\boldsymbol{q}} f(\omega(\boldsymbol{q})). \tag{3.20}$$

If the function $f(\omega(\boldsymbol{q}))$ is isotropic with a linear dispersion relation, i.e. $\omega(\boldsymbol{q}) = cq$, then the summation can be performed by using the Debye scheme. In this scheme the Brillouin zone summation is expressed as an integral over the Debye sphere of radius q_D

$$I = \int_0^{\omega_D} g_D(\omega) f(\omega) d\omega, \tag{3.21}$$

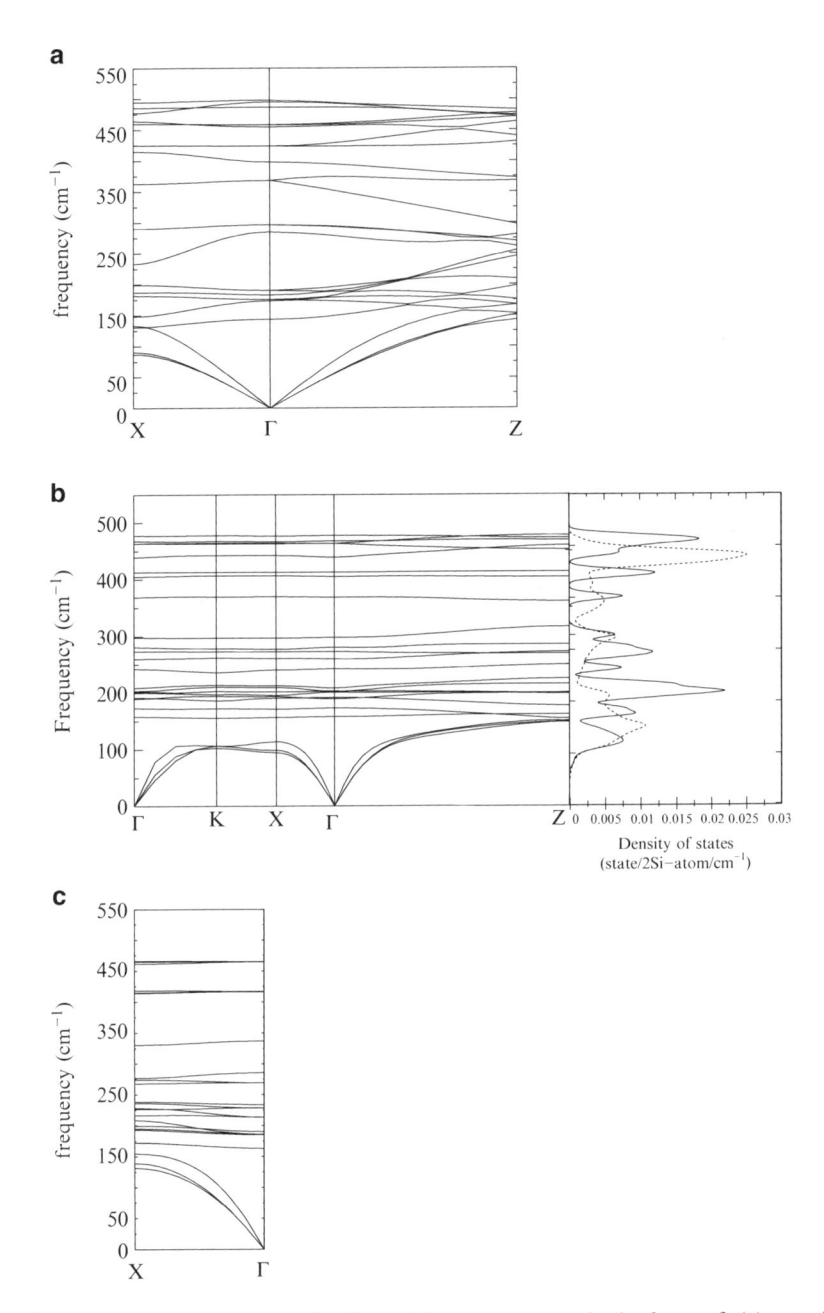

Fig. 3.4 Phonon dispersion curves for Si ultrathin nanostructures in the forms of: (**a**) nanoslab of thickness 0.543 nm, (**b**) rectangular nanowire of cross-section 0.543×0.543 nm, and (**c**) cubic nanodot of size 0.543×0.543×0.543 nm. $\Gamma - X$ is along the confinement direction. The *right-hand panel* in (**b**) shows a comparison of the density of states for the nanowire (*solid curves*) and bulk (*dashed curves*). Taken from [23]

where

$$g_D(\omega) = \frac{N_0 \Omega}{2\pi^2} \frac{\omega^2}{c^3}, \quad (3.22)$$

is the Debye density of states function and $\omega_D = cq_D$ is the Debye frequency. The simple isotropic continuum Debye method must be improved adequately to deal with non-cubic crystal structures, and dispersive phonon modes in acoustic as well as optical branches.

A general-purpose scheme for numerical evaluation of an integral of the type in Eq. (3.20) is based on the concept of "special q-points" [24, 25] inside the Brillouin zone for the crystal under consideration. Considering an appropriate selection of N special $\{q_i\}$ points and weight factors $\{W(q_i)\}$ associated with them, we can estimate the integral I as

$$I \simeq N_0 \sum_i^N f(\omega(q_i)) W(q_i). \quad (3.23)$$

For a given shape of Brillouin zone, different sets of special $\{q_i\}$ points can be generated. A set is considered more "efficient" if it provides an acceptable result for the integral with the least number of $\{q_i\}$ points [24].

3.3.3 Phonon Scattering Processes

Expressions for phonon scattering rates from various sources in bulk semiconductors can be derived by applying Fermi's Golden rule formula based on first-order time-dependent perturbation theory, and do exist in the literature (see, e.g. [3] and references therein). Here we simply reproduce some of the commonly used results.

3.3.3.1 Boundary Scattering

Purely diffusive phonon scattering rate from crystal boundaries can be expressed as

$$\tau_{bs}^{-1}(qs) = c_s/L_0. \quad (3.24)$$

Here L_0 denotes an effective boundary length, which depends on the geometrical shape of crystal. $L_0 = D$ for a crystal of cylindrical shape with circular cross-section of diameter D, and $L_0 = 1.12\, d$ for a square cross-section of side d. For polycrystalline solids L_0 is a measure of an effective grain size.

In reality, consideration of the *polish* quality of crystal surface must be made and an effective boundary length determined. This can be quite a difficult task, knowing especially that the precise nature of surface roughness and structure of grown samples is usually ill-defined. A phenomenological model [2] allows us to define an effectively longer boundary length L by incorporating the surface *polish* quality in terms of the *specularity* factor p,

$$L = \frac{(1+p)}{(1-p)} L_0. \quad (3.25)$$

The limiting cases $p = 0$ and $p = 1$ are purely diffusive (or Casimir) scattering and purely specular boundary scattering (or reflection), respectively. For a given p factor, each phonon is reflected $1/(1-p)$ times before being diffusively scattered.

3.3.3.2 Isotopic Mass-Defect Scattering

Consider a crystal with an average mass per unit cell as $\bar{M}$ and let f_i be the fraction of the unit cell containing the isotopic mass M_i. The scattering rate of a phonon mode qs due to isotopic mass defects is [3, 13]

$$\tau_{md}^{-1}(qs) = \frac{\Gamma_{md}\pi}{6N_0} \omega^2(qs) g(\omega(qs)), \quad (3.26)$$

where $g(\omega)$ is the density of states and Γ_{md} is the isotopic mass parameter. Within the isotropic continuum approximation, the expression in Eq. (3.26) reads

$$\tau_{md}^{-1}(qs) = \frac{\Gamma_{md}\Omega}{4\pi\bar{c}^3} \omega^4(qs),$$

$$= A_{md} \omega^4(qs), \quad (3.27)$$

with $\bar{c}$ as the average acoustic phonon speed. For a single-species crystal the isotopic mass parameter is evaluated as

$$\Gamma_{md} = \sum_i f_i (\Delta M_i / \bar{M})^2, \quad (3.28)$$

with $\Delta M_i = M_i - \bar{M}$, and M as the mass of atoms in the unit cell of volume Ω. Clearly, for a monatomic crystal, with one atom per unit cell, M is the mass of a single atom and Ω represents the atomic volume. For a composite material with molecular formula $A_x B_y C_z \ldots$ we can express Γ_{md} as [26]

$$\Gamma_{md} = \frac{x}{(x+y+z+\ldots)} \left(\frac{M_A}{\bar{M}}\right)^2 \Gamma(A)$$

$$+ \frac{y}{(x+y+z+\ldots)} \left(\frac{M_B}{\bar{M}}\right)^2 \Gamma(B)$$

$$+ \frac{z}{(x+y+z+\ldots)} \left(\frac{M_C}{\bar{M}}\right)^2 \Gamma(C)$$
$$+ \ldots, \qquad (3.29)$$

with

$$\Gamma(A) = \sum_i f_i \left(\frac{\Delta M_i(A)}{\bar{M}_A}\right)^2, \qquad (3.30)$$

etc. and

$$\bar{M} = \frac{xM_A + yM_B + zM_C + \ldots}{x + y + z + \ldots}. \qquad (3.31)$$

The above expression can be used for a superlattice structure $(AB)_n(CD)_m$ by considering it as a compound with formula $A_n B_n C_m D_m$.

3.3.3.3 Interface Scatterings

Low-dimensional solids, such as superlattices, and nanowires or nanodots embedded in a host matrix, can provide additional defect-related phonon scattering mechanisms: interface mass-mixing scattering (IMS) due to diffusion or mixing of atoms across interfaces, and interface dislocation scattering (IDS) which results from dislocations or missing bonds present at interfaces. The IMS scattering rate from a periodic distribution of nanodots in a host matrix has been studied by treating the embedded material as a small perturbation to the host material [27,28]. Here we briefly discuss the IMS for superlattice structures with planar (two-dimensional) interfaces [29].

(i) Mass-mixing scattering:

Let us consider a periodic un-reconstructed superlattice A_m/B_n with N_0 unit cells, and each unit cell containing m atomic layers of material A and n layers of material B. Assuming that interface mass mixing takes place within J layers on either side of an interface, for dealing with the theory of IMS the perturbed crystal Hamiltonian can be expressed as

$$H'(\text{IMS}) = \frac{1}{2} \sum_{i=m-J}^{m} (M_i |\mathbf{v_i}|^2 - M_A |\mathbf{v_A}|^2) + \frac{1}{2} \sum_{i=m+1}^{m+J} (M_i |\mathbf{v_i}|^2 - M_B |\mathbf{v_B}|^2). \quad (3.32)$$

where M_i is the mass of the ith atom, and $\mathbf{v}_i = d\mathbf{u_i}/dt$ with $\mathbf{u_i}$ being the relative displacement of the ith with respect to its neighbours. Application of Fermi's golden rule and utilisation of some simplifying assumptions lead to the following expression for the relaxation of a phonon mode $\boldsymbol{q}s$ [29]

$$\tau_{\text{IMS}}^{-1}(\mathbf{q}s) = \frac{\alpha\pi}{2N_0(n+m)^2} \int d\omega(\mathbf{q}'s')g(\omega(\mathbf{q}'s'))\omega(\mathbf{q}s)\omega(\mathbf{q}'s')$$
$$\times \frac{\bar{n}(\mathbf{q}'s')+1}{\bar{n}(\mathbf{q}s)+1}\delta(\omega(\mathbf{q}s)-\omega(\mathbf{q}'s'))\left[\left(1-\frac{e_A e'_A}{e_B e'_B}\right)^2 + \left(1-\frac{e_B e'_B}{e_A e'_A}\right)^2\right],$$
(3.33)

where α is regarded as a parameter related to the amount of mixing at the interface, $g(\omega(\mathbf{q}s))$ is the density of states, and e_B/e_A is the ratio of the amplitudes of eigenvectors in materials B and A along the superlattice growth direction.

(ii) Interface dislocation scattering:

Superlattices composed of lattice-mismatched layers are known to be characterised by the presence of dislocations at interfaces. Phonon scattering rate by such dislocations cannot adequately be described by the traditional theory for bulk solids. While the phonon scattering rate by dislocations in bulk has traditionally been derived by treating solids as elastic continuum (see, e.g., [2]), an atomic-scale theory is required when dealing with scattering from interface dislocations. Let us consider an interface dislocation as a series of randomly missing bonds located near the interface within a unit cell. With such a consideration we can write the perturbed crystal Hamiltonian as

$$H'(\text{IDS}) = \frac{1}{2}\sum_{i=m-l}^{m}(K_0|u_i|^2 - K_A|u_A|^2) + \frac{1}{2}\sum_{i=m+1}^{m+l}(K_0|u_i|^2 - K_B|u_B|^2), \quad (3.34)$$

where u_i as the relative displacement between two neighbouring atoms and l is number of atomic layers with broken bonds on either side of an interface, $K_A(K_B)$ represents the inter-atomic spring constant in the layer A(B), and K_0 represents a spring constant in the dislocation region, i.e. has a value equal or close to zero, for broken or missing bonds. From the application of Fermi's golden rule the following expression can be derived for phonon-IDS relaxation rate [29]

$$\tau_{\text{IDS}}^{-1}(\mathbf{q}s) = \frac{\pi\omega_0^4}{4N_0}\frac{\alpha'}{(n+m)^2}\int d\omega(\mathbf{q}'s')\frac{g(\omega(\mathbf{q}'s'))}{\omega(\mathbf{q}s)\omega(\mathbf{q}'s')}\frac{\bar{n}(\mathbf{q}'s')+1}{\bar{n}(\mathbf{q}s)+1}$$
$$\times \delta(\omega(\mathbf{q}s)-\omega(\mathbf{q}'s'))\left[1+\left(\frac{e_A e'_A}{e_B e'_B}\right)^2 + 1+\left(\frac{e_B e'_B}{e_A e'_A}\right)^2\right], (3.35)$$

where α' is a measure of dislocation concentration and ω_0 can be approximated as the highest zone-centre frequency.

3 Theories of Phonon Transport in Bulk and Nanostructed Solids

The following simplified expression for the amplitude ratio e_B/e_A, required for both IMS and IDS, can be derived by treating a superlattice as a diatomic linear chain along the growth direction

$$\frac{e_B}{e_A} = \frac{\left[\frac{1}{M_0} - \Delta\left(\frac{1}{M}\right)\right]\cos(l_z q_z)}{\left[\left(\frac{1}{M_0}\right)^2 \cos^2(l_z q_z) + \left(\Delta\left(\frac{1}{M}\right)\right)^2 \sin^2(l_z q_z)\right]^{\frac{1}{2}} - \Delta\left(\frac{1}{M}\right)}, \quad (3.36)$$

where $2M_0 = 1/M_A + 1/M_B$, and $\Delta(1/M) = (1/M_A - 1/M_B)/2$, and l_z is the superlattice period.

3.3.3.4 Anharmonic Scattering

Anharmonic interatomic potential is present in all crystals at finite temperatures. Obtaining an expression for crystal anharmonic potential, with its temperature dependence, from first principles is an enormous task and has not yet been achieved satisfactorily. A workable form of cubic anharmonic crystal potential can be expressed, by treating a crystal as an isotropic anharmonic elastic continuum, as [2, 3]

$$V_3 = \frac{1}{3!}\sqrt{\frac{\hbar^3}{2\rho N_0 \Omega}} \frac{\gamma(T)}{\bar{c}} \sum_{qsq's'q''s''} \sqrt{\omega(qs)\,\omega(q's')\,\omega(q''s'')}$$

$$\times (a_{qs}^\dagger - a_{-qs})(a_{q's'}^\dagger - a_{-q's'})(a_{q''s''}^\dagger - a_{-q''s''})\,\delta_{q+q'+q'',G}, \quad (3.37)$$

where $\gamma(T)$ as a mode-average but temperature-dependent Grüneisen's constant, **G** is a reciprocal lattice vector, and $a_{qs}^\dagger$, a_{qs}, etc. are the phonon creation and annihilation operators, respectively, and other symbols have previously been defined. It should be remarked that the above form of the anharmonic potential assumes that the phonon spectrum consists of only acoustic branches. Recently, a form of isotropic elastic anharmonic potential has been modelled [30] that considers acoustic as well as optical phonon branches. Here we will, however, neither present that form of the Hamiltonian nor any results obtained from its applications.

Application of Fermi's golden rule leads to the following single-mode relaxation time for a phonon mode qs due to the anharmonic three-phonon interactions in a bulk single crystal (see [3] for details):

$$\tau^{-1}(qs)\big|_{\text{bulk}} = \frac{\pi \hbar \rho^2 \gamma^2}{N_0 \Omega \bar{c}^2} \sum_{q's',q''s'',G} B(qs, q's', q''s''), \quad (3.38)$$

with

$$B(qs, q's', q''s'') = \omega(qs)\omega(q's')\omega(q''s'')\bar{n}(q's')$$
$$\times \left\{ \left[\frac{(\bar{n}(q''s'')+1)}{\bar{n}(qs)+1} \delta(\omega(qs)+\omega(q's') - \omega(q''s''))\delta_{q+q'.q''+G} \right] \right.$$
$$\left. + \left[\frac{1}{2} \frac{\bar{n}(q''s'')}{\bar{n}(qs)} \delta(\omega(qs) - \omega(q's') - \omega(q''s''))\delta_{q+G.q'+q''} \right] \right\} \quad (3.39)$$

The processes described by the first and second terms in Eq. (3.39) may be referred to as Class 1 and Class 2 events, governed by the momentum and energy conservation conditions:

Class 1 events : $\quad q + q' = q'' + G; \quad \omega + \omega' = \omega'',$ (3.40)

Class 2 events : $\quad q + G = q' + q''; \quad \omega = \omega' + \omega''.$ (3.41)

For each class, an event is called Normal (N) if it involves wavevectors of all participating phonons within the central Brillouin zone. If a reciprocal lattice vector G is required to meet the momentum conservation condition, the event is called Umklapp (U). These processes are schematically illustrated for a class 1 event in panels (a) and (b) of Fig. 3.5.

Callaway [14] took into account the momentum conserving condition of N-processes to derive an effective phonon relaxation time τ_C, which can be considered as a modification of the single-mode relaxation time τ_{smrt}. Within the isotropic continuum approximation, the following expression can be derived [31]

$$\tau_C = \tau_{smrt}(1 + \beta/\tau_N), \quad (3.42)$$

with τ_N as the relaxation time due to N processes, and

$$\beta = \frac{q}{\omega(qs)c_s(q)} \frac{\langle \omega c q \tau \tau_N^{-1} \rangle}{\langle q^2 \tau_N^{-1}(1 - \tau \tau_N^{-1}) \rangle}, \quad (3.43)$$

where the following notation has been used

$$< f > = \sum_{qs} f(qs)\bar{n}(qs)(\bar{n}(qs) + 1). \quad (3.44)$$

The above theory can be adopted for low-dimensional systems, but with modifications. Let us consider a superlattice structure as an example. Apart from making the obvious changes in the material density (ρ_{sl}), Grüneisen's constant (γ_{sl}) and average acoustic phonon speed ($\bar{c}_{sl}$), two further considerations must be made. (i) The shortest reciprocal lattice vector in the superlattice growth direction will in

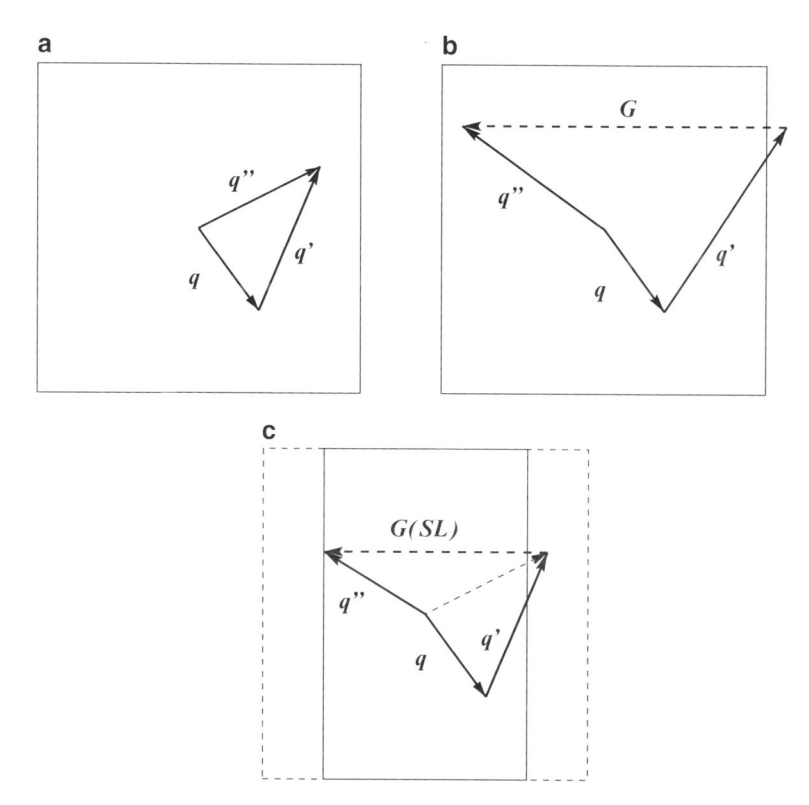

3 Theories of Phonon Transport in Bulk and Nanostructed Solids

Fig. 3.5 Schematic illustration of a Class 1 three-phonon scattering process in a single-crystal solid: (**a**) an N process and (**b**) an U process. (**c**) shows how the bulk N process shown in (**a**) becomes a "mini-U" process in a superlattice structure corresponding to a superlattice reciprocal lattice vector $G(\text{SL})$. For clarity, in (**c**) the bulk Brillouin zone is drawn by *dashed lines*, and the bulk q'' is shown by the *short-dashed vector*

general be much shorter than the shortest reciprocal lattice vector in the constituent materials. A simple illustration is provided in panel (c) of Fig. 3.5. This means that some of the bulk N-processes will turn into the superlattice mini-U processes. (ii) The presence of two materials in an A/B superlattice structure has to be incorporated in deriving an appropriate form of the anharmonic crystal potential V_3. An attempt in this regard was made by Ren and Dow [32]. Using a scheme similar to Ren and Dow, a suitable modification of the expression in Eq. (3.38) for an A/B superlattice structure can be presented as

$$\tau^{-1}(qs)\,|_{\text{superlattice}} = \frac{\pi\hbar\rho_{\text{sl}}^2\gamma_{\text{sl}}^2}{N_0\Omega\bar{c}_{\text{sl}}^2} \sum_{q's',q''s'',G_{\text{sl}}} B(qs,q's',q''s'')F_{\text{sl}}(qs,q's',q''s''),$$
(3.45)

where ρ_{sl} is superlattice density (i.e. the weighted average density of the two materials), $\bar{c}_{\text{sl}}$ is the average acoustic velocity in the superlattice, γ_{sl} is the Grüneisen constant for the joint system, G_{sl} represents a reciprocal lattice vector for the superlattice structure, and the term F_{sl} arises due to the presence of two materials in the system. Within the diatomic linear chain approximation, the term F_{sl} can be expressed as

$$F_{sl}(qs, q's', q''s'') = \frac{1}{64} \left\{ \frac{1}{2\rho_A^{\frac{3}{2}}} \left[1 + \frac{\rho_A^{\frac{1}{2}}}{\rho_B^{\frac{1}{2}}} \left(\frac{e_B}{e_A} + \frac{e_B'}{e_A'} + \frac{e_B''}{e_A''} \right) + \frac{\rho_A}{\rho_B} \left(\frac{e_B e_B'}{e_A e_A'} + \frac{e_B' e_B''}{e_A' e_A''} + \frac{e_B e_B''}{e_A e_A''} \right) \right. \right.$$

$$\left. + \frac{\rho_A^{\frac{3}{2}}}{\rho_B^{\frac{3}{2}}} \left(\frac{e_B e_B' e_B''}{e_A e_A' e_A''} \right) \right] + \frac{1}{2\rho_B^{\frac{3}{2}}} \left[1 + \frac{\rho_B^{\frac{1}{2}}}{\rho_A^{\frac{1}{2}}} \left(\frac{e_A}{e_B} + \frac{e_A'}{e_B'} + \frac{e_A''}{e_B''} \right) + \frac{\rho_B}{\rho_A} \left(\frac{e_A e_A'}{e_B e_B'} + \frac{e_A' e_A''}{e_B' e_B''} + \frac{e_A e_A''}{e_B e_B''} \right) \right.$$

$$\left. \left. + \frac{\rho_B^{\frac{3}{2}}}{\rho_A^{\frac{3}{2}}} \left(\frac{e_A e_A' e_A''}{e_B e_B' e_B''} \right) \right] \right\}^2. \tag{3.46}$$

where $e_A' \equiv e_A(q's')$, $e_A'' \equiv e_A''(q''s'')$, etc., ρ_A and ρ_B are the densities of materials A and B, respectively, and the expression for amplitude ratio e_B/e_A is given in Eq. (3.36).

3.3.3.5 Scattering from Donor Electrons

We will consider the scattering of phonons with donor electrons in doped semiconductors. In a doped semiconductor we may consider a local displacement field $u(r)$ produced by longitudinal acoustic phonons, causing an energy change of the form $E_d = C\Delta(r) = C_1 \nabla \cdot u(r) = C_2 \hat{q} \cdot e_{qs}$, with C_1 and C_2 as some parameters. The matrix element of the deformation potential E_d can be evaluated by expressing the donor electron wave function as a Bloch function. Application of Fermi's golden rule results in the following expression for the relaxation rate of a phonon mode qs (see [33, 34] for details)

$$\tau^{-1}(qs) = \frac{m^{*2} E_d^2 k_B T}{2\pi \rho c_L \hbar^4} \left[z - \ln \left(\frac{1 + \exp(\xi - \xi_0 + z^2/16\xi + z/2)}{1 + \exp(\xi - \xi_0 + z^2/16\xi - z/2)} \right) \right], \tag{3.47}$$

with c_L is the speed of longitudinal acoustic phonons, $z = \hbar\omega/k_B T$, $\xi = m^* c_L^2/2k_B T$ and $\xi_0 = \zeta/k_B T$. This expression can be reduced to the following form in the case of a heavily doped, degenerate, semiconductor with $\zeta > E_0$ and $\zeta - E_0 \gg k_B T$, where $E_0 = \hbar^2 \omega^2/8m^* c_L^2 + \frac{1}{2} m^* c_L^2 - \hbar\omega/2$,

$$\tau^{-1}(qs) = \frac{m^{*2} E_d^2}{2\pi \rho \hbar^3 c_L} \omega. \tag{3.48}$$

3 Theories of Phonon Transport in Bulk and Nanostructured Solids

For moderately doped semiconductors, Eq. (3.47) can be reduced to the following form [35]

$$\tau^{-1}(qs) = \frac{n_c E_d^2 \omega}{\rho c_L^2 k_B T} \sqrt{\frac{\pi m^* c_L^2}{2 k_B T}} \exp\left(\frac{-m^* c_L^2}{2 k_B T}\right), \tag{3.49}$$

where n_c is the carrier concentration.

3.4 Thermal Conductivity Results

Numerical calculations of thermal properties, such as lattice specific heat and lattice thermal conductivity (κ), require summation, or integration, of functions of phonon wave vectors over the entire Brillouin zone (BZ). In Sect. 3.3.2 we have mentioned two approaches for carrying out BZ summation. In this section we will present results of numerical calculations of κ for bulk and low-dimensional semiconductors. Some of the results have been obtained by using Debye's isotropic continuum method. Some other results have been carried out by using full lattice dynamical results for phonon dispersion curves, phonon velocities, and a realistic BZ summation technique.

Traditionally calculations of thermal properties have been carried out within Debye's isotropic continuum scheme. Within this scheme Callaway's expression for the lattice thermal conductivity can be written as follows:

$$\kappa_C = \frac{\hbar^2 q_D^5}{6\pi^2 k_B T^2}$$

$$\times \left[\sum_s c_s^4 \int_0^1 dx\, x^4 \tau \bar{n}(\bar{n}+1) + \frac{\{\sum_s c_s^2 \int_0^1 dx\, x^4 \tau \tau_N^{-1} \bar{n}(\bar{n}+1)\}^2}{\sum_s \int_0^1 dx\, x^4 \tau_N^{-1}(1 - \tau \tau_N^{-1}) \bar{n}(\bar{n}+1)} \right],$$

$$= \kappa_D + \kappa_{N-drift}, \tag{3.50}$$

where $x = q/q_D$ with q_D representing the Debye radius. The first and second terms in the above equation represent the single-mode relaxation time result κ_{smrt} (the Debye term κ_D) and the N-drift term $\kappa_{N-drift}$, respectively. The total phonon relaxation time is the sum of contributions to τ^{-1}, obtained within the isotropic continuum scheme, from all scattering processes relevant to the system under study. Expressions for commonly required processes are given in Sect. 3.3.3. Using the anharmonic crystal potential presented in Sect. 3.3.3.4 and employing Debye's isotropic continuum scheme, the anharmonic phonon relaxation time contributed by three-phonon processes can be expressed as follows (see [3] for details):

$$\tau_{qs}^{-1}(3\text{ ph}) = \frac{\hbar q_D^5 \gamma^2}{4\pi \rho \bar{c}^2} \sum_{s's''\epsilon}$$

$$\left[\int dx' x'^2 x''_+ \{1 - \epsilon + \epsilon(Cx + Dx')\} \frac{\bar{n}_{q's'}(\bar{n}''_+ + 1)}{(\bar{n}_{qs} + 1)} \right.$$

$$\left. + \frac{1}{2} \int dx' x'^2 x''_- \{1 - \epsilon + \epsilon(Cx - Dx')\} \frac{\bar{n}_{q's'} \bar{n}''_-}{\bar{n}_{qs}} \right].$$

Here $x' = q'/q_D$, $x''_\pm = Cx \pm Dx'$ and $\bar{n}''_\pm = \bar{n}(x''_\pm)$, $C = c_s/c_{s''}$, $D = c_{s'}/c_{s''}$, $\epsilon = 1$ for momentum conserving (normal, or N) processes, and $\epsilon = -1$ for momentum non-conserving (Umklapp, or U) processes. The first and second terms in the above equation are contributed by class 1 events $qs + q's' \to q''s''$ and class 2 events $qs \to q's' + q''s''$, respectively. The integration limits on the variable x', imposed by the energy and momentum conservation conditions in Eqs. (3.40) and (3.41), can be derived straightforwardly and are given below.

Class 1 events:

$$0 \leq x \leq 1$$

$$0, \frac{(1-C)x}{(1+D)} \leq x' \leq \frac{(1+C)x}{(1-D)}, \frac{(1-Cx)}{D}, 1 \quad N \text{ processes},$$

$$0, \frac{(2-(1+C)x)}{(1+D)} \leq x' \leq \frac{(1-Cx)}{D}, 1 \quad U \text{ processes.} \quad (3.51)$$

Class 2 events:
N processes:

$$0 \leq x \leq 1$$

$$0, \frac{(C-1)x}{D+1}, \frac{(Cx-1)}{D} \leq x' \leq \frac{(C+1)x}{D+1}, \frac{(C-1)x}{D-1}, 1 \quad (3.52)$$

U processes:

$$\frac{2}{1+C} \leq x \leq 1$$

$$0, \frac{2-(1+C)x}{1-D}, \frac{Cx-1}{D}, \frac{(C+1)x-2}{D+1} \leq x' \leq \frac{(C+1)x-2}{D-1}, 1. \quad (3.53)$$

3.4.1 Bulk Semiconductors

Figure 3.6 shows the temperature variation of the lattice thermal conductivity of bulk Si, Ge, and GaAs. The theory employed for numerical calculations was based on the

3 Theories of Phonon Transport in Bulk and Nanostructed Solids

Fig. 3.6 Thermal conductivity results for bulk semiconductors: (**a**) Si, (**b**) natural Ge, and (**c**) GaAs. *Solid curves* are obtained from theoretical calculations. *Symbols* represent experimentally measured values: Si [36]; Ge (natural sample of Geballe and Hull [37]); GaAs [26]. Reproduced from Srivastava [38]

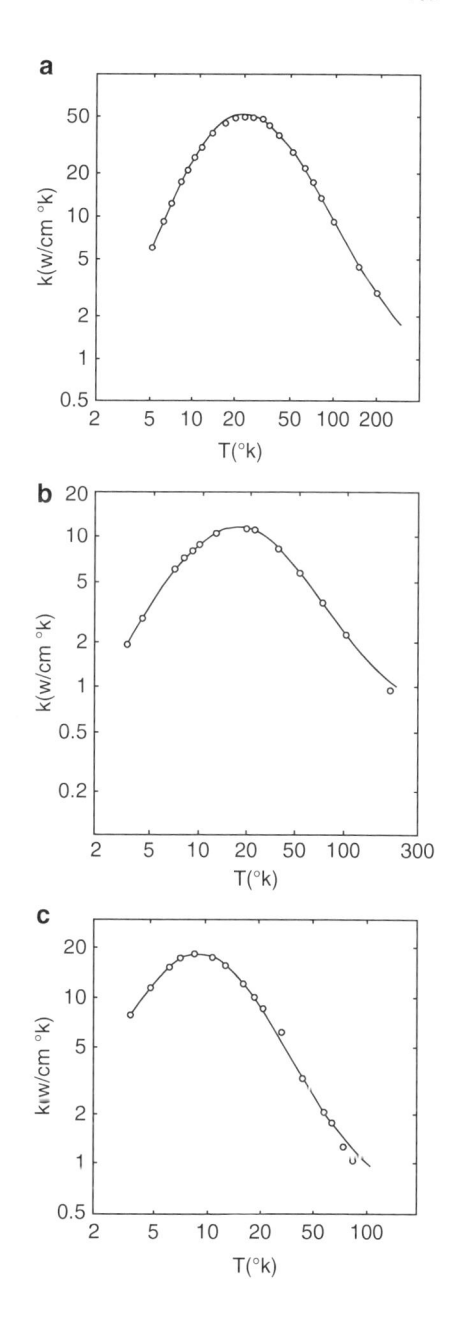

Srivastava's model [15] for an effective phonon relaxation time, *viz.* a modified form of Callaway's model [14]. The numerical results, obtained by employing Debye's isotropic continuum method and using reasonable values of fitting parameters, agree well with experimentally measured results. In general, with increase in temperature, κ increases as T^3 at low temperatures when the phonon-boundary scattering dominates, reaches a maximum in the temperature range 10–30 K due to the combined effect of defect and anharmonic scatterings of phonons, and decreases as $1/T$ at high temperatures when three-phonon scattering processes dominate. The maximum value of κ is generally determined by the average atomic mass and the level of sample purity. The maximum value of κ is approximately 5,000, 1,300, and 2,500 W/m-K and occurs at approximately 20, 20, and 15 K for Si, natural Ge, and GaAs, respectively. The room temperature values of κ are 130, 58, and 55 W/m-K for Si, Ge, and GaAs, respectively. Of these three semiconductors, Si is categorised as a "high thermal conductivity material" (a material with room-temperature conductivity larger than 100 W/m-K).

3.4.2 GaAs Suspended Nanobeams

Fon et al. [39] fabricated suspended GaAs nanobeams and made the first direct measurement of the thermal conductance in such structures. They considered a total of four samples: undoped nanobeam, doped nanobeam, a 6-beam device (4 doped nanobeams and 2 undoped nanobeams), and a 4-beam device (4 doped nanobeams). The nanobeams were of rectangular shape with cross-sectional dimensions $d_1 = 100$ nm and $d_2 = 250$ nm. Results for bulk and the four nanobeams of GaAs, obtained from numerical evaluation of the conductivity expression in Eq. (3.50) and the phonon anharmonic relaxation time expression in Eq. (3.51), are presented in Fig. 3.7. The parameters used in the calculations for bulk, undoped nanobeam, doped nanobeam, 6-beam device, and 4-beam device, respectively, were: 0.73 cm, 0.227 μm, 0.21 μm, 0.50 μm, and 0.37 μm for the effective boundary length; 0.0, 0.5, 0.4, 0.5, and 0.4 for the surface specularity parameter p; 0.0, 0.0, 5.0 × 10^{24} m^{-3}, 0.0, and 5.0 × 10^{24} m^{-3} for the donor electron concentration; 4.36 × 10^{-42} s^3, 46.76 × 10^{-42} s^3, 436.27 × 10^{-42} s^3, 31.2 × 10^{-42} s^3, and 311.7 × 10^{-42} s^3 for the mass defect parameter A_{md} in Eq. (3.27). The Grüneisen constant was set to 1.8 for all the samples. We will describe the role of various phonon scattering processes in explaining the experimental results for bulk GaAs [26] and the nanobeam structures [39] in the following paragraphs.

In bulk GaAs, phonon boundary scattering is only important below 10 K. In contrast, boundary scattering has a very strong influence on the thermal conductivity of the undoped nanobeam up to about 100 K, and it also controls the shape of the thermal conductivity curve up to about 300 K. In order to explain the low-temperature conductivity results of the nanobeams it was necessary to include specular phonon boundary scattering events. Our fitted value of the effective boundary length of 0.68 μm is consistent with the specularity factor $p = 0.5$. This suggests that each phonon is specularly reflected on an average of $1/(1 - p) \approx 2$ times before being diffusely scattered in these nanobeams. Our work also suggests that, compared to bulk, phonon scattering by point defects is much stronger

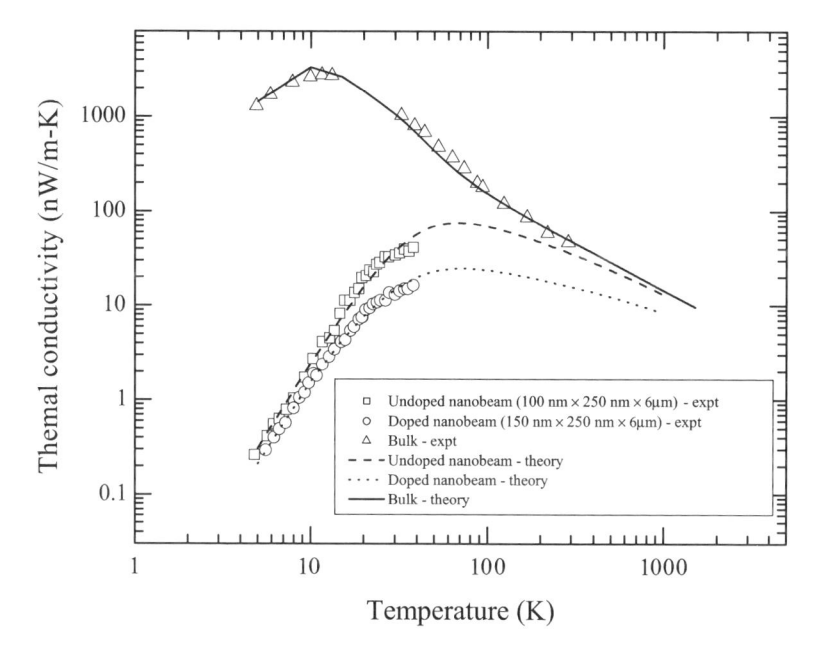

Fig. 3.7 Thermal conductivity of bulk GaAs (theory: *solid curve*; experiment [26]: *triangles*). Also shown are the thermal conductivity results for suspended GaAs nanobeams (theory: *dotted curves*; experiment [39]: *squares and circles*). Reproduced from [40]

in the nanobeams. The fitted values of the effective boundary length are much larger than the effective physical width $L_0 = 1.12\sqrt{d_1 d_2} = 0.177$ μm of the nanobeams fabricated by Fon et al. A similar conclusion was reached by Fon et al. in their attempt to theoretically explain the low-temperature conductivity results. Surprisingly, in our theoretical work we had to use a big range of effective boundary length (from 0.49 μm for doped beam to 1.5 μm for the six-beam device).

The doped beams fabricated by Fong et al. contain donor dopants with concentration 5×10^{18} cm^{-3} in the topmost 50 nm layers. We used Eq. (3.49) to account for the electron–phonon scattering in this sample. However, in order to successfully explain the conductivity results of the doped nanobeams, we had to consider the point defect scattering rate approximately 10 times stronger than that for the undoped beam. This suggests that doping of the mesoscopic beams was accompanied by structural disorder. The peak at around 10 K in the conductivity-temperature curve for the bulk sample has shifted to a much higher value of around 60 K for the nanobeams. The conductivity of the undoped nanobeam merges with the conductivity of bulk above the Debye temperature (345 K) of GaAs. However, the conductivity of the doped nanobeams continues to remain lower than that of the undoped nanobeam for all temperatures up to at least 1,000 K.

3.4.3 Si Nanowires

In the previous subsection we showed that the Debye–Callaway model of lattice thermal conductivity, developed for bulk solids, can be successfully applied to explain the conductivity results for GaAs beams of cross-sectional dimensions in the range of 100–250 nm. Phonon transport in thin quasi-one-dimensional structures, such as thin nanowires with cross-sectional dimensions smaller than 50 nm, is likely to be sensitive to both surface roughness and quantum confinement effects. There have been several theoretical attempts for explaining experimentally measured thermal conductivity results of Si nanowires, including the semi-classical Boltzmann-type approach [41], a semi-empirical molecular dynamics approach [42], and the quantum Landauer formalism [43]. All these models are capable of explaining the measured thermal conductivity [44] of Si nanowires of thicknesses (diameters) 115, 56, and 37 nm. However, it has been argued [43] that in order to explain the conductivity of the nanowire of diameter 22 nm it is important to employ a theoretical model that combines incoherent surface scattering for short-wavelength phonons with nearly ballistic long-wavelength phonons. However, in a recent work [45] we have argued that Callaway's relaxation-time model for thermal transport in bulk materials, described at the start of this section, can be successfully applied to explain the measurements for Si nanowires with cross-sectional dimensions down to 22 nm.

Calculations were made with the following parameters for Si nanowires of diameters 115, 56, 37, and 22 nm, respectively: effective boundary length values of 0.115, 0.056, 0.037, 0.010 μm; effective mass-defect scattering parameter A_{md} (see Eq. (3.27)) values of 45.0×10^{-46} s^3, 45.0×10^{-46} s^3, 250.0×10^{-46} s^3, 500.0×10^{-46} s^3; Grüneisen's constant values of 1.4, 1.7, 1.7, 1.7. Figure 3.8 clearly shows that there is reasonably good agreement between the presently calculated results and the experimental measurements for Si nanowires of all the four thicknesses studied by Li et al. [44]. For reproducing the experimental results for the nanowires of diameters 115 nm and 56 nm we had to use several times stronger mass-defect scattering parameter than what is needed for bulk Si. For the thinner nanowire of diameter 37 nm, the mass-defect parameter had to be increased five-fold compared to that for the thicker nanowires. For the thinnest nanowire (diameter 22 nm) we had to consider a very strong diffused surface (boundary) scattering, and a mass defect scattering parameter twice that for the 37 nm wire. The choice of significantly smaller effective boundary length for the 22 nm wire clearly indicates that surface of this wire is quite rough. We also note that for nanowires of all thickness, we had to use a stronger anharmonicity factor (Grüneisen's constant γ) than is needed for bulk Si (usually $\gamma = 0.8$ is adequate). Following the discussion in the theory section, we re-iterate that the level of successful agreement between theory and experiment for the conductivity achieved for the 22 nm nanowire is not expected to continue to much thinner nanowires.

Three clear trends can be noticed. Firstly, the maximum of the conductivity generally shifts to higher temperatures with decrease in the nanowire thickness. This is due to the joint effect of strong boundary and strong mass defect scatterings

3 Theories of Phonon Transport in Bulk and Nanostructed Solids 107

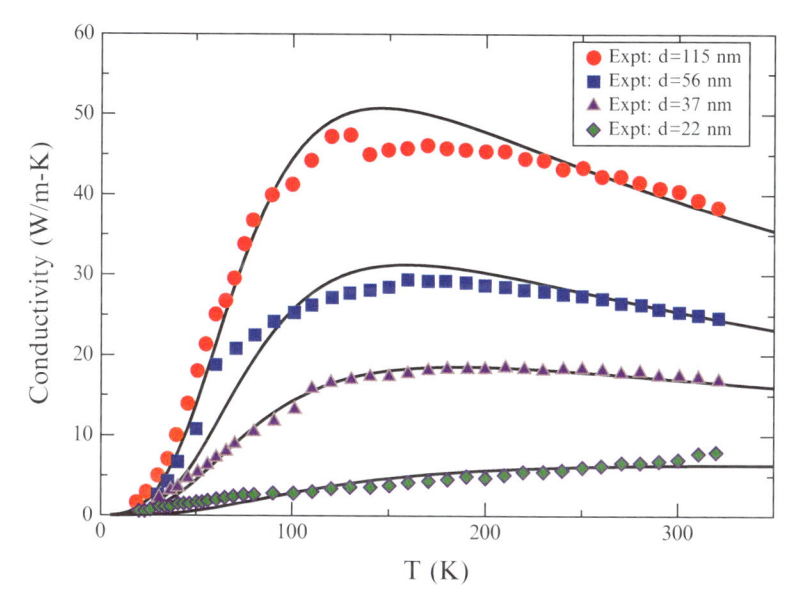

Fig. 3.8 Thermal conductivity of (**a**) Si bulk and (**b**) Si nanowires: theory (*lines*); experiment [44] (*symbols*). Taken from [45] (NWs)

of phonons. Secondly, for temperatures up to well over the room temperature the conductivity of nanowires has been reduced by 2–3 orders of magnitude with respect to the bulk values. A similar conclusion has been reached from another theoretical investigation [42]. Thirdly, at a given (low) temperature, the conductivity decreases with the decrease in nanowire thickness. A definite relationship for the decrease in the conductivity with nanowire thickness is difficult to establish, as the surface and interior qualities are not guaranteed to be maintained during nanowire fabrication.

3.4.4 Si/Ge Superlattices

The lattice thermal conductivity of superlattices has been reported to be at least two orders of magnitude lower than that of constituent bulk materials [46, 47]. What mechanisms govern the low thermal conductivity of superlattices is not generally well understood. Let us consider a superlattice of repeat period thickness $d = d_1 + d_2$, containing two components of thicknesses d_1 and d_2. Let us further consider that the Landau–Peirls–Ziman condition in Eq. (3.18) is satisfied. This can be interpreted as the superlattice sample size L being larger than the phonon mean free path Λ. We note that the sample size L contains several multiples of the superlattice unit cell size d, so that $L = Nd$ with N ranging from about 10–1,000 for typically grown samples. With this understanding, one can distinguish three regimes of phonon transport along the superlattice growth direction [48]:

(i) For superlattices with $d \gg \Lambda$, the thermal conductivity κ can be fairly well expressed as

$$1/\kappa = (d_1/\kappa_1 + d_2/\kappa_2)/(d_1 + d_2), \quad (3.54)$$

where κ_i is the conductivity of the ith bulk material;

(ii) For superlattices with $d \sim \Lambda$,

$$1/\kappa = (d_1/\kappa_1 + d_2/\kappa_2 + 2W_K)/(d_1 + d_2), \quad (3.55)$$

where W_K is the Kapitza resistance of the superlattice interface;

(iii) For $d \ll \Lambda$, it will be important to consider the superlattice as a single new material and a detailed consideration of phonon group velocities, phonon density of states, and phonon lifetime will be required for thermal conductivity calculations.

Here we will discuss the phonon conductivity of thin Si/Ge superlattices (i.e. for case (iii): $d \ll \Lambda$). As discussed in Sect. 3.3.1.2, the phonon dispersion relations in thin Si/Ge superlattices are significantly different from those in bulk Si or Ge. These changes lead to reduction in the speeds of acoustic phonons and changes in the density of states. Although such changes can be manipulated to some extent by altering the superlattice period and layer thicknesses, the resulting changes would not be enough to explain the reduced thermal conductivity of the superlattices. With the help of the numerically obtained results for phonon group velocities, density of states and phonon relaxation times, we provide a plausible explanation for this by considering numerical results for two specific choices of Si/Ge superlattices, Si(19)/Ge(5)[001] and Si(72)/Ge(30)[001], as fabricated and studied by Lee et al. [46].

Equation (3.19) was used to express the components of the conductivity tensor, with axes along [$\bar{1}$10], [110], and [001]. Numerical results for phonon dispersion curves and density of states were obtained by employing an enhanced adiabatic bond charge model [18]. Phonon scattering rates due to point mass defects, interface mass-mixing (IMS), interface dislocations (IDS), and anharmonic interactions were calculated using Eqs. (3.26), (3.33), (3.35), and (3.45), respectively. The amplitude ratio e_B/e_A was calculated using Eq. (3.36). The required BZ summations were carried out using the special q-points scheme as described in Sect. 3.3.2. However, the IMS and IDS parameters α and α' had to be treated as adjustable parameters. This became necessary due to the fact that no information is available for the amount of mass mixing and the nature and concentration of interface dislocations, except that these features are always present during the growth of Si/Ge systems [49].

The numerically calculated results for the lattice thermal conductivity of the two superlattices, along the growth direction, are presented in Fig. 3.9. Also presented are the experimentally measured results, obtained by Lee et al. [46]. In order to fit

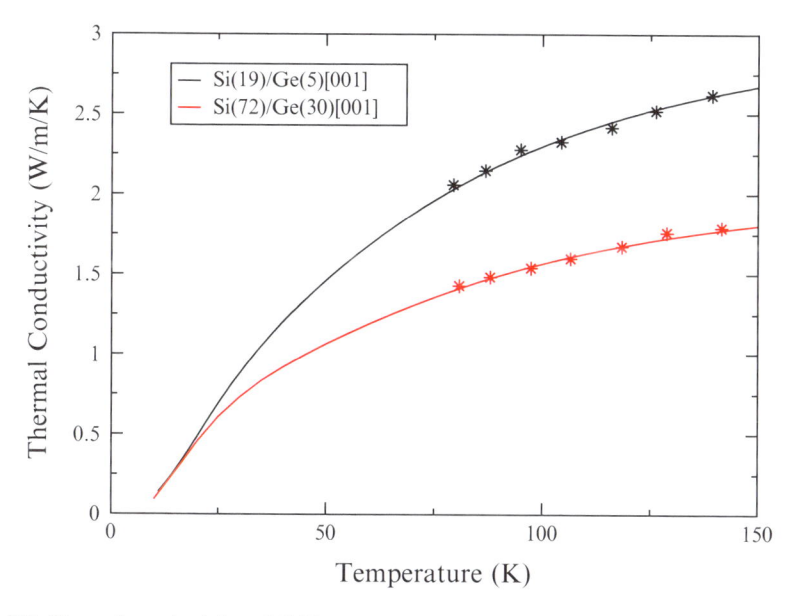

3 Theories of Phonon Transport in Bulk and Nanostructed Solids

Fig. 3.9 Thermal conductivity of Si/Ge superlattices along [001], the growth direction: theory [29] (*lines*), experiment [46] (*symbols*). Reproduced from [29]

the experimentally measured numerical results and the temperature variation of the conductivity up to 150 K, the role of phonon anharmonic interactions was found to be unimportant. While in the boundary scattering region there is essentially no difference in the conductivity of the two samples, at higher temperatures the lower conductivity in the (72,30) superlattice is due to the dominant IDS mechanism. Indeed, it was found that for the thinner superlattice IMS is the dominant scattering mechanism (with $\alpha = 550$) and the contribution from IDS is negligible ($\alpha' = 0$). For the thicker superlattice both IMS and IDS mechanisms are significant, but IDS dominates ($\alpha \simeq 10^7$ and $\alpha' \simeq 10^{-4}$).

The thermal conductivity of both the Si/Ge superlattices studied here is much lower compared to the conductivities of bulk Si or Ge. This is generally true of both the peak value of the conductivity and of the value at any temperature. This can be ascertained from an examination of the bulk results presented in Fig. 3.6 and the superlattice results presented in Fig. 3.9. At 100 K, the conductivity of the superlattices is three orders of magnitude smaller than the average of the conductivities of the bulk materials. It is interesting to note that at 100 K, the conductivity of the Si nanowires (cf. Fig. 3.8) is also three orders of magnitude smaller than the bulk conductivity in Si. However, the physical reasons for the reduction in the conductivity values in the superlattice and nanowire structures are different. The reduction in the conductivity of the Si nanowires is due to the phonon boundary scattering mechanism. In contrast, the reduction in the conductivity of the Si/Ge superlattices is due to a combined effects of reduced phonon velocity, stronger anharmonic interactions (consistent with larger periodicity), and phonon interface scattering mechanisms.

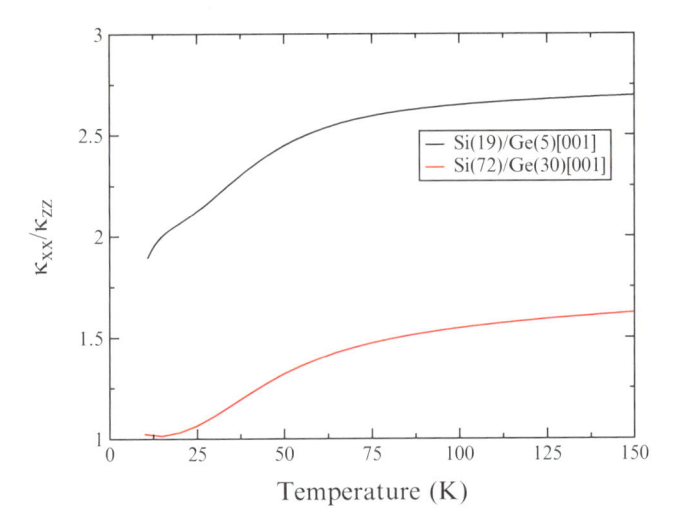

Fig. 3.10 The thermal conductivity ratio κ_{xx}/κ_{zz} for two Si/Ge superlattices as a function of temperature

While the conductivity in the bulk materials is essentially isotropic, it shows a clear anisotropic behaviour in the superlattice structure. Figure 3.10 shows the numerically obtained values of the conductivity in the superlattice plane (κ_{xx}) and along the superlattice growth direction (κ_{zz}). At low temperatures, the ratio κ_{xx}/κ_{zz} tends to unity for the thicker superlattice, but remains larger than unity for the thinner superlattice. This difference in the low-temperature ratio κ_{xx}/κ_{zz} arises due to different behaviours of the IMS and IDS mechanisms in thinner and thicker superlattices. Only low-lying phonon modes are appreciably populated at low temperatures. Also, as discussed earlier, zone-edge modes are more strongly scattered than zone-centre modes. For the thinner superlattice, with less zone-folding, the zone edge modes are not populated until higher temperatures. In contrast, in the thicker superlattice, there is a larger amount of zone folding, lending to population of a larger number of phonon modes both at the zone centre and at the zone edge. Thus, both the IMS and IDS mechanisms become more anisotropic for the thinner superlattice than for the thicker superlattice.

3.5 Concept of Phonon Engineering of Thermoelectric Materials

Thermoelectricity (TE) is the process of generating either electricity from heat engines or heating devices from electricity. Examples of modern TE applications include portable refrigerators, beverage coolers, electronic component coolers, infrared sensing, etc. Possible future applications of TE devices include efficient conversion of waste heat (e.g. from waste and during powering of vehicles) into usable energy, improving efficiency of photovoltaic cells, etc. Thermoelectric materials have been investigated for several decades due to their energy efficiency. The efficiency of thermoelectric materials is defined by the figure of merit quantity ZT given by $ZT = S^2\sigma T/\kappa$, where S is the Seebeck coefficient (a measure of conversion of temperature difference into electricity), σ is the electrical conductivity, and κ is the thermal conductivity. Larger values of ZT require high S, high σ, and low κ. After decades of research it has been established that alloyed semiconductors with high carrier concentration are the most efficient TE bulk materials.

The thermal conductivity κ of semiconducting materials can be contributed by carriers (donor electrons or acceptor holes in doped samples), lattice (or phonons), and electron–hole pairs (bipolar contribution in intrinsic semiconductors): $\kappa = \kappa_{el} + \kappa_{ph} + \kappa_{bp}$. An increase in S normally implies a decrease in σ because of carrier density considerations, and an increase in σ normally implies an increase in κ_{el} (as given by the Wiedemann–Franz law). The bipolar contribution κ_{bp} increases rapidly above temperatures corresponding to thermal energy larger than the semiconductor band gap. Thus, it is very difficult to increase ZT in typical bulk semiconductor TE materials. Clearly, for high values of ZT, we require materials which are characterised by less efficient scattering of carriers (to increase σ) and efficient scattering of phonons (to reduce κ). From research over the past several decades it has been found that semiconductor bulk alloys such as SiGe and BeTeSe, and PbTeS are amongst the most promising TE materials. In fact, SiGe are good high-temperature TE materials [50] and bismuth chalcogenides (e.g. BiTeSe) are good low-temperature TE materials [51]. Good reviews of the current status can be found in [52–54]. For most bulk materials the room-temperature ZT has been found to be less than 1. However, there is a report of $ZT = 2$ at $T = 800$ K for $AgPb_mSbTe_{2+m}$ [55].

Several theoretical and computational attempts have been made to obtain numerically accurate estimates of the TE coefficients σ, S, κ_{el}, κ_{bp}, and κ_{ph}. In order to compute σ, S, and κ_{el} at different temperatures it is important first to compute Fermi level both in extrinsic (carrier controlled) and intrinsic (host controlled) temperature ranges [56]. With that information being available, the nearly free electron theory can reasonably well be applied to compute σ, S, and κ_{el}. Useful expressions can be found in [57, 58]. The coefficient κ_{bp} is usually computed using the Price theory [59]. The theory discussed in this chapter can be used to compute κ_{ph}. Interested readers are referred to [30, 60–62] for details.

In recent years efforts have started to overcome the challenge of developing materials with $ZT > 3$. Such efforts have concentrated on creating new semiconducting materials. Two primary approaches are being considered: formation of complex crystal structures or fabrication of reduced-dimensional (especially nanostructured) materials. Reduced-dimensional systems can be categorised as 2D structures (i.e. thin films), 1D structures (i.e. nanowires), or 0D structures (i.e. quantum dots). Nanostructures can also be fabricated using two or more materials (known generally as nanocomposites), such as superlattices of alternating layers of two materials, an array of nanowires of one material embedded in another host, and an array of nanodots of one material embedded in the host of another material. Theoretical modelling suggests [58] that a Bi_2Te_3 quantum-well structure has the potential to increase ZT by an order of magnitude over the bulk value. Although Si is a poor thermoelectric material, arrays of rough Si nanowires of diameter in the range 20–300 nm are found to exhibit $ZT = 0.6$ at room temperature [63]. Recently, it has been found that stacks of thin films of Bi_2Te_3 exhibit enhanced ZT [64]. In particular, it has been theoretically suggested that quintuple atomic layers of Bi_2Te_3 can exhibit $ZT = 7.15$ [65].

It is exciting to think that in general fabrication of reduced-dimensional structures, in particular nanocomposite structures, can be tailored to exhibit much reduced lattice (phonon) conductivity, thus leading to an enhancement in ZT. However, a detailed and accurate investigation of an enhancement in ZT of nanocomposites will require knowledge of phonon dispersion relations and phonon scattering processes and their relative strengths. In particular, for nanowires and nanodots phonon scattering at rough boundaries is likely to play a dominant role. For superlattices, as discussed earlier in this chapter, interface scattering and enhanced anharmonic interaction due to the onset of mini-Umklapp processes and the dual mass term will play a dominant role in reducing the phonon lifetime and thus the thermal conductivity. The theoretical developments in Sect. 3.3.3.3 have indicated how to deal with phonon interface scatterings. Due to our poor understanding of temperature-dependent crystal anharmonic forces and the resultant phonon–phonon scattering strength, accurate calculations of phonon conductivity is, at least at present, essentially a very difficult problem even for single crystal semiconductors. In Sect. 3.3.3.4 in this chapter we have indicated how the concept of phonon anharmonic interactions in bulk can be extended to the case of a superlattice. These ideas can be modified and extended to deal with phonon scattering rates in nanocomposites in general. With the help of accurate determination of phonon lifetimes and phonon conductivity κ_{ph} for various types and sizes of nanocomposites it would be possible to develop the concept of phonon engineering for efficient TE materials.

3.6 Summary

In this chapter we have reviewed some of the existing theories of phonon transport in bulk and nanostructured solids. Particular attention has been paid to a detailed description of the essential steps required in the derivation and numerical evaluation of the thermal conductivity within the single-mode and an effective-mode phonon relaxation time approximations. Lattice thermal conductivity results have been presented for bulk Si, Ge, and GaAs, for suspended nanobeams of GaAs, for Si nanowires, and for Si/Ge superlattices.

It has been shown that the effect of reduction in dimensionality of a material results in significant reduction in its lattice thermal conductivity. Reduction of up to three orders of magnitude has been noted for thin Si nanowires (quasi one-dimensional structures) and thin Si/Ge superlattices (quasi two-dimensional structures). It has been explained that the physical reasons behind the same amount of reduction in the conductivity due to the Si/Ge superlattice and Si nanowire formations are different. Another significant effect of reduction in dimensionality is to change the isotropic nature of the conductivity into a tensor quantity. This has been illustrated from numerical calculations for Si/Ge superlattices. It has been explained that the anisotropic nature of the conductivity is more pronounced in thinner superlattices.

Finally, it has been pointed out that a huge reduction in the lattice thermal conductivity can be achieved by the formation of nanostructured semiconductors, resulting into the possibility of significant enhancement in thermoelectric figure of merit. Such an enhancement can be achieved, within reasons, by employing the concept of phonon engineering of nanocomposite semiconductors: i.e. by reducing the velocities and lifetimes of thermally active phonon modes by fabricating nanostructures of different shapes and sizes.

Acknowledgements I wish to thank my past and present Ph. D. students and postdoctoral fellows for their contribution towards the development of the subject matter presented here. Special thanks to Ceyda Yelgel for careful reading of the manuscript. This work has been supported by EPSRC (UK) through the grant number EP/H046690.

References

1. Kubo R (1957) J Phys Soc Jpn 12:570
2. Ziman JM (1960) Electrons and phonons. Clarendon, Oxford
3. Srivastava GP (1990) The physics of phonons. Adam Hilger, Bristol (now Taylor and Francis Group)
4. McGaughey AJH, Kaviany M (2004) Phys Rev B 69:094303
5. Kaburaki H, Yip S, Kimizuka H (2007) J Appl Phys 102:043514
6. Huang B-L, Kaviany M (2008) Phys Rev B 77:125209
7. Ziman JM (1969) Elements of advanced quantum thoery. Cambridge University Press, Cambridge
8. Guyer RA, Krumhansl JA (1966) Phys Rev 148:766
9. Benin D (1970) Phys Rev B 1:2777
10. Srivastava GP (1976) J Phys C Solid State Phys 9:3037
11. Srivastava GP (1976) J Phys C Solid State Phys 10.1843
12. Arthurs AM (1970) Complementary variational principles. Clarendon, Oxford
13. Carruthers P (1961) Rev Mod Phys 33:92
14. Callaway J (1959) Phys Rev 113:1046
15. Srivastava GP (1976) Phil Mag 34:795
16. Weber W (1977) Phys Rev B 15:4789
17. Tütüncü HM, Srivastava GP (1996) Phys Rev B 53:15675
18. Hepplestone SP, Srivastava GP (2008) Phys Rev Lett 101:105502
19. Baroni S, de Gironcoli S, Dal Corso A, Giannozzi P (2001) Rev Mod Phys 73:515
20. Dolling G (1963) Inelastic scattering of neutrons in solids and liquids, vol I. IAEA, Vienna, p 37
21. Nilsson G, Nelin G (1972) Phys Rev B 6:3777

22. Ezzahri Y, Grauby S, Rampnoux JM, Michel H, Pernot G, Claeys W, Dilhaire S, Rossignol C, Zeng G, Shakouri A (2007) Phys Rev B 75:195309
23. Hepplestone SP, Srivastava GP (2006) Nanotechnology 17:3288
24. Chadi DJ, Cohen ML (1973) Phys Rev B 8:5747
25. Monkhorst HJ, Pack JD (1976) Phys Rev B 13:5189
26. Holland MG (1964) Phys Rev 134:A471
27. Kim W, Majumdar A (2006) J Appl Phys 99:084306
28. Gillet J-N, Chalopin Y, Volz S (2009) J Heat Transfer 131:043206
29. Hepplestone SP, Srivastava GP (2010) Phys Rev B 82:144303
30. Thomas IO, Srivastava GP (2012) Phys Rev B 86:045205
31. Parrott JE (1971) Phys Status Solid B 48:K159
32. Ren SY, Dow JD (1982) Phys Rev B 25:3750
33. Ziman JM (1956) Phil Mag 1:191
34. Ziman JM (1957) Phil Mag 2:292
35. Parrott JE (1979) Rev Int Hautes Tem Refract 16:393
36. Holland MG, Neuringer LJ (1962) In: Proc. Int. congr. on the physics of semiconductors, exeter, Institute of Physics, London, p 475
37. Geballle TH, Hull GW (1958) Phys Rev 110:773
38. Srivastava GP (1980) J Phys Chem Solid 41:357
39. Fon W, Schwab KC, Worlock JM, Roukes ML (2002) Phys Rev B 66:045302
40. Barman S, Srivastava GP (2006) Phys Rev B 73:205308
41. Walkauskas SG, Broido DA, Kempa K, Reinecke TL (1999) J Appl Phys 85:2579
42. Volz SG, Chen G (1999) Appl Phys Lett 75:2056
43. Murphy PG, Moore JE (2007) Phys Rev B 76:155313
44. Li D, Wu Y, Kim P, Shi L, Yang P, Majumdar A (2003) Appl Phys Lett 83:2934
45. Srivastava GP (2009) Mat Res Soc Symp Proc 1172:T08-07
46. Lee SM, Cahill DG, Vekatasubramanian R (1997) Appl Phys Lett 70:2957
47. Cahill DG, Ford WK, Goodson KE, Mahan GD, Majumdar A, Maris HJ, Merlin R, Phillpot SR (2003) J Appl Phys 93:793
48. Capinski WS, Maris HJ, Ruf T, Cardona M, Ploog K, Katzer DS (1999) Phys Rev B 59:8105
49. Mo Y-W, Savage DE, Swartzentruber BS, Lagally MG (1990) Phys Rev Lett. 65:1020; Ma T, Tu H, Shao B, Liu A, Hu G (2006) Mater Sci Semicond Process 9:49
50. Dismukes JP, et al (1964) J Appl Phys 35:2899; Heddins HR, Parrott JE (1976) J Phys C Solid State Phys 9:1263
51. Goldsmid HJ (1964) Thermoelectric refrigeration. Plenum, New York; Goldsmid HJ (1986) Electronic refrigeration. Pion, London
52. Venkatasubramanium R, Siivola E, Colpitts T, O'Quinn B (2001) Nature 413:697
53. Majumdar A (2004) Science 303:777
54. Minnich AJ, Dresselhaus MS, Ren ZF, Chen G (2009) Energy Environ Sci 2:466
55. Hsu KF, Loo S, Guo F, Chen W, Dyck JS, Uher C, Hogan T, Polychroniadis EK, Kanatzidis MG (2004) Science 303:818
56. McKilvey JP (1966) Solid state and semiconductor physics (International edition). Harper & Row, New York-Evanston-London, and John Weatherhill, Tokyo
57. Drabble JR, Goldsmit HJ (1961) Thermal conduction in semiconductors. Pergamon Press, Oxford, pp 115–117
58. Hicks LD, Dresselhaus MS (1993) Phys Rev B 47:12727
59. Price PJ (1955) Phil Mag 46:1252
60. Vining CB (1991) J Appl Phys 69:331
61. Minnich AJ, Lee H, Wang XW, Joshi G, Dresselhaus MS, Ren ZF, Chen G, Vashaee D (2009) Phys Rev B 80:155327
62. Yelgel ÖC, Srivastava GP (2012) Phys Rev B 85:125207
63. Hochbaum A, Chen R, Deigado RD, Liang W, Garnett EC, Najarian M, Majumdar A, Yang P (2008) Nature 451:163
64. Goyal V, Teweldebrhan D, Balandin AA (2010) Appl Phys Lett 97:133117
65. Zahid F, Lake R (2010) Appl Phys Lett 97:212102

Chapter 4
First-Principles Determination of Phonon Lifetimes, Mean Free Paths, and Thermal Conductivities in Crystalline Materials: Pure Silicon and Germanium

Jivtesh Garg, Nicola Bonini, and Nicola Marzari

Abstract The thermal properties of insulating, crystalline materials are essentially determined by their phonon dispersions, the finite-temperature excitations of their phonon populations—treated as a Bose–Einstein gas of harmonic oscillators—and the lifetimes of these excitations. The conceptual foundations of this picture are now a well-established cornerstone in the theory of solids. However, only in recent years our theoretical and algorithmic capabilities have reached the point where we can now determine all these quantities from first-principles, i.e. from a quantum-mechanical description of the system at hand without any empirical input. Such advances have been largely due to the development of density-functional perturbation theory that allows to calculate second- and third-order perturbations of a system of interacting electrons with a cost that is independent of the wavelength of the perturbation. Here we present an extensive case study for the phonon dispersions, phonon lifetimes, phonon mean free paths, and thermal conductivities for isotopically pure silicon and germanium, showing excellent agreement with experimental results, where available, and providing much needed microscopic insight in the fundamental atomistic processes giving rise to thermal conductivity in crystals.

J. Garg (✉)
Department of Mechanical Engineering, Massachusetts Institute of Technology, Cambridge, MA, USA
e-mail: jivtesh@mit.edu

N. Bonini
Department of Physics, King's College, London, UK
e-mail: nicola.bonini@kcl.ac.uk

N. Marzari
Theory and Simulation of Materials (THEOS), École Polytechnique Fédérale de Lausanne, Lausanne, Switzerland
e-mail: nicola.marzari@epfl.ch

Thermal conductivity is a fundamental transport property that plays a vital role in many applications. In semiconductors and insulators, heat is conducted by lattice vibrations. Understanding and quantifying the interactions between these vibrations is critical to accurately predict thermal transport properties. A theoretical approach to calculate lattice thermal conductivity in these materials would facilitate our understanding of heat dissipation in microelectronics and nanoelectronics as well as assist in design of high efficiency thermoelectrics for both refrigeration and power generation applications. At temperature above a few tens of degrees Kelvin the lattice thermal conductivity of semiconductors is usually dominated by three-phonon scattering, arising from the anharmonicity of the interatomic potential. Anharmonic phonon scattering is an intrinsic resistive process and does not require the presence of defects, impurities or grain boundaries in the material. In 1929 Peierls [24] first formulated a microscopic description of the intrinsic lattice thermal conductivity of semiconductors and insulators through what has become known as the phonon Boltzmann equation (PBE). The equation involves the unknown perturbed population of a phonon mode and balances the perturbation due to the temperature gradient to the change in phonon population due to scattering. The perturbed phonon populations can be obtained by solving the PBE. While the framework to describe thermal conductivity is well known, the development of a predictive theoretical approach to calculate thermal conductivity has been hindered by the significant complexity inherent in describing (a) interatomic forces between atoms (IFCs) and (b) the inelastic phonon–phonon scattering processes.

The first issue can be addressed by using density-functional perturbation theory [1, 6a, 14] to obtain interatomic force constants. Use of DFPT has been shown in the past to yield accurate IFCs, which have led to the prediction of material properties in good agreement with measured values. The challenge involved in the second issue lies in the dependence of inelastic phonon–phonon scattering rates upon the unknown perturbed phonon populations of the phonon modes in the Brillouin zone. A tremendous simplification is, however, achieved in the calculation of thermal conductivity in bulk semiconductors by using the single-mode relaxation time (SMRT) approximation [29]. In this approximation it is assumed that only the phonon mode under consideration is out of equilibrium and relaxes to its equilibrium state, while all other modes remain in their equilibrium states. This allows the three-phonon scattering rate to be expressed in terms of only the unknown population of that mode and a phonon relaxation time which is completely known. This further allows the Boltzmann equation to be solved for the unknown population. The thermal conductivity in this approximation is determined by fundamental properties such as phonon frequencies, group velocities, phonon populations and phonon relaxation times. Recently we have used this approximation with interatomic force constants computed from DFPT to explore the thermal transport properties of silicon-germanium alloys [11] and superlattices [12] as well as of free-standing graphene [2].

Going beyond the use of single-mode relaxation time approximation, the PBE in its linearized form can be solved exactly by using a self-consistent iterative procedure first developed by Omini and Sparavigna [22]. Broido et al. [4] used such

4 First-Principles Determination of Phonon Lifetimes...

an exact solution of the phonon Boltzmann equation along with the interatomic force constants derived from density-functional perturbation theory [7] to compute the thermal conductivity of isotopically pure silicon and obtained excellent agreement with measured values. Recently, a novel approach to solve the PBE using a variational formulation and conjugate-gradient minimization has been introduced [10], showing excellent computational performance.

In this chapter, the theory for first-principles thermal conductivity calculations is presented along with its implementation for the paradigmatic case of isotopically pure Si and Ge. The results are compared against the experimentally measured values, allowing for a benchmarking of the approach.

4.1 Theory of Thermal Conductivity

In this section we present the theory of thermal conductivity following Ref. [27]. The potential energy V of a crystal, in which the unit cell is characterized by the vector l and the atomic positions in each unit cell are described by the vector b, can be expanded in a Taylor series in powers of the atomic displacements $u(lb)$; we show here the expansion up to third-order:

$$V = V_0 + \sum_{lb\alpha} \frac{\partial V}{\partial u_\alpha(lb)}\bigg|_0 u_\alpha(lb)$$

$$+ \frac{1}{2} \sum_{lb,l'b'} \sum_{\alpha\beta} \frac{\partial^2 V}{\partial u_\alpha(lb)\partial u_\beta(l'b')}\bigg|_0 u_\alpha(lb)u_\beta(l'b')$$

$$+ \frac{1}{3!} \sum_{lb,l'b',l''b''} \sum_{\alpha\beta\gamma} \frac{\partial^3 V}{\partial u_\alpha(lb)\partial u_\beta(l'b')\partial u_\gamma(l''b'')}\bigg|_0 u_\alpha(lb)u_\beta(l'b')u_\gamma(l''b'').$$

(4.1)

The second and third-order derivatives of the energy with respect to atomic displacements yield the second- and third-order interatomic force constants given by Eqs. (4.2) and (4.3), respectively.

$$\Phi_{\alpha\beta}(lb,l'b') = \frac{\partial^2 V}{\partial u_\alpha(lb)\partial u_\beta(l'b')}\bigg|_0 \qquad (4.2)$$

$$\Phi_{\alpha\beta\gamma}(lb,l'b',l''b'') = \frac{\partial^3 V}{\partial u_\alpha(lb)\partial u_\beta(l'b')\partial u_\gamma(l''b'')}\bigg|_0. \qquad (4.3)$$

The second-order force constants are used to compute the dynamical matrix $D_{\alpha\beta}(bb'|q)$ given by,

$$D_{\alpha\beta}(bb'|q) = \frac{1}{\sqrt{m_b m_{b'}}} \sum_{l'} \Phi_{\alpha\beta}(0b,l'b') exp(iq.l'), \qquad (4.4)$$

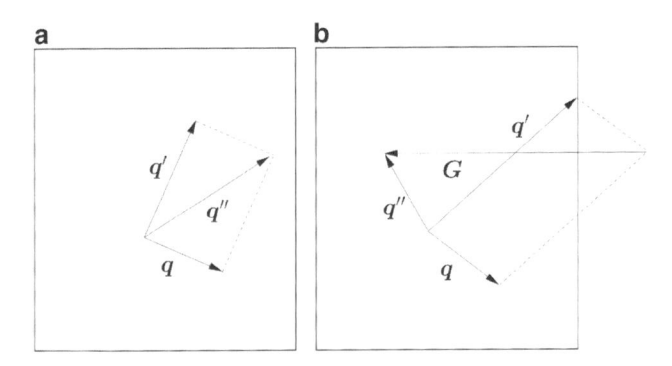

Fig. 4.1 (a) Normal and (b) Umklapp processes—for class 1 events or the "coalescence processes"

where q is the wave-vector and m denotes the mass of an atom at a particular site in the crystal. Diagonalizing the dynamical matrix yields the phonon frequencies ω.

The third-order terms lead to the scattering of phonons through intrinsic three-phonon scattering processes. These processes conserve momentum such that $G = q + q' + q''$, where G is the reciprocal lattice vector, q, q' and q'' are the wave-vectors of the three phonon modes involved in scattering. The above constraint leads to two types of scattering processes depending upon the magnitude of the reciprocal lattice vector G.

1. Normal processes: For these types of processes, $G = 0$. For example, when a phonon q scatters by absorbing another phonon q' to yield phonon q'' the momentum conservation for this process can be written as $-q - q' + q'' = 0$. This is shown in Fig. 4.1a. These processes preserve the direction of energy flow, since the resulting phonon is in the same direction as the combining phonon modes. As a result such processes do not contribute towards thermal resistance in a material.
2. The second type of processes are characterized by $G \neq 0$. Here the momentum conservation for the process indicated above would be $q + q' = q'' + G$. This is shown in Fig. 4.1b. As can be seen, in this process the direction of energy flow is reversed. These types of scattering processes were given the name "Umklapp" by Peierls [25], and they give rise to thermal resistance in a material.

The scattering processes can be further classified as class 1 events or "coalescence processes" where a phonon mode (qs) scatters by absorbing another mode $(q's')$ to yield a third phonon mode $(q''s'')$ and class 2 events or "decay processes" where a phonon mode (qs) scatters by decaying into two phonon modes $(q's')$ and $(q''s'')$. These two classes of events are shown in Fig. 4.2a,b, respectively.

Both processes satisfy momentum and energy conservation: For class 1 events, these are:

$$q + q' = q'', \quad \hbar\omega(qs) + \hbar\omega(q's') = \hbar\omega(q''s''), \tag{4.5}$$

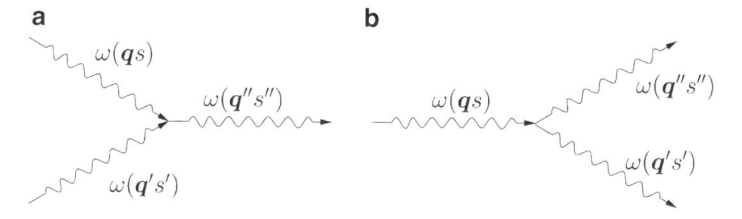

4 First-Principles Determination of Phonon Lifetimes... 119

Fig. 4.2 (**a**) Class 1 and (**b**) Class 2 events associated with three-phonon scattering

while for class 2 events we have

$$q = q' + q'', \quad \hbar\omega(qs) = \hbar\omega(q's') + \hbar\omega(q''s''). \tag{4.6}$$

The scattering rate for a three-phonon scattering process can be computed using Fermi's golden rule [8, 9]:

$$P_i^f(3ph) = \frac{2\pi}{\hbar} |\langle f|V_3|i\rangle|^2 \delta(E_f - E_i). \tag{4.7}$$

In the above equation i and f denote the initial and final state, V_3 is the three phonon coupling potential and is related to the crystal anharmonicity expressed by Eq. (4.3), and $\delta(E_f - E_i)$ denotes energy conservation between the initial and final states. The net scattering rate of a phonon mode is the sum of scattering rates due to the class 1 and class 2 events and, for a small deviation from equilibrium, is given by

$$-\frac{\partial n_{qs}}{\partial t}\bigg|_{\text{scatt}} = \sum_{q's',s''} \bigg[\tilde{P}^+_{qs,q's'q''s''}(\Psi^s_q + \Psi^{s'}_{q'} - \Psi^{s''}_{q''})$$

$$+ \frac{1}{2}\tilde{P}^-_{qsq's'.q''s''}(\Psi^s_q - \Psi^{s'}_{q'} - \Psi^{s''}_{q''})\bigg]. \tag{4.8}$$

In the above $\tilde{P}^{\pm}_{qs,q's'q''s''}$ are the intrinsic scattering rates for the class 1 and class 2 events and are given by,

$$\tilde{P}^{\pm}_{qs,q's'q''s''} = 2\pi \bar{n}_{qs} \left(\bar{n}_{q's'} + \frac{1}{2} \mp \frac{1}{2}\right)(\bar{n}_{q''s''} + 1)|\tilde{V}_3(-qs, \mp q's', q''s'')|^2$$

$$\times \delta(\omega(qs) \pm \omega(q's') - \omega(q''s'')), \tag{4.9}$$

and Ψ_{qs} is a first order perturbation that relates the perturbed population n_{qs} to the equilibrium Bose–Einstein distribution $\bar{n}_{qs}$ through

$$n_{qs} = \bar{n}_{qs} - \frac{k_B T}{\hbar} \frac{\partial \bar{n}_{qs}}{\partial \omega(qs)} \Psi_{qs}, \tag{4.10}$$

leading to $n_{qs} = \bar{n}_{qs} + \bar{n}_{qs}(\bar{n}_{qs} + 1)\Psi_{qs}$.

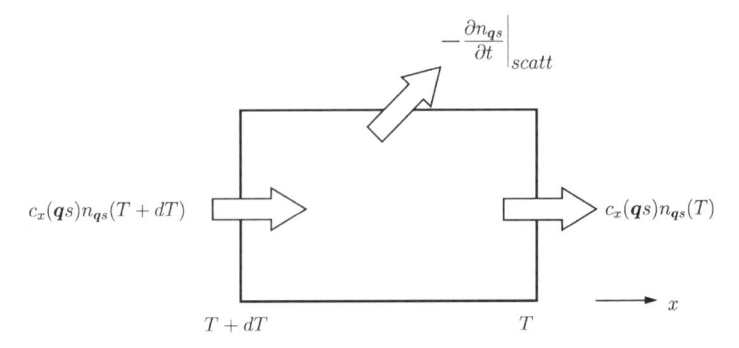

Fig. 4.3 Perturbation in phonon population due to temperature gradient balanced by the change due to scattering induced by the anharmonicity of the interatomic potential

$\tilde{V}_3(qs, q's', q''s'')$ is the three-phonon anharmonic coupling given by

$$\tilde{V}_3(qs, q's', q''s'') = \left(\frac{\hbar}{8N_0\omega(qs)\omega(q's')\omega(q''s'')}\right)^{1/2} \sum_{b,b',b''\alpha\beta\gamma} \Phi_{\alpha\beta\gamma}(qb, q'b', q''b'')$$
$$\times \frac{e_\alpha(b|qs)}{\sqrt{m_b}} \frac{e_\beta(b'|q's')}{\sqrt{m_{b'}}} \frac{e_\gamma(b''|q''s'')}{\sqrt{m_{b''}}}, \quad (4.11)$$

where e are vibration eigenvectors, m are the atomic masses and $\Phi_{\alpha\beta\gamma}(qb, q'b', q''b'')$ are the Fourier transformed third-order interatomic force constants given by,

$$\Phi_{\alpha\beta\gamma}(qb, q'b', q''b'') = \sum_{l',l''} \Phi_{\alpha\beta\gamma}(ob, l'b', l''b'') e^{iq'l'} e^{iq''l''}. \quad (4.12)$$

The scattering rate given by Eq. (4.8) when combined with the temperature gradient induced phonon diffusion yields the phonon Boltzmann equation (PBE) [22, 27, 29] for the perturbed phonon populations (see Fig. 4.3):

$$-c(qs)\cdot\nabla T\left(\frac{\partial n_{qs}}{\partial T}\right) + \frac{\partial n_{qs}}{\partial t}\bigg|_{scatt} = 0, \quad (4.13)$$

where n_{qs} and $c(qs)$ are the perturbed phonon population and group velocity, respectively, of mode qs. Assuming that the perturbation from equilibrium is small, the temperature gradient of the perturbed phonon population can be replaced with the temperature gradient of the equilibrium phonon population, $\partial n_{qs}/\partial T \approx \partial \bar{n}_{qs}/\partial T$.

Equations (4.8) and (4.13) show that the PBE for the unknown Ψ_q^s is coupled together with the unknown phonon populations of all other modes ($\Psi_{q'}^{s'}$, $\Psi_{q''}^{s''}$) all over the Brillouin zone. The complexity involved in solving the phonon Boltzmann equation (PBE) based on the three-phonon processes lies in the dependence of the distribution function n_{qs} on the occupation of all other states, allowed by energy and momentum conservation. The scattering rate of a mode when the entire system relaxes to equilibrium would in general not be the same as when all other modes are in equilibrium. In the first situation the PBEs of all the different modes qs are coupled together and have to be solved simultaneously in a self-consistent way [3, 22]. The second situation corresponds to the single mode relaxation time approximation [29]. In this approximation, the PBE is solved for n_{qs} by assuming that $\Psi_{q'}^{s'}, \Psi_{q''}^{s''}$ are zero, where $q's'$ and $q''s''$ are modes involved in the scattering of mode qs. We first calculate the thermal conductivity in the single mode relaxation time approximation and later compare it with the result obtained from the full self-consistent solution of the Boltzmann transport equation.

In the single-mode relaxation time approximation, by setting $\Psi_{q'}^{s'}, \Psi_{q''}^{s''} = 0$ Eq. (4.8) can be rewritten as,

$$-\frac{\partial n_{qs}}{\partial t}\bigg|_{scatt} = \bar{n}_{qs}(\bar{n}_{qs}+1)\Psi_q^s \pi \sum_{q's',s''} |\tilde{V}_3(-qs,q's',q''s'')|^2$$
$$\times \big[2(\bar{n}_{q's'} - \bar{n}_{q''s''})\delta(\omega(qs) + \omega(q's') - \omega(q''s''))$$
$$+ (1 + \bar{n}_{q's'} + \bar{n}_{q''s''})\delta(\omega(qs) - \omega(q's') - \omega(q''s''))\big]. \quad (4.14)$$

The above scattering rate can also be written using the perturbation in the phonon population and a relaxation time as

$$-\frac{\partial n_{qs}}{\partial t}\bigg|_{scatt} = \frac{n_{qs} - \bar{n}_{qs}}{\tau_{qs}}, \quad (4.15)$$

where τ_{qs} is the phonon relaxation time and is given by the following expression:

$$\frac{1}{\tau_{qs}} = 2\Gamma_{qs} = \pi \sum_{q's',s''} |\tilde{V}_3(-qs,q's',q''s'')|^2$$
$$\times \big[2(\bar{n}_{q's'} - \bar{n}_{q''s''})\delta(\omega(qs) + \omega(q's') - \omega(q''s''))$$
$$+ (1 + \bar{n}_{q's'} + \bar{n}_{q''s''})\delta(\omega(qs) - \omega(q's') - \omega(q''s''))\big]. \quad (4.16)$$

$2\Gamma_{qs}$ is the full linewidth at half maximum (FWHM).

The heat flux Q in the presence of a temperature gradient can be written in terms of phonon energies $\hbar\omega(qs)$, perturbed phonon populations n_{qs} and phonon group velocities $c(qs)$ as

$$Q_\alpha = \frac{1}{N_0 \Omega} \sum_{qs} \hbar\omega(qs) c_\alpha(qs) n_{qs} = -k_{\alpha\beta} |\nabla T|_\beta, \quad (4.17)$$

where Ω is the volume of the unit-cell. In the above thermal conductivity is a tensor whose components $k_{\alpha\beta}$ give the direction of heat flux along a direction α for a temperature gradient along direction β.

Finally solving for the perturbed phonon populations using Eqs. (4.13), (4.15) and (4.16) and substituting them in the expression for heat flux yield the thermal conductivity,

$$k_{\alpha\beta} = \frac{\hbar^2}{N_0 \Omega k_B T^2} \sum_{qs} c_\alpha(qs) c_\beta(qs) \omega^2(qs) n_{qs}(n_{qs}+1) \tau_{qs}. \qquad (4.18)$$

4.2 Implementation

In order to compute thermal conductivity in the single mode relaxation time approximation, the only inputs required are the second-order and the third-order interatomic force constants (IFCs).

The steps involved in the thermal conductivity calculation are outlined below:

1. The second-order interatomic force constants (IFCs) $\Phi_{\alpha\beta}(ob, hb')$ are obtained. The second-order IFCs allow computation of the dynamical matrix $D_{\alpha\beta}(bb'|q)$, whose eigenvalues yield the phonon frequencies and the dispersion, from which the phonon group velocities and Bose–Einstein populations can be computed. The second-order force constants yield the second derivative of energy with respect to two atomic displacements; as the distance between the atoms increases, these force constants diminish in magnitude. In order to ensure an accurate estimate of dynamical matrix, these force constants have to be obtained in real space on a large enough supercell such that the force constants have decayed to negligibly small values. We find that a supercell size of $10 \times 10 \times 10$ is enough to ensure this. However, direct calculation of force constants in real space is computationally expensive as it requires using a supercell with thousands of atoms. Instead, first the Brillouin zone is discretized into $10 \times 10 \times 10$ grid of q points; then the force constants in q space $\Phi_{\alpha\beta}(bb'|q)$ are computed for q belonging to this grid, using density-functional perturbation theory in reciprocal space. This calculation is computationally much cheaper as it involves using the primitive fcc unit cell with only two atoms. The force constants in q space, $\Phi_{\alpha\beta}(bb'|q)$, are then inverse Fourier transformed to obtain the force constants in real space:

$$\Phi_{\alpha\beta}(ob, hb') = \sum_q \Phi_{\alpha\beta}(bb'|q) exp(iq.h), \qquad (4.19)$$

where q in the above equation now belongs to the $10 \times 10 \times 10$ grid in the first Brillouin zone, and h is the lattice vector of a unit cell on a $10 \times 10 \times 10$ supercell in real space. The dynamical matrix at any arbitrary q' can now be obtained by a simple Fourier transform of these real space force constants

$$D_{\alpha\beta}(bb'|-q') = \frac{1}{\sqrt{m_b m_{b'}}} \sum_h \Phi_{\alpha\beta}(ob, hb') exp(-iq'.h). \qquad (4.20)$$

2. Next, the third-order interatomic force constants $\Phi_{\alpha\beta\gamma}(lb, l'b', l''b'')$ are computed. The third-order IFCs are used to compute the three-phonon scattering matrix elements, which along with the phonon frequencies and populations yield the phonon relaxation times. As in the case of second-order force constants, the third-order force constants decay with the distance between atoms and have to be computed on a large enough supercell such that for the farthest atoms in the supercell, these force constants have diminished to negligible values. In the case of second-order force constants this calculation was performed indirectly, the force constants were first obtained in q space and then inverse Fourier transformed to get them in real space. To repeat the same process for the third-order force constants, force constants in q space $\Phi_{\alpha\beta\gamma}(qb, q'b', q''b'')$ need to be determined for q, q' and q'' belonging to a chosen grid in the first Brillouin zone. As in the case of second-order force constants, this approach would reduce computational cost as it would allow using the primitive fcc unit cell with only two atoms. However DFPT as implemented in the Quantum-ESPRESSO package [13] currently only allows the above calculation for $q = 0$, $q'' = -q'$. Knowledge of the force constants $\Phi_{\alpha\beta\gamma}(0b, q'b', -q'b'')$ on the two atom unit cell is only sufficient to compute the linewidth of the phonon mode at Γ (since momentum conservation $q + q' + q'' = 0$ leads to $q'' = -q'$ for $q = 0$).

Due to the above limitation the third-order force constants $\Phi_{\alpha\beta\gamma}(lb, l'b', l''b'')$ are computed at Γ using density-functional perturbation theory on larger supercells sized $2 \times 2 \times 2$ and $3 \times 3 \times 3$ containing 16 and 54 atoms, respectively. This makes the calculation computationally expensive. By comparing the phonon linewidths obtained from force constants on a $2 \times 2 \times 2$ versus $3 \times 3 \times 3$ supercell, an estimate of the convergence of the phonon linewidth is obtained. We find that the force constants obtained on a $3 \times 3 \times 3$ supercell are adequate for the present calculation and lead to only a small error in the estimate of the phonon linewidth (inverse of phonon relaxation time). The three-phonon matrix elements are then computed using these force constants through Eq. (4.11), which are then used to compute phonon relaxation times using Eq. (4.16).

3. To compute the thermal conductivity, the first Brillouin zone is discretized into a grid of q points, and the thermal conductivity is computed using Eq. (4.18). At any q in the grid, the phonon frequencies are computed using the second-order force constants obtained in step 1, and the phonon group velocities are computed from the derivative of the phonon dispersion $\partial \omega / \partial q$, using the central difference technique

$$c(qs) = \frac{\partial \omega(qs)}{\partial q} = \frac{\omega(q + \Delta q, s) - \omega(q - \Delta q, s)}{2\Delta q}. \quad (4.21)$$

Finally, the phonon population is computed using the Bose–Einstein distribution,

$$\bar{n}_{qs} = \frac{1}{e^{\hbar \omega(qs)/(k_B T)} - 1}. \quad (4.22)$$

4. To compute the relaxation time of any phonon mode q, the Brillouin zone is again discretized into a grid of q'. The relaxation time is then computed by evaluating the sum in Eq. (4.16). The convergence of the computed relaxation time with respect to the size of the q' grid is studied. It is found that for a grid of size $30 \times 30 \times 30$, relaxation times are sufficiently converged. Also to compute the relaxation time, the delta function for the energy conservation in Eq. (4.16) is replaced by a Gaussian

$$\delta(\omega) = \frac{1}{\sqrt{\pi}\epsilon} exp(-(\omega/\epsilon)^2); \qquad (4.23)$$

a width of $\epsilon = 2.5\,\mathrm{cm}^{-1}$ along with a q' grid of size $30 \times 30 \times 30$ is found to lead to reasonably converged relaxation times.

5. Finally, the convergence of the computed thermal conductivity with respect to the size of the q grid in the first Brillouin zone is studied. The converged result is taken to be the thermal conductivity in the single mode relaxation time approximation.

For all density-functional perturbation theory calculations an $8 \times 8 \times 8$ Monkhorst–Pack [21] mesh is used to sample electronic states in the Brillouin zone and an energy cutoff of 20 Ry is used for the plane-wave expansion. Convergence of all quantities with respect to these parameters is carefully tested. First-principles calculations within density-functional theory are carried out using the PWscf and PHonon codes of the Quantum-ESPRESSO distribution [13] with norm-conserving pseudopotentials based on the approach of von Barth and Car [5].

4.3 Phonon Linewidth of the Zone Center Optical Mode in Si and Ge

Calculation of the linewidth of zone center optical phonon modes can be performed more accurately than that of any arbitrary q. This is due to the fact that for computing the linewidth of the zone center optical phonons, the elements $\Phi_{\alpha\beta\gamma}(qb, q'b', q''b'')$ need to be known only for $q = 0, q'' = -q'$, i.e. only $\Phi_{\alpha\beta\gamma}(0b, q'b', -q'b'')$ need to be known. As previously discussed, these can be obtained directly from density-functional perturbation theory as implemented in Quantum-ESPRESSO for q' belonging to grids even larger than $3 \times 3 \times 3$, and relatively cheaply.

Thus, to study the convergence of linewidth of the zone center optical mode, we obtain directly $\Phi_{\alpha\beta\gamma}(0b, q'b', -q'b'')$ using DFPT on grids of size $2 \times 2 \times 2$, $3 \times 3 \times 3$, and $4 \times 4 \times 4$. The theoretical linewidth are compared against the experimentally measured values in Fig. 4.4. It can be seen that the linewidths computed using $\Phi_{\alpha\beta\gamma}(0b, q'b', -q'b'')$ obtained on a $4 \times 4 \times 4$ initial grid agree very well with both the first-principles calculation performed by Lang et al. [19] and the experimental values. Interestingly however, the linewidths computed even with the $\Phi_{\alpha\beta\gamma}(0b, q'b', -q'b'')$ obtained on a $3 \times 3 \times 3$ initial grid agree well with

Fig. 4.4 Comparison of the phonon linewidth of the zone center optical mode in Si^{28} computed using the anharmonic force constants interpolated from three different initial q' grids in the Brillouin zone, $2 \times 2 \times 2$, $3 \times 3 \times 3$ and $4 \times 4 \times 4$. The computed values are compared against the computed values obtained by Lang et al. [19] and the experimentally measured values obtained by Menéndez and Cardona [20]

Fig. 4.5 Comparison of the phonon linewidth of the transverse acoustic modes along Γ-X $(0,0,\lambda)$ in Si^{28} at 300 K computed using anharmonic force constants obtained on two different supercells, $2 \times 2 \times 2$ and $3 \times 3 \times 3$

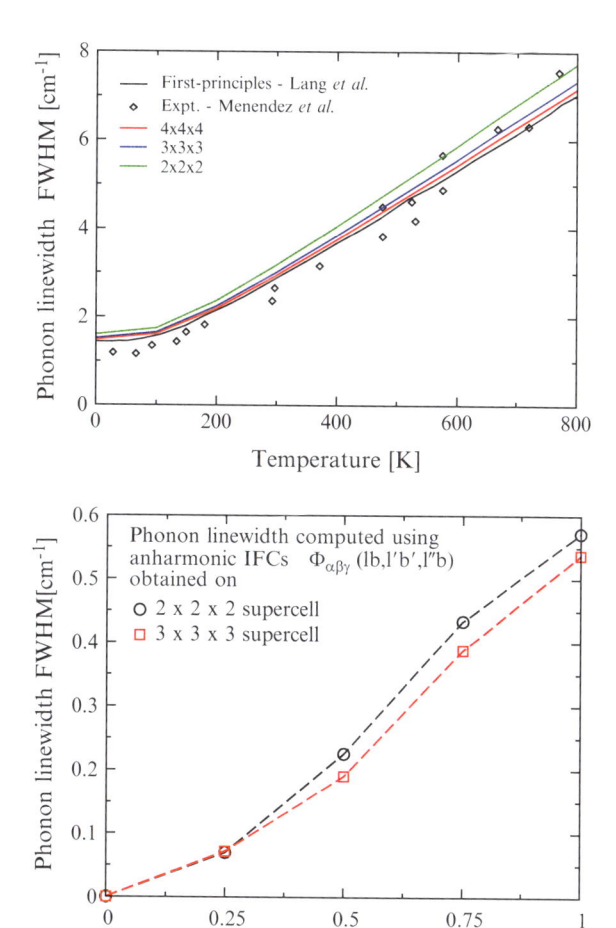

experimentally measured values, with only a small error. This confirms that the anharmonic force constants decay rapidly in real space and thus vary smoothly in q space, allowing accurate interpolations based on even a $3 \times 3 \times 3$ grid.

4.4 Phonon Lifetimes in Si and Ge

As indicated earlier, the third-order force constants $\Phi_{\alpha\beta\gamma}(ob, h'b', h''b'')$ needed to compute phonon linewidths were obtained on two different supercells, $2 \times 2 \times 2$ and $3 \times 3 \times 3$. The phonon linewidth of phonon modes along the direction Γ-X was computed using these two different sets of interatomic force constants. The results for the TA modes at 300 K are presented in Fig. 4.5. The difference between the two linewidths is only about 6.3% for the TA mode at $(0, 0, 1.0)$; however, it is much larger, about 18%, at $(0, 0, 0.5)$. Although the change in linewidth can be expected

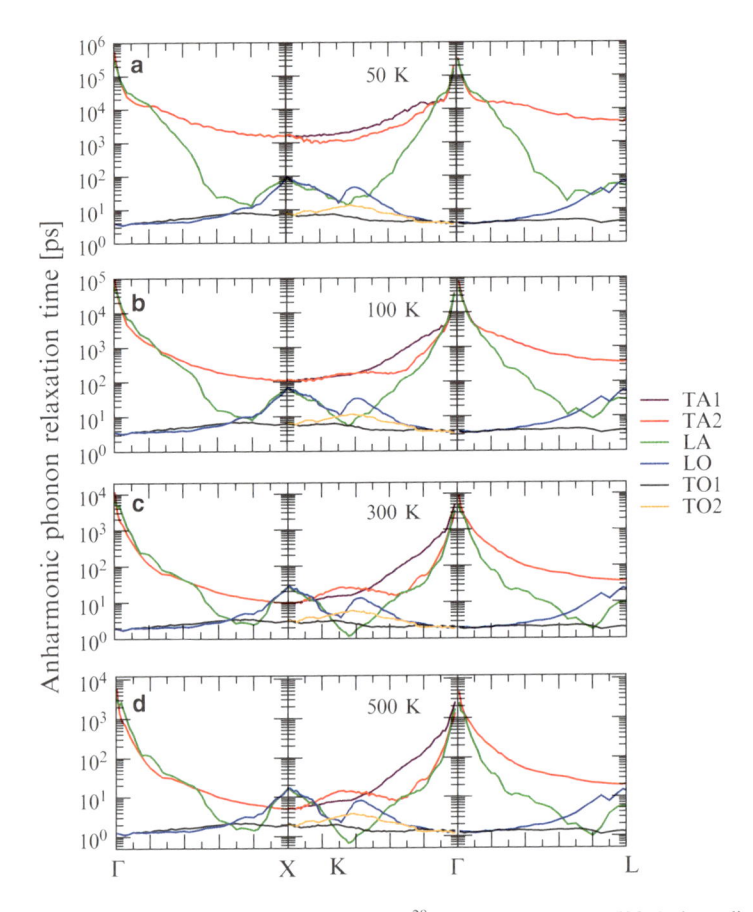

Fig. 4.6 Anharmonic phonon relaxation times in Si^{28} at 50, 100, 300 and 500 K along directions of high symmetry

to be smaller in going from a $3 \times 3 \times 3$ to a $4 \times 4 \times 4$ supercell, there is certainly a small error introduced in the thermal conductivity calculation due to the inability to compute the third-order IFCs on a supercell larger than $3 \times 3 \times 3$.

Phonon relaxation times to compute the thermal conductivity were thus computed using the third-order IFCs obtained on a $3 \times 3 \times 3$ supercell. In Fig. 4.6 the anharmonic relaxation times in silicon are presented along directions of high symmetry at 50, 100, 300, and 500 K, respectively.

4.5 Thermal Conductivity of Si and Ge

In this section, the heat carrying ability of the different modes is compared. The total thermal conductivity is computed and compared against experimentally measured values.

Fig. 4.7 (a) Convergence of the computed thermal conductivity of Si^{28} with the size of q grid in first Brillouin zone (b) Comparison between the computed thermal conductivity in the single-mode relaxation time approximation and the experimentally measured values for Si^{28} and Ge^{70}. The discrepancy is about 15% at 300K. Experimentally measured values are from [17] (Si^{28}, *solid squares*) and [23] (Ge^{70}, *solid triangles*)

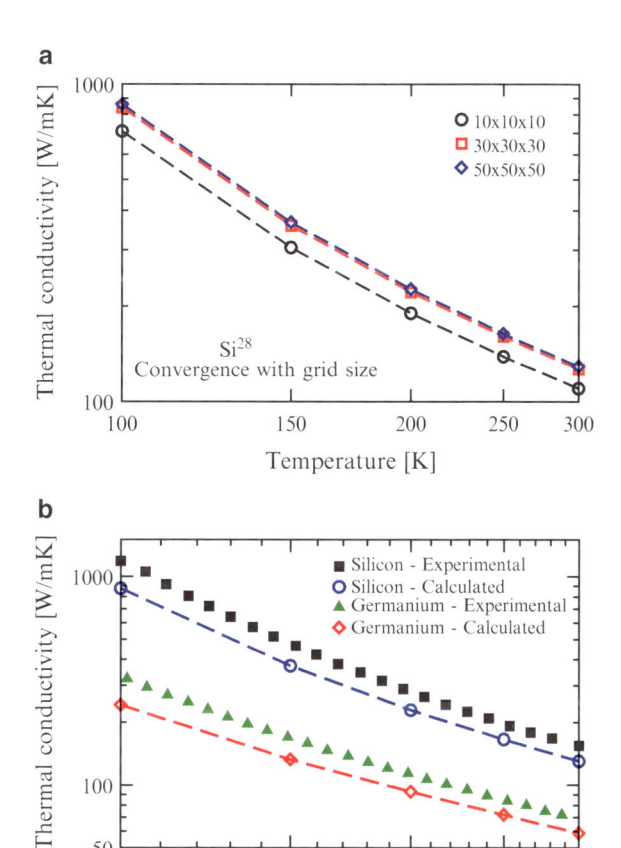

As indicated before, to compute the thermal conductivity the Brillouin zone is discretized in a grid of q wave-vectors. For each vibration mode qs in the grid the phonon frequencies, group velocities, populations and relaxation times are computed. The thermal conductivity is then computed by using Eq. (4.18). We first study the convergence of the computed thermal conductivity with respect to the size of the q grid. Figure 4.7a shows the thermal conductivity of Si^{28} computed using three different grid sizes of $10 \times 10 \times 10$, $30 \times 30 \times 30$ and $50 \times 50 \times 50$, respectively. The computed value converges for a grid size of $30 \times 30 \times 30$.

In Fig. 4.7b we compare the converged computed thermal conductivity of Si^{28} and Ge^{70} in the single mode relaxation time approximation with experimentally measured values. The computed values in the single mode relaxation time approximation agree well with the experimentally measured values [17, 23] both qualitatively and quantitatively, the disagreement being about 16% for Si^{28} and 14% for Ge^{70} at 300 K.

4.5.1 Contribution of TA and LA Modes to Thermal Conductivity

The good agreement between computed and measured values allows these results to be used for a more detailed understanding of the parameters controlling thermal transport in bulk semiconductor materials. One of the issues that has been strongly debated is the relative contribution of longitudinal and transverse acoustic modes in conducting heat in Silicon and Germanium. Hamilton and Parott [15] solved the Boltzmann transport equation by using a variational approach using a trial function; using a linear phonon dispersion, they showed that in Germanium transverse acoustic modes conduct about 80–90% of the heat, while the contribution of longitudinal phonons to thermal conductivity is less than 20%. Their work led to the idea that transverse acoustic modes play the dominant role in thermal conduction. Savvides and Goldsmid [26] used the results of Hamilton and Parott to explain their experimental results. However, Ju and Goodson [18] measured thermal conductivity of silicon thin films and through modelling explained the results by assuming that LA modes were the dominant heat carriers. More recently Henry and Chen [16] performed molecular dynamics simulations using an environment-dependent interatomic potential (EDIP) to study thermal transport in silicon. They found that LA phonons contributed roughly 45% to thermal conductivity while TA modes conducted about 50% of the heat. Clearly there is a large scatter in the values reported for the relative importance of TA and LA modes in conducting heat. The main reason for this disagreement is that while empirical potentials have been partially successful in capturing the second-order vibration properties such as the phonon dispersion correctly, their use to predict the anharmonic behaviour is largely unsuccessful. Empirical potentials are almost never fitted to any properties related to third-order derivatives (such as Gruneisen parameter) and therefore cannot be expected to yield the correct third-order behaviour.

In Fig. 4.8 we compare the different modes along Γ-L line in terms of their phonon frequencies, group velocities, populations and relaxation times, i.e all the ingredients necessary to compute thermal conductivities. The populations and relaxation times are presented at 300K: it can be seen right away that optical modes have much smaller group velocities, phonon populations and relaxation times, compared to the acoustic phonons. Optical phonons can thus be expected to have only a small contribution to the thermal conductivity. Among the acoustic modes, while transverse acoustic modes have lower frequencies and group velocities compared to LA modes, they have higher populations and relaxation times; Actually the relaxation times of TA modes are higher than those of the LA modes, almost by an order of magnitude in certain parts of the Brillouin zone.

In Fig. 4.9a we compare the heat carrying ability of the different modes along the direction Γ-L(λ,λ,λ). It can be seen that lower frequencies and group velocities of the TA modes are compensated by their higher relaxation times, leading to the heat conduction of each TA mode being comparable to that of the LA modes.

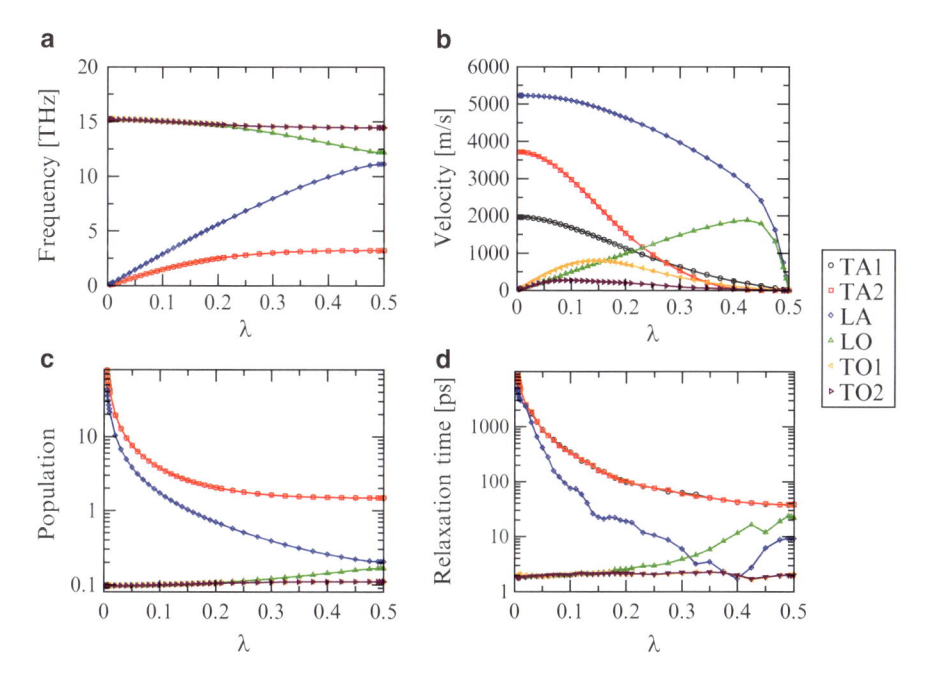

Fig. 4.8 (**a**) Phonon frequencies, (**b**) group velocities, (**c**) Bose–Einstein populations and (**d**) relaxation times—in Si^{28} at 300K for different modes along the Γ-L direction $(\lambda, \lambda, \lambda)$

In Fig. 4.9b the total contribution to the thermal conductivity of the TA and LA modes is compared. In silicon, at room temperature, it is found that the TA modes conduct about 63% of the heat and LA modes conduct about 32%, the remaining 5% being conducted by optical phonons. These values are significantly different from results reported above [15, 16, 18, 26]. First-principles calculations can thus provide more accurate understanding of parameters controlling thermal transport.

Figure 4.10a,b show the frequency dependence of the thermal conductivity in Si^{28} and Ge^{70}, respectively: in Si^{28}, even though acoustic modes extend in frequencies to more than 10 THz, only modes up to about 6 THz contribute to the thermal conductivity. A decrease in relaxation times and phonon group velocities with increase in frequency diminishes the heat conduction ability of higher frequency phonons. The small jump in thermal conductivity at 12 THz occurs due to the contribution of longitudinal optical modes. Similar trends can be seen for Ge^{70}.

4.5.2 Phonon Mean Free Path Dependence

The dependence of thermal conductivity on phonon mean free path is presented in Fig. 4.11. The phonon mean free path of a mode qs is taken to be the product of its relaxation time τ_{qs} and the magnitude of its group velocity $|c(qs)|$.

Fig. 4.9 (a) Heat conduction ability of different modes along Γ-L (λ, λ, λ) at 300 K in Si[28] (b) Comparison of the contribution of the transverse and longitudinal acoustic modes to total thermal conductivity in Si[28]. At 300 K, TA modes contribute almost 63% and LA modes contribute 32% of the thermal conductivity

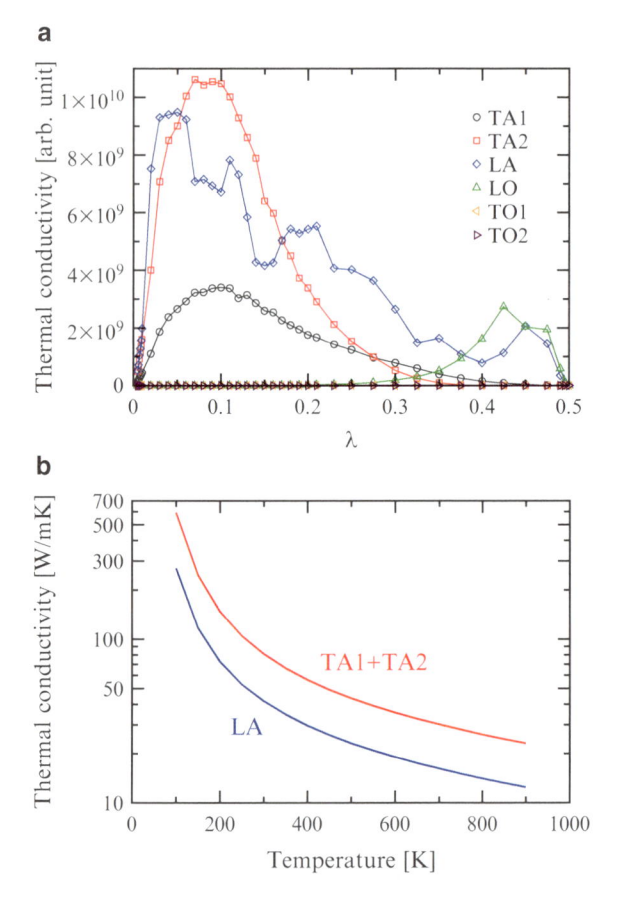

Peak contribution to thermal conductivity at room temperature comes primarily from phonons of relatively small mean free path, about 50 nm in Si[28] and about 40 nm in Ge[70]. As the temperature is lowered, an increase in relaxation times shifts the peak contribution to higher mean free paths. Even though the contribution to thermal conductivity drops significantly with the increase in mean free path, the long tail ensures that these large mean free path phonons still contribute significantly to the total thermal conductivity.

Figure 4.12 shows the accumulation of thermal conductivity as a function of phonon mean free path. In Si[28] significant heat is conducted by phonons of long mean free paths. This provides avenues to lower thermal conductivity through nanostructuring. Indeed room temperature thermal conductivity of polycrystalline silicon was measured to be almost an order of magnitude lower than that of single-crystal silicon [28].

Fig. 4.10 Frequency dependence of the thermal conductivity in (**a**) Si^{28} and (**b**) Ge^{70}

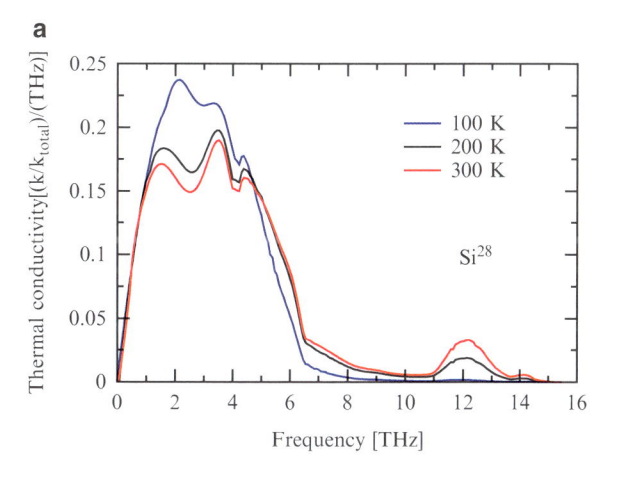

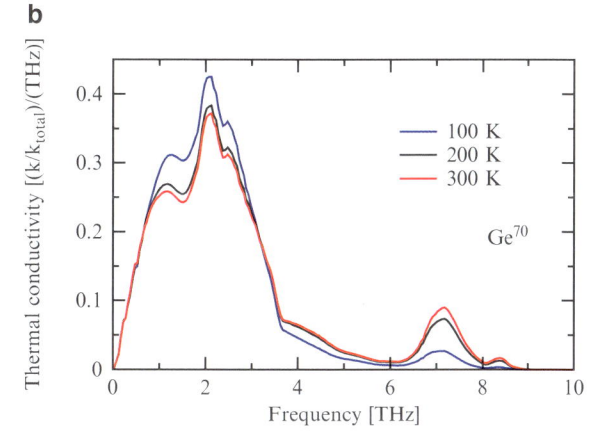

4.6 Full Iterative Solution

In the previous sections the thermal conductivity was computed by solving the Boltzmann transport equation in the single-mode relaxation time approximation. However, the Boltzmann transport equation can be solved exactly using a self consistent iterative solution [3, 22]. In this section, following [3], the full iterative solution is implemented, and the methodology presented.

Rewriting the phonon Boltzmann equation (PBE) as

$$-c(qs).\nabla T \left(\frac{\partial \bar{n}_{qs}}{\partial T} \right)$$
$$= \sum_{q's',s''} \left[\tilde{P}^+_{qs,q's'q''s''}(\Psi^s_q + \Psi^{s'}_{q'} - \Psi^{s''}_{q''}) + \frac{1}{2} \tilde{P}^-_{qsq's',q''s''}(\Psi^s_q - \Psi^{s'}_{q'} - \Psi^{s''}_{q''}) \right],$$

(4.24)

Fig. 4.11 Dependence of the thermal conductivity on phonon mean free paths in (**a**) Si^{28}. Peak contribution to thermal conductivity at 300 K comes from phonons of mean free path around 50 nm. However, as shown by the long tail present even higher mean free path phonons make significant contributions to thermal conductivity. (**b**) Same for Ge^{70}

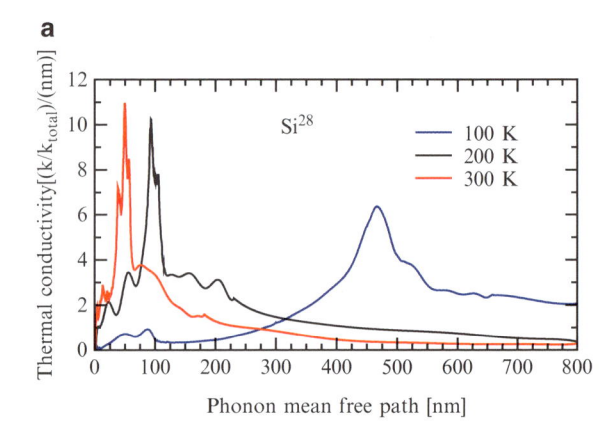

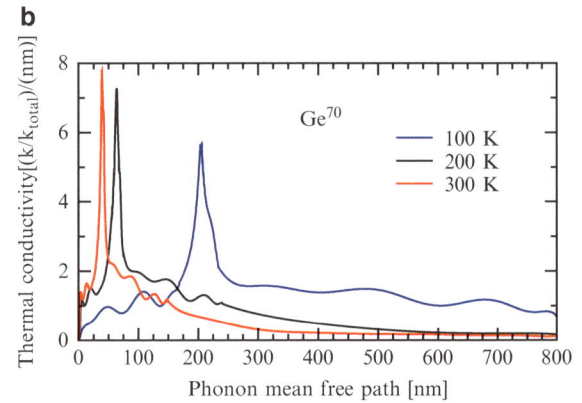

using the shorthand λ for the vibration mode ($\boldsymbol{q}s$) and defining $\Psi_{\boldsymbol{q}}^s = \sum_\alpha F_{\lambda\alpha} (\partial T/\partial x_\alpha)$ the PBE can be rewritten as

$$-c_\alpha(\lambda)\frac{\hbar\omega(\lambda)\bar{n}_\lambda(\bar{n}_\lambda+1)}{k_B T^2} = \sum_{\lambda',\lambda''} \left[\tilde{P}^+_{\lambda,\lambda'\lambda''}(F_{\lambda\alpha} + F_{\lambda'\alpha} - F_{\lambda''\alpha}) \right.$$
$$\left. + \frac{1}{2}\tilde{P}^-_{\lambda\lambda',\lambda''}(F_{\lambda\alpha} - F_{\lambda'\alpha} - F_{\lambda''\alpha}) \right] \quad (4.25)$$

where it is understood that the sum over λ'' only involves the sum over the mode s''. Defining

$$Q_\lambda = \sum_{\lambda',\lambda''} \left[\tilde{P}^+_{\lambda,\lambda'\lambda''} + \frac{1}{2}\tilde{P}^-_{\lambda\lambda',\lambda''} \right] \quad (4.26)$$

and

$$-c_\alpha(\lambda)\frac{\hbar\omega(\lambda)\bar{n}_\lambda(\bar{n}_\lambda+1)}{k_B T^2} = F^0_{\lambda\alpha} Q_\lambda \quad (4.27)$$

Fig. 4.12 Thermal conductivity accumulation with phonon mean free path in (**a**) Si[28]. Long mean-free-path phonons play a significant role in heat conduction in Silicon. This provides avenues to lower thermal conductivity through nanostructuring. (**b**) Same for Ge[70]

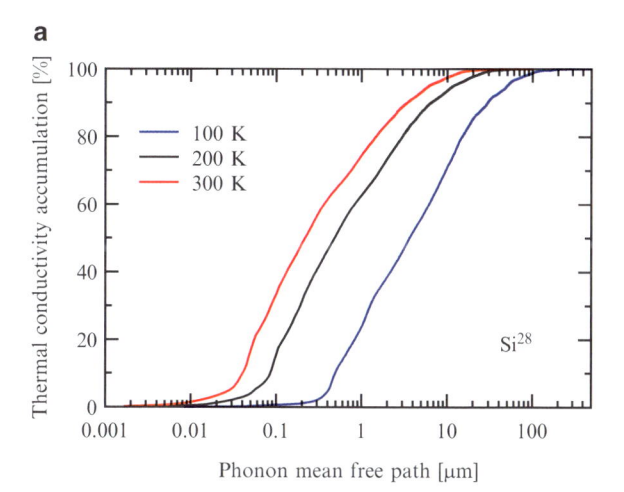

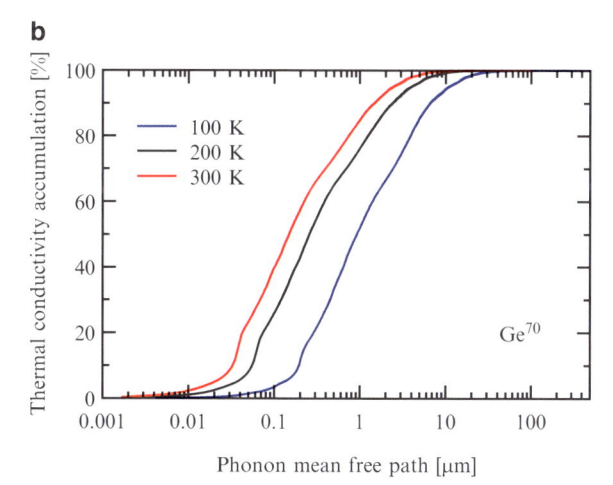

the PBE can be rewritten as

$$F^0_{\lambda\alpha} Q_\lambda = F_{\lambda\alpha} Q_\lambda - \sum_{\lambda',\lambda''} \left[\tilde{P}^+_{\lambda,\lambda'\lambda''}(F_{\lambda''\alpha} - F_{\lambda'\alpha}) + \frac{1}{2} \tilde{P}^-_{\lambda\lambda',\lambda''}(F_{\lambda'\alpha} + F_{\lambda''\alpha}) \right], \quad (4.28)$$

and further rewritten as:

$$F_{\lambda\alpha} = F^0_{\lambda\alpha} + \frac{1}{Q_\lambda} \sum_{\lambda',\lambda''} \left[\tilde{P}^+_{\lambda,\lambda'\lambda''}(F_{\lambda''\alpha} - F_{\lambda'\alpha}) + \frac{1}{2} \tilde{P}^-_{\lambda\lambda',\lambda''}(F_{\lambda'\alpha} + F_{\lambda''\alpha}) \right]. \quad (4.29)$$

Fig. 4.13 Thermal conductivity of Si^{28} and Ge^{70} computed using the full self-consistent solution of the phonon Boltzmann transport equation. Experimental values are from [17] (Si^{28}) and [23] (Ge^{70})

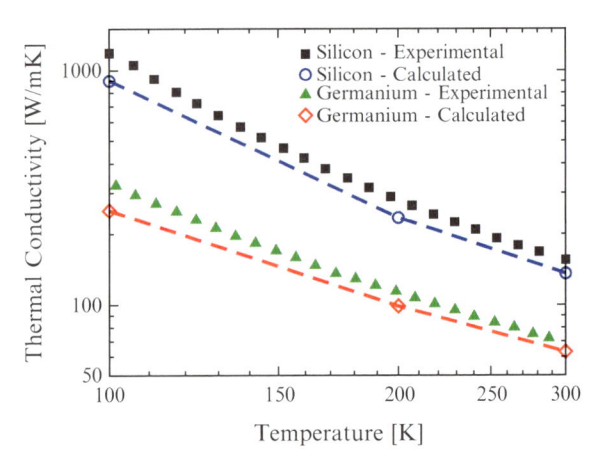

The above equation has to be solved for $F_{\lambda\alpha}$ for all the modes λ on a chosen grid in the first Brillouin zone. The iterative solution starts by assuming that the second term on the right-hand side is zero. This gives the zeroth order solution $F_{\lambda\alpha} = F^0_{\lambda\alpha}$. For the next iteration, the required values of $F_{\lambda'\alpha}$ and $F_{\lambda''\alpha}$ are taken from the zeroth order solution. Substituting these into the second term on the right-hand side yields the first-order solution $F^1_{\lambda\alpha}$. Continuing this process yields the converged values of $F_{\lambda\alpha}$.

The thermal conductivity after solving exactly the PBE is obtained as

$$k_{\alpha\beta} = \frac{1}{N_0\Omega} \sum_\lambda \hbar\omega(\lambda) c_\alpha(\lambda) \bar{n}_\lambda (\bar{n}_\lambda + 1) F_{\lambda\beta}. \quad (4.30)$$

The computed thermal conductivity after self-consistently solving the PBE is compared against experimental values in Fig. 4.13. It can be seen that solving the PBE exactly leads to a marginally better agreement with experimentally measured values as compared to the use of the SMRT approximation. However for pure Si^{28} and Ge^{70} the difference between the two results is small. Furthermore, it can be seen that there is still a small disagreement between the computed and the experimentally measured values. This is primarily due to the fact that the anharmonic force constants $\Phi_{\alpha\beta\gamma}(ob, h'b', h''b'')$ were obtained on a $3 \times 3 \times 3$ supercell. As shown in Fig. 4.5, there is a small difference in the linewidths computed using $\Phi_{\alpha\beta\gamma}(ob, h'b', h''b'')$ obtained on a $2 \times 2 \times 2$ versus a $3 \times 3 \times 3$ supercell. Thus the computation of linewidths is not fully converged with respect to the size of the supercell on which the third-order anharmonic force constants were obtained. Computation of the force constants on a supercell larger than $3 \times 3 \times 3$ was found computationally too expensive, thus somewhat limiting the accuracy of the final computed thermal conductivity.

4.7 Conclusions

The application of first-principles approaches to the thermal transport properties of crystals is just beginning, driven by the development of third-order density-functional perturbation theory [6a–6c]. Current efforts in extending its reciprocal space implementation to any generic triplet of phonon wave-vectors, and in efficient solution of the Boltzmann equation, envision a broad applicability of these techniques to ever more complex materials, with major implication for micro- and nano-electronics (especially for novel two-dimensional and layered materials) or, in a completely different context, to earth and planetary science, to understand thermal transport under extreme conditions. Among the remaining challenges we can list the extension to polar materials (where non-analytic terms in third-order derivatives, arising from the coupling with macroscopic fields, need special care) and the assessment of the relevance of higher-order processes in materials at the desired conditions.

References

1. Baroni S, Giannozzi P, Testa A (1987) Phys Rev Lett 58:1861
2. Bonini N, Garg J, Marzari N (2012) Nano Lett 12:2673
3. Broido DA, Ward A, Mingo N (2005) Phys Rev B 72:14,308
4. Broido DA, Malorny M, Birner G, Mingo N, Stewart DA (2007) Appl Phys Lett 91:231,922
5. Dal Corso A, Baroni S, Resta R, de Gironcoli S (1993) Phys Rev B 47:3588
6a. Debernardi A, Baroni S, Molinari E (1995) Phys Rev Lett 75:1819
6b. Lazzeri M, de Gironcoli S (2002) Phys Rev B 65:245402
6c. Paulatto L, Mauri F, Lazzzeri M (2013) Phys Rev B 87:214303
7. Deinzer G, Birner G, Strauch D (2003) Phys Rev B 67:144,304
8. Dirac PAM (1927) Proc Roy Soc (London) A114:243
9. Fermi E (1950) Nuclear physics. University of Chicago Press, Chicago
10. Fugallo G, Lazzeri M, Paulatto L, Mauri F (2013) Phys Rev B 88:045430
11. Garg J, Bonini N, Kozinsky B, Marzari N (2011a) Phys Rev Lett 106:45,901
12. Garg J, Bonini N, Marzari N (2011b) Nano Lett 11:5135
13. Giannozzi P, Baroni S, Bonini N, Calandra M, Car R, Cavazzoni C, Ceresoli D, Chiarotti GL, Cococcioni M, Dabo I, Dal Corso A, de Gironcoli S, Fabris S, Fratesi G, Gebauer R, Gertsmann U, Gougoussis C, Kokalj A, Lazzeri M, Martin-Samos L, Marzari N, Mauri F, Mazzarello R, Paolini S, Pasquarello A, Paulatto L, Sbraccia C, Scandolo S, Sclauzero G, Seitsonen AP, Smogunov A, Umari P, Wentzcovitch RM (2009) J Phys Conden Matter 21:395,502
14. Gonze X (1995) Phys Rev A 52:1086
15. Hamilton R, Parrot J (1969) Phys Rev 178:1284
16. Henry AS, Chen G (2008) J Comput Theor Nanosci 5:1
17. Inyushkin A, Taldenkov A, Cibin AM, Gusev AV, Pohl HJ (2004) Phys Status Solid C 1:2995
18. Ju Y, Goodson K (1999) Appl Phys Lett 74:3005
19. Lang G, Karch K, Schmitt M, Pavone P, Mayer A, Wehner R, Strauch D (1999) Phys Rev B 59:6182
20. Menendez J, Cardona M (1984) Phys Rev B 29:2051
21. Monkhorst HJ, Pack JD (1976) Phys Rev B 13:5188

22. Omini M, Sparavigna A (1996) Phys Rev B 53:9064
23. Ozhogin VI, Inyushkin AV, Taldenkov AN, Tikhomirov AV, Popov GE (1996) JETP Lett 63:1996
24. Peierls R (1929) Ann Phys 3:1055
25. Peierls R (1955) Quantum theory of solids. Clarendon Press, Oxford
26. Savvides N, Goldsmid H (1973) J Phys C Solid State Phys 6:1701
27. Srivastava GP (1990) The physics of phonons. Taylor and Francis Group, New York
28. Uma S, McConnell AD, Ashegi M, Kurabayashi K, Goodson KE (2001) Int J Thermophysics 22:605
29. Ziman JM (1960) Electrons and phonons. Oxford University Press, London

Chapter 5
Ab Initio Thermal Transport

N. Mingo, D.A. Stewart, D.A. Broido, L. Lindsay, and W. Li

Abstract Ab initio (or first principles) approaches are able to predict materials properties without the use of any adjustable parameters. This chapter presents some of our recently developed techniques for the ab initio evaluation of the lattice thermal conductivity of crystalline bulk materials and alloys, and nanoscale materials including embedded nanoparticle composites.

5.1 Introduction

The lattice thermal conductivity is a fundamental thermal transport parameter that determines the utility of materials for specific thermal management applications. Accurate theoretical modeling of the lattice thermal conductivity is extremely important for modern science and technology. The applicability of such a theory would be ubiquitous: It is essential to numerous fields including microelectronics cooling, efficient thermoelectric refrigeration and power generation, and even planetary science.

N. Mingo • W. Li
CEA-Grenoble, 17 rue des Martyrs, 38054 Grenoble, France
e-mail: natalio.mingo@cea.fr; Wu.LI@cea.fr

D.A. Stewart
Cornell Nanoscale Facility, 250 Duffield Hall, Cornell University, Ithaca, NY 14853, USA
e-mail: derek.stewart@cornell.edu

D.A. Broido (✉)
Department of Physics, 335 Higgins Hall, Boston College, 140 Commonwealth Ave, Chestnut Hill, MA 02467, USA
e-mail: broido@bc.edu

L. Lindsay
Naval Research Laboratory, Washington, DC 20375, USA
e-mail: lucas.lindsay.ctr@nrl.navy.mil

For decades, the development of an accurate, predictive theory of the lattice thermal conductivity of materials has remained a long-standing and yet unsolved problem [1]. Instead, most previous theoretical approaches to describe the lattice thermal conductivity in bulk and nanomaterials have resorted to the use of relaxation time approximations (RTAs), which require several fitting parameters [2–8] and so provide little predictive power, or molecular dynamics (MD), which omit any quantum mechanical description [9–13].

With advances in theoretical formalisms [14, 15] combined with dramatic increases in computing power in recent years, it has now become possible to address the lattice thermal conductivity problem using ab initio approaches. Such approaches' predictive power stems from the fact that they require no adjustable input parameters and only need a material's crystal structure as their starting point.

During the past several years, our group has initiated a comprehensive ab initio-based effort to describe phonon thermal transport in bulk and nanostructured materials [16–21]. The approach involves two steps: (1) First principles calculation of the harmonic and if necessary the anharmonic interatomic force constants of a material; (2) Use of the force constants from (1) in a self-consistent solution of atomistic, quantum mechanical phonon transport equations. For bulk crystalline materials, this involves an exact numerical solution of the Boltzmann transport equation (BTE) for phonons. For nanostructured materials, such as nanotubes with defects, a Non-Equilibrium Green's Function (NEGF) approach has been implemented. Recently, we have used our parameter-free approach to calculate the lattice thermal conductivity for bulk silicon, germanium, and diamond and demonstrated strikingly good agreement with measured thermal conductivity data for natural and isotopically enriched samples [15,20]. We have also obtained results for simple and nanostructured alloys [22], as well as for silicon and diamond nanowires [23], which are discussed in the examples.

In the following pages, we summarize our own efforts in advancing a first principles approach to thermal transport in materials. Considerable work has been published recently by a number of researchers and the field is rapidly developing. We apologize in advance to any whose contributions have been inadvertently left out.

5.2 Theory of Lattice Thermal Transport

In this section we will discuss the general formalism to atomistically compute the lattice thermal conductivity of bulk materials. Aspects specifically related to the *ab initio* calculation of phonon dispersions and phonon interactions are treated in the next section. Under *bulk materials* we include any systems that are homogeneous at scales larger than the phonon mean free paths. This not only includes single crystals but also encompasses disordered alloys or solid solutions. All these systems can be investigated via the space independent Boltzmann Transport Equation. In addition, some nanocomposites, such as *nanoparticle embedded in alloy thermoelectric* (NEAT) materials, are also well suited for this kind of description, when the nanoparticle concentration is relatively low.

5.2.1 The Linearized Boltzmann Transport Equation for Phonons

A microscopic description of lattice thermal conduction in bulk materials was first formulated by Peierls in 1929 [24, 25], who wrote down a Boltzmann transport equation (BTE) for phonons that included three-phonon scattering. This intrinsic scattering process is difficult to treat because it is (i) inelastic, (ii) requires accurate descriptions of both harmonic and anharmonic interatomic forces, and (iii) depends on precise determination of the phase space and matrix elements for three-phonon scattering. This prompted Ziman to write three decades after Peierls' original publication [2, p. 298]:

> It is the author's belief that progress towards a more accurate evaluation of the lattice conductivity can only be made [...] by the use of more complicated [variational] trial functions. The Boltzmann equation is so exceedingly complex that it seems hopeless to expect to generate a solution from it directly.

As a consequence, relaxation time approximations (RTAs) were widely used as well as occasional variational solutions [2]. It is well known that the RTA is in principle not valid for describing inelastic scattering. Furthermore, both RTA and variational approaches introduce *ad hoc* adjustable parameters and so lack predictive capability. Ziman's prediction held true until the mid-1990's when Omini and Sparavigna [26, 27] pioneered a general iterative solution to the phonon BTE and applied it to several materials [28–30] using simple models to describe the interatomic forces. Subsequent calculations of the lattice thermal conductivity based on an exact solution of the phonon BTE used empirical interatomic potentials (EIPs) to obtain the interatomic forces with mixed success [31,32]. However, the fundamental problem with such approaches is that EIPs are only available for a handful of well-studied materials, and these are typically not engineered to accurately describe lattice dynamical and thermal properties. The next step towards a fully predictive approach, free of adjustable parameters, requires ab initio determination of the interatomic forces

5.2.1.1 Formulation of the BTE in a Computational Framework

Here we present the BTE in a form that allows for its practical implementation and numerical solution within an ab initio scheme.

First of all, let us define some shorthand notation that will be used throughout the rest of the chapter. A traveling wave will be uniquely labeled by its phonon branch index, α, and wavevector, $\mathbf{q}$, shortened as the single symbol

$$\lambda \equiv (\alpha, \mathbf{q}). \qquad (5.1)$$

We will often encounter a combined summation over branches and integration over the first Brillouin zone, which will be denoted as

$$\sum_{\lambda'} F_{\lambda,\lambda'} \equiv \sum_{\alpha'} \int_{BZ} F_{(\alpha,\mathbf{q}),(\alpha',\mathbf{q}')} \frac{d\mathbf{q}'}{V_{BZ}}. \qquad (5.2)$$

Here, V_{BZ} is the "volume" of the Brillouin zone. As we will see later, various numerical techniques are available to conveniently perform Brillouin zone integrations. We will also encounter a variant of this integral, in the case of three-phonon processes, as

$$\sum_{\lambda'\lambda''}^{\pm} F_{\lambda,\lambda',\lambda''} \equiv \sum_{\lambda'\alpha''} F_{\lambda,\lambda',\lambda\pm\lambda'} \equiv \sum_{\alpha'} \sum_{\alpha''} \int_{BZ} F_{(\alpha,\mathbf{q}),(\alpha',\mathbf{q}'),(\alpha'',\mathbf{q}\pm\mathbf{q}')} \frac{d\mathbf{q}'}{V_{BZ}}. \qquad (5.3)$$

With these notations in hand, we proceed to write down the linearized BTE and explain how to solve it. The BTE is frequently written in terms of the non-equilibrium distribution, n_λ. For small temperature gradients, ∇T, this can be linearized so that: $n_\lambda = n_{0\lambda} + n_{1\lambda}$, where $n_{0\lambda} = 1/(\exp(\hbar\omega_\lambda/k_B T) - 1)$ is the equilibrium (Bose) distribution function, and the deviation from equilibrium, $n_{1\lambda}$, is of order ∇T. It is convenient to express $n_{1\lambda}$ in terms of the self-consistent lifetimes, τ_λ:

$$n_{1\lambda} = n_{0\lambda}(n_{0\lambda} + 1) \frac{v_\lambda^z \hbar \omega_\lambda}{k_B T^2} \tau_\lambda \left(-\frac{dT}{dz}\right), \qquad (5.4)$$

where ω_λ and v_λ^z are, respectively, the phonon frequency and z-component of group velocity in mode λ. The thermal gradient is taken along the z direction. In contrast with τ^0, the self-consistent phonon lifetimes depend on the direction of the applied thermal gradient. For convenience, we will omit the z subscript for the self consistent phonon lifetimes throughout this chapter.

The linearized Boltzmann Transport Equation can then be written in general form as

$$\tau_\lambda = \tau_\lambda^0 + \tau_\lambda^0 \Delta_\lambda. \qquad (5.5)$$

The unknowns to be determined are the list of all the τ_λ. Δ_λ on the right-hand side is a linear function of the ensemble of τ_λ's, and τ_λ^0 are the inhomogeneous terms, which are related to the scattering rates, as described below, and so are known quantities. Therefore, this is a linear equation which can be solved by iteration. Before discussing its solution, let us explicitly write down all these quantities in terms of magnitudes that we can calculate.

First, the inhomogeneous term contains harmonic and anharmonic scattering probability contributions:

$$1/\tau_\lambda^0 \equiv \sum_{\lambda'\lambda''}^{+} \Gamma_{\lambda\lambda'\lambda''}^{+} + \sum_{\lambda'\lambda''}^{-} \frac{1}{2} \Gamma_{\lambda\lambda'\lambda''}^{-} + \sum_{\lambda'} \Gamma_{\lambda\lambda'}. \quad (5.6)$$

The first two terms correspond to three-phonon processes, in which one phonon is absorbed or emitted, respectively. The third term is linked to the harmonic processes, and it thus only involves two phonons of exactly the same frequency. The same scattering probabilities appear in the homogeneous term,

$$\Delta_\lambda \equiv \sum_{\lambda'\lambda''}^{+} \Gamma_{\lambda\lambda'\lambda''}^{+}(\xi_{\lambda\lambda''}\tau_{\lambda''} - \xi_{\lambda\lambda'}\tau_{\lambda'})$$

$$+ \sum_{\lambda'\lambda''}^{-} \frac{1}{2} \Gamma_{\lambda\lambda'\lambda''}^{-}(\xi_{\lambda\lambda''}\tau_{\lambda''} + \xi_{\lambda\lambda'}\tau_{\lambda'})$$

$$+ \sum_{\lambda'} \Gamma_{\lambda\lambda'}\xi_{\lambda\lambda'}\tau_{\lambda'}. \quad (5.7)$$

In this term, the group velocities, v_λ, and frequencies, ω_λ, of the phonons are also involved through the quantity $\xi_{\lambda\lambda'} \equiv \frac{\omega'_\lambda v'_\lambda}{\omega_\lambda v_\lambda}$.

Thus, one only needs to calculate all the scattering amplitudes Γ's between states λ using a suitable discretization of the reciprocal space. Afterwards, the integrals over the Brillouin zone need to be performed, and the BTE solved by iteration. This is discussed in the next subsections.

5.2.1.2 Three-Phonon Scattering Amplitudes

The different scattering amplitudes Γ are properties of the lattice interatomic force constants. In the rest of this section we will denote the equilibrium Bose–Einstein distribution n_0 simply by n, omitting the subindex 0. Similarly $\omega' \equiv \omega_{\lambda'}$, etc. The transition probability for three-phonon processes is [31]

$$\Gamma_{\lambda\lambda'\lambda''}^{\pm} = \frac{\hbar\pi}{4} \left\{ \begin{array}{c} n' - n'' \\ n' + n'' + 1 \end{array} \right\} \frac{\delta(\omega \pm \omega' - \omega'')}{\omega\omega'\omega''} |V_{\lambda\lambda'\lambda''}^{\pm}|^2, \quad (5.8)$$

where the upper (lower) row in curly brackets goes with the + (−) sign.

In Eq. (5.8), $\lambda'' \equiv \{\alpha'', \mathbf{q}''\}$ where, using momentum conservation

$$\mathbf{q}'' = \mathbf{q} \pm \mathbf{q}' + \mathbf{K} \quad (5.9)$$

and **K** is a reciprocal lattice vector that is zero for Normal processes and nonzero for Umklapp processes [2]. The scattering matrix elements $V_{\lambda\lambda'\lambda''}$ only depend on the eigenfunctions of the three phonons involved, and the third order interatomic force constants $\Phi_{ijk} = \frac{\partial^3 E}{\partial u_i \partial u_j \partial u_k}$:

$$V_{\lambda\lambda'\lambda''} = \sum_{i \in u.c.} \sum_{j,k} \Phi_{ijk} \frac{e_\lambda(i) e_{\pm\lambda'}(j) e_{-\lambda''}(k)}{\sqrt{M_i M_j M_k}}. \tag{5.10}$$

Indexes i, j, and k denote individual atomic degrees of freedom. In the sums, i runs through just one unit cell, which we shall call the central unit cell; j and k run over all degrees of freedom in those cells that interact with the central unit cell. Equation (5.8) differs from Eq. (4) in [31] by $1/N$, where N is the number of unit cells in the crystal. The difference is due to the way the integral in Eq. (5.3) is defined. In general, a certain cutoff radius needs to be defined such that cells further apart will be considered as non-interacting. As we will see in the next section, this imposes a need to slightly modify the values of the ab initio calculated Interatomic Force Constants (IFC's). The $e_\lambda(i) \equiv \langle i|\lambda\rangle$ denotes the value of the eigenfunction corresponding to quantum numbers λ, evaluated at the degree of freedom i. These eigenfunctions are defined to be normalized to 1 inside the unit cell. We have also introduced the notation $-\lambda \equiv \{\alpha, -\mathbf{q}\}$.

5.2.1.3 Impurity Scattering Amplitudes

Traditionally, impurity scattering has been approximated using Rayleigh's law. However, when aiming at an ab initio description, it is important to derive the scattering rates from atomistic expressions instead. The elastic scattering amplitude due to a random distribution of independent scatterers in a homogeneous medium is

$$\Gamma_{\lambda\lambda'} \equiv \sum_p f^p \Gamma^p_{\lambda\lambda'} \tag{5.11}$$

with

$$\Gamma^p_{\lambda\lambda'} = \frac{\Omega\pi}{2\omega^2} \frac{1}{V_p} |\langle\lambda|\mathbf{T_p}(\omega^2)|\lambda'\rangle|^2 \delta(\omega - \omega'), \tag{5.12}$$

where f^p is the volume fraction of scatterers of type p, V_p is the scatterer's volume, Ω is the volume into which the wave functions $|\lambda\rangle$ are normalized, and $\mathbf{T_p}(\omega^2)$ is the T matrix associated with the scatterer of type p [33].

In the case of alloys it is convenient to adopt a virtual crystal approximation (VCA) model for the medium, where the interatomic force constants and atomic masses of the pure crystals are averaged according to their relative concentrations in the alloy. For example, in a binary alloy of species A and B, the total alloy scattering for bulk A_xB_{1-x} is given by the concentration weighted sum of the scattering probabilities of an A impurity in the VCA medium, and a B impurity in the VCA medium: $\Gamma_{\lambda\lambda'}^{AB} = x\Gamma_{\lambda\lambda'}^{A} + (1-x)\Gamma_{\lambda\lambda'}^{B}$, where $\Gamma_{\lambda\lambda'}^{A}$ is defined as in Eq. (5.12) for $p = A$, and similarly for $\Gamma_{\lambda\lambda'}^{B}$.

One can derive useful atomistic expressions for the total scattering rates due to impurity scattering. To do so, the main quantity to compute is the scattering cross section of the impurities, σ, which relates to the scattering rate as

$$\tau_\lambda^{-1} = \frac{f}{V_{\text{imp}}} \sigma_\lambda |\mathbf{v}_\lambda|, \qquad (5.13)$$

where f is the volume fraction of impurities, V_{imp} is the volume of a single impurity, and $\mathbf{v}_\lambda$ is the phonon group velocity. The exact expression for the total scattering cross section was given in [33]:

$$\sigma = \frac{2\Omega \Im\{\langle\lambda|\hat{T}|\lambda\rangle\}}{|\nabla_\mathbf{q}\omega_\lambda^2|}, \qquad (5.14)$$

where Ω is the volume into which the $|\lambda\rangle$ are normalized, and $\Im\{\}$ means the imaginary part. The real space T matrix associated with the impurity is the same size as the perturbation induced by the impurity, $\hat{V}$. We can now concentrate on the case of isotope or mass difference impurities and derive a useful approximation. The perturbation in this case is a frequency-dependent diagonal matrix, nonzero only on the degrees of freedom associated with the impurity atoms, with elements

$$V = \omega^2 \frac{\delta M}{M}, \qquad (5.15)$$

where M is the host atomic mass and δM is the difference between the impurity and host atomic masses. We then use the expansion of the T matrix in powers of V:

$$\hat{T} \simeq \hat{V} + \hat{V}\hat{g}\hat{V}, \qquad (5.16)$$

where $\hat{g}(\omega^2)$ is the perfect crystal resolvent, or retarded Green's function. Now we use the relation between the resolvent and the matrix phonon density of states, $\hat{\rho}_{\text{ph}}$:

$$\Im\hat{g}/\pi = \frac{\hat{\rho}_{\text{ph}}}{2\omega_\lambda}. \qquad (5.17)$$

In the case of a single isotope impurity, the volume of a single impurity equals V_{at}, the atomic volume in the lattice. This yields

$$\Omega\Im\{\langle\lambda|\hat{T}|\lambda\rangle\} \simeq \frac{\pi V_{\text{at}}^2}{6\omega_\lambda} \rho_\lambda \left(\frac{\delta M}{M}\right)^2 \omega_\lambda^4, \qquad (5.18)$$

where $V_{at}^2 \rho_\lambda / 3 \equiv \Omega \sum_{i,j \in \text{impurity}} \langle \lambda | i \rangle \langle i | \hat{\rho}_{ph} | j \rangle \langle j | \lambda \rangle$, and the local indexes i and j run through the three atomic degrees of freedom of the impurity. For an isotropic cubic crystal, ρ_λ thus defined is equal to the total phonon density of states, $\rho(\omega_\lambda)$. Also we have

$$|\nabla_\mathbf{q} \omega_\lambda^2| = 2\omega_\lambda |\nabla_\mathbf{q} \omega_\lambda| = 2\omega_\lambda |\mathbf{v}_\lambda|. \tag{5.19}$$

Substituting all this in Eq. (5.13) yields

$$\tau^{-1} = f \frac{\pi}{6} \rho_{ph}(\omega) V_{at} \left(\frac{\delta M}{M} \right)^2 \omega^2. \tag{5.20}$$

This approximation was first derived using perturbation theory by Tamura [34], who rigorously showed that it is correct for cubic crystals even including anisotropy. The expression is very advantageous because it only requires calculating the ab initio phonon density of states, which can be easily done using standard tetrahedron integration. From this equation it is easy to derive the Rayleigh approximation by simply replacing the density of states with its low frequency analytical expression. This is left as an exercise to the reader.

5.2.1.4 Defining Grids, and Performing Integrals

Numerically solving the integrals in Eqs. (5.2) and (5.3) requires some sort of discretization of the Brillouin zone. However, the problem is not so simple, because the function to integrate in general involves a delta function of the frequencies, resulting from the conservation of energy. Thus we are faced with integrals of the type $\sum W_{\lambda\lambda'} \delta(\omega - \omega')$. Obviously, making sums over one single grid of $\mathbf{q}$ points would not work, because nearly all the elements would be zero due to the delta function. To solve this there are various options: (1) explicitly finding the points out of the grid that fulfill energy conservation (*Gaussian quadrature* method), (2) using integration methods based on *interpolation*, or (3) using a *Gaussian smearing* of the delta functions.

In our work we have extensively employed the first approach, using an initial Gaussian quadrature grid [31]. This approach is numerically challenging since, for adequate quadrature grids, the phase space for three-phonon processes typically contains millions of scattering events that satisfy the conservation conditions. The second set of approaches is best represented by the Gilat–Raubenheimer or tetrahedron methods [35, 36]. In this type of approach, only one q-grid is defined at the beginning of the calculation. All the eigenfrequencies ω_λ and group velocities $\mathbf{v}_\lambda$ are also calculated for the q-points in the grid in the first step. The method reduces the integrals to a summation over $\mathbf{q}$ in the grid, of the integrand value at each point times an associated weight $w_{\alpha',\mathbf{q}'}(\omega)$ automatically computed by the algorithm:

$$\sum_{\alpha'} \int F_{\alpha',\mathbf{q}'} \delta(\omega - \omega') \frac{d\mathbf{q}'}{V_{BZ}} \simeq \sum_{\alpha'} \sum_{\mathbf{q}' \in grid} w_{\alpha',\mathbf{q}'}(\omega) F_{\alpha',\mathbf{q}'} \tag{5.21}$$

We have only performed preliminary tests on the tetrahedron approach, which suggest that the integration and iterative schemes converge.

The third method, i.e. Gaussian smearing, is arguably the most practical of the three. Although it requires a denser grid, this method is easier to implement. It is also much more robust, especially for complex unit cells or abundant branch crossings.[23] We have compared the Gaussian quadrature, and Gaussian smearing methods, obtaining very similar results. The delta function $\delta(\omega_\lambda - W)$, with $W = \omega_{\lambda'}$ in Eq. (5.2) and $W = \pm\omega_{\lambda'} + \omega_{\lambda''}$ in Eq. (5.3), can be considered as the limit of the Gaussian function,

$$g(\omega_\lambda - W) = \frac{1}{\sqrt{\pi}\sigma} e^{\frac{-(\omega_\lambda - W)^2}{\sigma^2}}. \tag{5.22}$$

when the smearing parameter $\sigma \to 0$. The integration is approximated by the numerical summation over the finite number of discretized $\mathbf{q}'$ points, and thus one has to use a finite smearing σ. σ should be chosen such that two Gaussian functions at two neighboring W overlap. The overlapping condition sets a criterion for σ, which means σ is of the order of ΔW, where ΔW is the spacing of W. ΔW can be estimated as

$$\Delta W = \left|\frac{\partial W}{\partial \mathbf{q}'}\right| |\Delta q'| = |\mathbf{v}_{\lambda'}||\Delta q'|, \tag{5.23}$$

for Eq. (5.2) [37] and

$$\Delta W = \left|\frac{\partial W}{\partial \mathbf{q}'}\right| |\Delta q'| = |\mathbf{v}_{\lambda'} - \mathbf{v}_{\lambda''}||\Delta q'|, \tag{5.24}$$

for Eq. (5.3), where $|\Delta q'|$ is simply the spacing of the sampling $\mathbf{q}$ points in the Brillouin zone.

5.2.1.5 Iterative Solution of the BTE

Once the transition rates have been obtained, we are ready to iteratively solve the BTE. We take $\tau_\lambda^{(0)} = \tau_\lambda^0$ to be the starting (zeroth order) solution. Then, $\tau_\lambda^{(1)} = \tau_\lambda^0 + \tau_\lambda^0 \Delta_\lambda^{(0)}$, where $\Delta_\lambda^{(0)}$ is evaluated using τ_λ^0. For the nth iteration:

$$\tau_\lambda^{(n+1)} = \tau_\lambda^0 + \tau_\lambda^0 \Delta_\lambda^{(n)}. \tag{5.25}$$

Quantity Δ in general contains the factor $v_\mathbf{q}^z / v_{\mathbf{q}'}^z$, leading to infinities when the group velocity in the z direction is zero. To avoid this, in the practical implementation it is possible not to use $\tau_{\alpha,\mathbf{q}}$, but $F_{\alpha,\mathbf{q}} = \tau_{\alpha,\mathbf{q}}(v_{\alpha,\mathbf{q}}^z \hbar \omega_{\alpha,\mathbf{q}}/T)$ as the quantity to solve for in the BTE.

Figure 5.1 illustrates the convergence procedure for the thermal conductivity, κ, upon iteration (see next section).

Fig. 5.1 Scaled thermal conductivity as a function of iteration obtained in the full solution of the phonon BTE

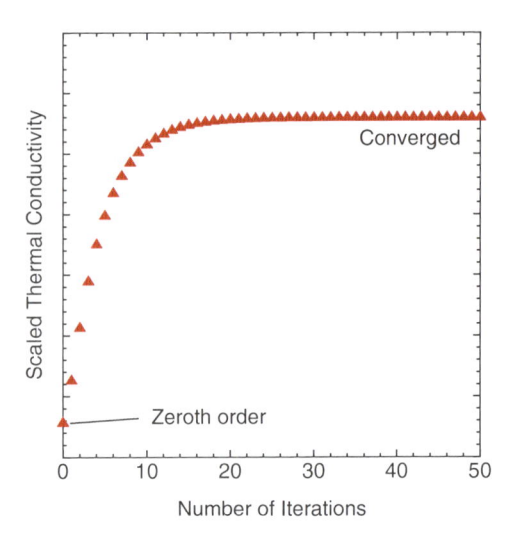

Note that κ increases with iteration until it converges to a higher value. The increase is a consequence of the fact that the zeroth order calculation is equivalent to the relaxation time approximation in which Normal and Umklapp processes are both treated as resistive. However, only the Umklapp processes provide thermal resistance. The iterative procedure corrects for this yielding the exact numerical solution to the BTE, which necessarily gives a higher κ.

5.2.1.6 Thermal Conductivity

Once the phonon lifetimes are known from solving the BTE, the thermal conductivity is simply an integral over the Brillouin zone involving these lifetimes:

$$\kappa = \frac{1}{k_B T^2} \sum_\alpha \int n_{0\lambda}(n_{0\lambda}+1)(\hbar\omega_\lambda)^2 (v_\lambda^z)^2 \tau_\lambda \frac{d\mathbf{q}}{(2\pi)^3}. \tag{5.26}$$

This integral can be performed directly on the **q** grid. The lattice thermal conductivity can also be written as an integral over frequency which may prove useful in some cases,

$$\kappa = \frac{1}{k_B T^2} \int n_0(1+n_0)(\hbar\omega)^2 \Sigma^{ph}(\omega) d\omega, \tag{5.27}$$

where $n_0 \equiv \frac{1}{e^{\hbar\omega/k_B T}-1}$, and

$$\Sigma^{ph}(\omega) = \frac{1}{8\pi^3} \sum_\alpha \int_{BZ} \tau_{\alpha,\mathbf{q}} (v_{\alpha,\mathbf{q}}^z)^2 \delta(\omega - \omega_{\alpha,\mathbf{q}}) d\mathbf{q} \tag{5.28}$$

5 Ab Initio Thermal Transport

This integral over the Brillouin zone can then be evaluated using integration methods based on either interpolation or Gaussian smearing. Special attention needs to be paid in those cases to ensure that interpolation does not introduce errors, especially at low frequency.

5.2.2 Ab initio Computation of Phonon Dispersions and Scattering Rates

Previous sections have discussed the problem of calculating the thermal conductivity of a system for which the harmonic and anharmonic force constants are known. In the following section, we will describe the basic approach to compute these quantities so that they can be properly employed in thermal transport calculations.

5.2.2.1 Ab Initio Computation of Interatomic Force Constants and Their Derivatives

A holy grail of computational materials research is the ability to predict all relevant material properties based solely on atomic positions and their nuclear charges. Such a virtual laboratory would provide an efficient test-bed to understanding existing materials and also design new materials with desired properties. These computational approaches are typically termed *first principles* or *ab initio* which translated from Latin means "from the beginning."

While a number of approaches have been proposed to reduce the complexity of solving a many-body Schrodinger equation, currently density functional theory is the only approach that comes close to making this vision a practical reality for materials. The main tenet of this approach is that the necessary information to find the ground state energy of an N electron system is encapsulated within the electron charge density, a function of only three space variables [38]. Density functional theory recasts the description of a complex many-body system into an equivalent set of quasiparticle equations where many-body effects are folded into an approximate exchange-correlation functional [39]. These equations can be solved self-consistently to determine the ground state potential and electron density of the system under study. This provides a powerful predictive tool for studying systems ranging from molecules to crystalline structures. It has been remarkably accurate at determining cohesive properties of crystals such as equilibrium lattice constant, bulk modulus, and elastic constants. In addition, it is an integral tool for understanding magnetic properties in alloys, magnetic multilayers, and dilute magnetic semiconductors. The physical properties that density functional theory predicts poorly (e.g., band gap) are also well known. Within the last fifteen years, density functional formalism has also been expanded so that electronic transport in bulk and nanoscale systems can be described with reasonable accuracy. In addition, the phonon dispersion of materials can now be determined with little more computational effort than that required to calculate the ground state electronic structure. For a more thorough description of density functional theory and applications, several excellent references [40, 41] are available to the interested reader.

While much effort has focused on predicting the mechanical and electrical properties of materials using ab initio techniques, the ability to accurately predict thermal transport from first principles has, until recently, been a significant gap in our knowledge. In this section, we will show how the harmonic and anharmonic terms necessary to predict thermal conductivity can be calculated using existing first principle techniques.

Periodic Systems. The equilibrium or ground-state configuration for a system can be found either by determining the energy minimum or by finding the lattice constants and atomic configuration where the hydrostatic pressure on the unit cell is zero and the sum of the forces acting on each atom is zero. Based on the Hellmann–Feynman theorem [42, 43], the force, F_i, acting on the ith atom is given by the expectation value of the derivative of the total crystal Hamiltonian with respect to the ion position:

$$F_i = -\left\langle \Psi(R) \left| \frac{\partial H(R)}{\partial R_i} \right| \Psi(R) \right\rangle \tag{5.29}$$

For a periodic lattice of atoms, the only portions of the crystal Hamiltonian that depend on the ion positions are the ion–electron interaction term, $V_R(r)$, and the ion–ion interaction term, $E_{\text{ion}}(R)$. This allows us to express the force on a given atom in terms of a component that is responding to the surrounding electron charge density, $n_R(r)$, and another term that accounts for ion–ion interactions.

$$F_i = -\int n_R(r) \frac{\partial V_R(r)}{\partial R_i} dr - \frac{\partial E_{\text{ion}}(R)}{\partial R_i} \tag{5.30}$$

The interatomic force constant matrix, Φ_{total} [14] used to determine the phonon frequencies in the system is found by differentiating the force on a given ion, i, by the change in position of ion j.

$$\Phi_{R_i R_j}^{total} = -\frac{\partial F_i}{\partial R_j}$$

$$= \int \frac{\partial n_R(r)}{\partial R_j} \frac{\partial V_R(r)}{\partial R_i} dr + \int n_R(r) \frac{\partial^2 V_R(r)}{\partial R_i \partial R_j} dr + \frac{\partial^2 E_{\text{ion}}(R)}{\partial R_i \partial R_j} \tag{5.31}$$

The interatomic force constant matrix (IFC) above consists of separate electronic (first two integral terms on right) and ionic contributions (final term on right). The ionic contribution is equivalent to the second derivative of an Ewald sum and can be calculated directly [44]. The electronic portion is given by:

$$\Phi_{R_i,R_j}^{\text{elec}} = \int \frac{\partial n_R(r)}{\partial R_j} \frac{\partial V_R(r)}{\partial R_i} dr + \int n_R(r) \frac{\partial^2 V_R(r)}{\partial R_i \partial R_j} dr \tag{5.32}$$

This equation shows that two factors are critical for the determination of the electronic contributions to the IFC and ultimately the phonon properties of a material. The first is the calculation of the ground state electron charge density, $n_R(r)$. The second is the determination of the linear response of the ground state electron density, $\partial n_R(r)/\partial R_I$, to a change in the ion geometry. These quantities can be calculated directly within the density functional framework without resorting to fitting data from experiment.

While the first principle calculations readily determine the equilibrium electron charge density, $n_R(r)$, additional computational effort is required to determine $\partial n_R(r)/\partial R_I$ and the corresponding IFC. However, this is on the same order as the standard ground state energy calculation. Interatomic force constants for periodic structures are determined through the use of density functional perturbation theory [14, 45, 46]. In this approach, the Kohn–Sham equations for the charge density, self-consistent potential, and orbitals are linearized with respect to changes in wave function, density, and potential variations. For phonon calculations, the perturbation, δV_{ion}, is periodic with a wave vector $\mathbf{q}$. This perturbation results in a corresponding change in the electron charge density, $\delta n(r)$. Since the perturbation generated by a phonon is periodic with respect to the crystal lattice, we can Fourier transform the self-consistent equations for the first order corrections and consider the Fourier components, $\delta n(\mathbf{q} + \mathbf{K})$, of the change in electron density where $\mathbf{K}$ is any reciprocal lattice vector. In k-space, the problem can be addressed on a sufficiently dense k-point grid in the unit cell Brillouin zone and fast Fourier transforms (FFT) can be used to convert back to real space [47]. However, while this reciprocal space approach is powerful, it requires that the system is periodic. For non-periodic systems, optimized real space approaches must be used to determine phonon properties.

Since phonons can be defined in terms of $\mathbf{q}$ vectors within the first Brillouin zone, it is more efficient to solve for the interatomic force constants in a reciprocal space representation.

The position of any atom i within a crystal can be defined in terms of the specific unit cell, l, it is located in, its index κ and position τ_κ in the unit cell, and finally its displacement, u_s from the equilibrium position.

$$r_i = R_l + \tau_\kappa + u_\kappa(i) \tag{5.33}$$

The atomic displacements in reciprocal space are given by:

$$u_\kappa(\mathbf{q}) = \frac{1}{\sqrt{N}} \sum_R e^{-i\mathbf{q}\cdot\mathbf{R}} u_\kappa(\mathbf{R}) \tag{5.34}$$

The dynamical matrix or reciprocal space interatomic force constant matrix can be related to the real space IFC using:

$$\tilde{\Phi}_{\alpha,\beta}(\mathbf{q}) = \sum_{l'} \Phi_{\alpha,\beta}\left(0\kappa; l'\kappa'\right) e^{-i\mathbf{q}\cdot\mathbf{R}_{l'}} \tag{5.35}$$

where α and β indicate the respective Cartesian indexes (x,y,z) and $\mathbf{R}_{l'}$ indicates the position of the l'th unit cell.

The phonon eigenvectors, $e_{\alpha\kappa}$ and frequencies, ω, can be determined by solving the eigenvalue equation:

$$\sum_{\beta\kappa'} \frac{1}{\sqrt{M_\kappa M_{\kappa'}}} \tilde{\Phi}_{\alpha,\beta}(\mathbf{q}) e_{\beta\kappa'}(\mathbf{q}) = \omega^2 e_{\alpha\kappa}(\mathbf{q}) \qquad (5.36)$$

where M_κ and $M_{\kappa'}$ are the masses of atoms κ and κ', respectively. The density functional perturbation approach to determine harmonic force constants has been implemented in several widely available plane-wave codes, including Quantum-Espresso [48], Abinit [49], and CASTEP [50].

Anharmonic Effects in Crystals. To find the interatomic force constant matrix and phonon frequencies for a solid, we evaluate the change in the electron density due to a small perturbation with wavevector $\mathbf{q}$. In order to determine the change in the electron density, we self-consistently evaluate the first order derivatives of the self-consistent potential and the electron wavefunction as well. By taking advantage of the $2n + 1$ theorem, we can use this information to calculate third-order anharmonic terms in the energy expansion necessary for examining thermal conductivity. The $2n + 1$ theorem states that if the derivatives of the wave functions for a system up to order n are known, then it is possible to calculate the energy derivatives for the system up to order $2n + 1$. This theorem has been shown to hold within the density functional framework [51] and full expressions for the energy derivatives up to fourth order have been derived from information obtained using standard density functional calculations [52]. This results in significant savings in terms of computational time for periodic systems. Several works have calculated the anharmonic decay of optical phonons at the Γ point where the initial phonon wavevector $\mathbf{q}$ is zero [53–55]. The first calculation of anharmonic interatomic force constants for general pairs of q-vectors $(\mathbf{q}, \mathbf{q}')$ in the Brillouin zone was done by Deinzer et al. in 2003 [56].

The anharmonic force constants can be expressed as the third-order derivative of the energy in terms of the atomic displacements. It can be written in reciprocal space and the anharmonic terms can be expressed in terms of interactions between three different phonons with arbitrary $\mathbf{q}$ vectors, $\mathbf{q}$, $\mathbf{q}'$, and $\mathbf{q}''$ [56].

$$\Phi_{\alpha\beta\gamma}^{\kappa\kappa'\kappa''}(\mathbf{q}\mathbf{q}'\mathbf{q}'') = \frac{\partial^3 E}{\partial u_\alpha^\kappa(\mathbf{q}) u_\beta^{\kappa'}(\mathbf{q}') u_\gamma^{\kappa''}(\mathbf{q}'')} \qquad (5.37)$$

Given that the anharmonic IFCs are symmetrical with respect to the indexes $(\kappa, \alpha, \mathbf{q})$ [57], an anharmonic IFC can be expanded into six terms. From a physical perspective, these different mathematical terms represent the possible three phonon interactions where either two phonons combine to create a third phonon or else a

5 Ab Initio Thermal Transport

given phonon generates two phonons. To simplify the expression, we use $\mathbf{1} = (\kappa, \alpha, \mathbf{q})$, $\mathbf{2} = (\kappa', \beta, \mathbf{q'})$, and $\mathbf{3} = (\kappa'', \gamma, \mathbf{q''})$.

$$\Phi(\mathbf{1,2,3}) = \tilde{\Phi}(\mathbf{1,2,3}) + \tilde{\Phi}(\mathbf{2,3,1}) + \tilde{\Phi}(\mathbf{3,1,2})$$
$$+\tilde{\Phi}(\mathbf{1,3,2}) + \tilde{\Phi}(\mathbf{3,2,1}) + \tilde{\Phi}(\mathbf{2,1,3}) \qquad (5.38)$$

Working through a fair amount of algebra, a given term in this expansion for the anharmonic force constants such as, $\tilde{\Phi}^{\kappa\kappa'\kappa''}_{\alpha\beta\gamma}$, takes the following form [56]:

$$\tilde{\Phi}^{\kappa\kappa'\kappa''}_{\alpha\beta\gamma}(\mathbf{q,q',q''}) = \frac{\delta_{\mathbf{q}+\mathbf{q'}+\mathbf{q''},\mathbf{K}}\delta_{\kappa\kappa'}}{3} \sum_{v\mathbf{k}} \left\langle \psi_{v\mathbf{k}} \left| \frac{\partial^3 V_{\text{ext}}}{\partial u^\kappa_\alpha(0) \partial u^{\kappa'}_\beta(0) \partial u^{\kappa''}_\gamma(0)} \right| \psi_{v\mathbf{k}} \right\rangle$$
$$+ 2\delta_{\kappa\kappa'} \sum_{v\mathbf{k}} \left\langle \psi_{v\mathbf{k}} \left| \frac{\partial^2 V_{\text{ext}}}{\partial u^{\kappa'}_\beta(-\mathbf{q}) \partial u^{\kappa''}_\gamma(0)} \wp_c \right| \frac{\partial \psi_{v\mathbf{k}}}{\partial u^\kappa_\alpha(\mathbf{q})} \right\rangle$$
$$+ 2 \sum_{v\mathbf{k}} \left\langle \frac{\partial \psi_{v\mathbf{k}}}{\partial u^\kappa_\alpha(-\mathbf{q})} \left| \wp_c \frac{\partial V_{\text{KS}}}{\partial u^{\kappa'}_\beta(\mathbf{q'})} \wp_c \right| \frac{\partial \psi_{v\mathbf{k}}}{\partial u^{\kappa''}_\gamma(-\mathbf{q''})} \right\rangle$$
$$- 2 \sum_{vv'\mathbf{k}} \left\langle \frac{\partial \psi_{v\mathbf{k}}}{\partial u^\kappa_\alpha(-\mathbf{q})} \left| \wp_c \right| \frac{\partial \psi_{v'\mathbf{k}+\mathbf{q''}}}{\partial u^{\kappa'}_\beta(-\mathbf{q'})} \right\rangle \left\langle \psi_{v\mathbf{k}+\mathbf{q''}} \left| \frac{\partial V_{\text{KS}}}{\partial u^{\kappa''}_\gamma(\mathbf{q''})} \right| \psi_{v\mathbf{k}} \right\rangle$$
$$+ \frac{1}{6} \int d^3r d^3r' d^3r'' f^{\text{LDA}}_{\text{xc}}(r,r',r'') \frac{\partial n(r)}{\partial u^\kappa_\alpha(\mathbf{q})} \frac{\partial n(r')}{\partial u^{\kappa'}_\beta(\mathbf{q'})} \frac{\partial n(r'')}{\partial u^{\kappa''}_\gamma(\mathbf{q''})}$$

$$(5.39)$$

where $\psi_{v\mathbf{k}}$ is the Bloch orbital for the vth band at the k-vector k, $n(r)$ is the electron density at the point r, V_{ext} is the external potential, V_{KS} is the self-consistent Kohn–Sham potential, and P_C is a projector over the conduction band manifold [14]. The third-order derivative of the exchange and correlation energy with respect to the density in the local density approximation is given by $f^{\text{LDA}}_{\text{xc}}$. It is important to note that only the first derivatives of the wavefunctions, charge density, and self-consistent potential are required to solve the equation above in agreement with the $2n + 1$ theorem. Care must be taken in evaluating these terms since multiple integrations are required over q-vector shifted Brillouin zones. The calculation time can also be reduced by a careful analysis of the symmetries present in a particular crystal which can greatly reduce the number of independent anharmonic terms [57].

Interatomic Interactions for Non-periodic Systems. While IFCs for periodic systems can be calculated rapidly through the use of fast Fourier transforms, non-periodic systems present particular challenges since phonon dispersions must be calculated directly in real space. In structures with long-range interactions, this can result in time-consuming real space calculations for systems with large numbers of atoms. In these cases, a density functional approach that scales with the number of atoms in the system, otherwise known as an order-N approach, is crucial. Several order-N density functional codes are available or under development, such as Siesta [58], OpenMX [59], Conquest [60], and ONETEP [61]. In this work, we will focus primarily on Siesta since it has a specific mode to calculate harmonic force constants and also because we have used it in the past to examine thermal transport in nanostructures [17–20]. Siesta relies on a basis set that is strictly localized in space to construct sparse Hamiltonians. This helps provide calculations close to the speed of tight binding approaches, but with the accuracy of first-principles approaches. For non-periodic systems, we treat the region of interest (e.g., interface and surface) as an isolated cluster. We then calculate the perturbation to the electron density due to the phonon perturbation self-consistently in real space.

The real space interatomic force constant matrix can be calculated using the finite displacement method [62–64]. In this technique, the ground state electronic structure for a large supercell is determined using density functional calculations. After this, a single atom is shifted in the central unit cell in the $\pm x$, $\pm y$, and $\pm z$ directions and the resulting Hellmann–Feynman forces on the neighboring atoms are tabulated. This is repeated for all atoms in the central unit cell in order to build up the full interatomic force constant matrix. Symmetries in the system under study can reduce the number of atoms that need to be shifted in the unit cell. Convergence of phonon properties is ensured by monitoring the interatomic force constants for different orders of neighbors and increasing the size of the system until the interatomic force constants become insignificant for the nth order of atomic neighbors. Since IFCs are calculated using a finite difference approach based on the calculated forces, care must be taken to avoid numerical errors due to a poor choice of the size of the atomic shift or due to poor density functional convergence parameters (insufficient basis set, plane wave energy cutoff, etc). For example, in the case of Siesta, a mesh cutoff energy defines the fineness of the real space grid used to sample space-dependent properties in the cell, such as the potential. In order to calculate a reasonable phonon dispersion, the mesh cutoff energy must be large enough to ensure that the grid does not significantly break translational symmetry and cause the system to suffer from eggbox effects [65].

One of the primary benefits of the real space approach is that this technique can be used with any density functional code that can calculate forces on atoms. Several packages have been developed to calculate harmonic interatomic force constants and phonon dispersions based on this technique such as the Siesta Vibra package [65], PHON [66], PHONON [64], and Phonopy [67]. A few works recently have also calculated anharmonic force constants based on a real space technique where two atomic displacements are required [68–71]. In this case, the use of symmetries to reduce the number of terms considered is crucial for feasible calculations.

Before we move on, it is important to note some key differences between the density functional perturbation approach and the real space finite displacement method in polar materials. In this case, the long-range Coulomb interactions between the ions in the polar crystal results in a non-analytic term in the dynamical matrices that leads to the Lydanne–Sachs–Teller splitting of the longitudinal (LO) and transverse optical (TO) modes [72]. The non-analytic term depends on the

Born effective charge on each ion, Z^*, and also on the dielectric constant, ε_∞. Since density functional perturbation theory can also determine perturbations due to electric field, it is straightforward to evaluate these terms and the non-analytic portion of the dynamical matrix separately from the analytic portion [14, 73]. Lattice dynamics calculations of polar systems using DFPT accurately reproduce the optical splitting observed at the Γ point. Real space finite displacement approaches, however, cannot calculate this term directly [74, 75] and the necessary values must be supplied either from experimental data or a separate DFPT or Berry phase calculation. In many cases, the splitting of these branches is small and the overall effect on the thermal conductivity will be minimal. However, in thermoelectrics, such as PbTe, that displays a substantial LO-TO splitting [76, 77], this could lead to significant changes in scattering between heat carrying acoustic branches and low lying optical modes. The macroscopic electric field in polar systems will also give rise to non-analytic terms in the third and fourth order anharmonic force constants. G. Deinzer et al. [78] have derived expressions for these terms and examined how they will affect damping of long-wavelength optical modes in GaAs.

Finally, the two techniques can also be compared in terms of computational effort required to calculate harmonic force constants. Expanding on the discussion in S. Baroni et al. [14], for the finite displacement supercell technique, our supercell should be larger than the interaction radius, R_{IFC}, where IFCs between atoms are non-negligible. This defines a supercell with volume R_{IFC}^3. The number of atoms in our supercell will also be proportional to this volume, $N_{sc} \sim R_{IFC}^3$. For the real space approach, we need to shift all the unique atoms in the central unit cell, N_{uq}, in six directions. Standard density functional calculations scale as N^3 where N is the number of atoms in the system. So for the supercell calculations, the DFT calculations will scale as N_{sc}^3 which leads to R_{IFC}^9 scaling in terms of the interaction radius. Since we need to do this calculation $6N_{uq}$ times, the total scaling for the real space technique for a crystal is $6N_{uq}R_{IFC}^9$. If we use an order-N density functional approach, this scaling improves to $6N_{uq}R_{IFC}^3$. In addition, using an order-N approach like Siesta [58] on a one-dimensional structure (e.g., nanowire or nanoribbon), the computational cost is greatly reduced to $6N_{uq}R_{IFC}$.

Turning to the DFPT approach, in order to get accurate interatomic force constants, we need to sample the Brillouin zone with a dense mesh of q points where the spacing should be approximately $2\pi/R_{IFC}$. So in the Brillouin zone, this results in a 3D grid of roughly R_{IFC}^3 q points. For each column of the dynamical matrix at a given q, the computational cost of the calculation is N_{uq}^3 and this calculation must be done for all $3N_{uq}$ columns. Therefore, the cost to determine the phonon dispersion of a material using DFPT scales as $6N_{uq}^4 R_{IFC}^3$ for a standard density functional plane wave code. For a one-dimensional system, the scaling goes as $6N_{uq}^4 R_{IFC}$. So for bulk systems, the ratio of the computation time for the real space approach to the q-space approach, t_r/t_q, will scale as R_{IFC}^6/N_{uq}^3. Since $R_{IFC}^3 \sim N_{sc}$, this can also be expressed as N_{sc}^2/N_{uq}^3. For the case where the size of the supercell increases uniformly in all directions to including additional neighbors, then the number of unit cells included in the supercell follows the series $1, 27, 125, \ldots (2n+1)^3$ where $n = 0, 1, 2, 3, etc.$

In this case, the supercell of nth order contains $N_{sc} = (2n+1)^3 N_{cell}$ atoms. Note that N_{cell} is the total number of atoms in the unit cell and $N_{cell} \geq N_{uq}$. For this estimate, however, we assume $N_{cell} \approx N_{uq}$. The ratio of computation time using an N^3 real space DFT approach, $t_r^{N^3}$, and N^3 DFTP approach, $t_q^{N^3}$, becomes:

$$\frac{t_r^{N^3}}{t_q^{N^3}} = \frac{(2n+1)^6}{N_{uq}^2} \tag{5.40}$$

This shows that for bulk materials, unless the IFCs are very short range, the density functional perturbation approach is preferable. It is interesting to note that the use of order-N density functional approach for real space phonon calculations may tip the scaling in favor of a real space technique, given that there are currently no order-N DFPT approaches.

$$\frac{t_r^N}{t_q^{N^3}} = \frac{1}{N_{uq}^3} \tag{5.41}$$

However, it should be emphasized that while order-N approaches are available, their scaling time has a large prefactor. This can lead to the order-N approach being more costly than standard N^3 approaches for small scale calculations ($N < 100$). In addition, this discussion has not addressed how parallelization over atomic shifts for the real space technique or over $\mathbf{q}$ vectors for the DFPT approach may affect computational efficiency. Therefore, further empirical studies of the computation time for real space and DFPT approaches would be helpful in comparing the scaling of these approaches.

5.2.2.2 Symmetrization of Interatomic Force Constants in Complex Materials

An important aspect of ab initio phonon transport calculations is the issue of force constant symmetrization. This problem does not arise in electron transport calculations, but it is extremely important for phonon calculations.

Briefly stated, the problem is that the crystal must be invariant with respect to rigid translations and rotations of the system. This defines many constraint equations relating the harmonic and anharmonic IFCs, Φ_{ij} and Φ_{ijk}. While the translational invariance conditions relate IFCs of the same order, the rotational invariance conditions involve different order energy derivatives within the same equation (see for example [79, 80].)

For simplicity, we neglect the anharmonic terms, Φ_{ijk}, for now and consider only the second order derivatives of the potential energy, and assume that no external forces are acting on the atoms, so translational and rotational invariance may be expressed as

$$\sum_j \Phi_{ij} R_j^n = 0, \tag{5.42}$$

where R_j^n is the displacement of the jth degree of freedom upon one of the translations or rotations, labeled by $n = 1, \ldots, 6$. When calculating force constants from standard parameterized interatomic potentials, symmetry conditions are usually automatically satisfied, if the potentials are physically meaningful [81]. However, when one tries to do the same thing from ab initio calculations, this is generally not the case. For practical calculations where the interatomic force constants are calculated in real space within a supercell, it is also important to ensure that the supercell is large enough to include non-negligible interactions with atoms in neighboring cells.

As an illustration, Fig. 5.2 shows the calculated interatomic force constant magnitude as a function of distance from the atom shifted in a (8,0) boron nitride nanotube (32 atoms per unit cell). The nanotube is oriented along the z-axis and the force constants are shown for the case where the atom is shifted in the x direction and along the z direction for 3 and 5 unit supercells. The calculated IFC magnitude drops rapidly with distance. In this case, the IFCs at 8 Å away are roughly three orders of magnitude smaller than those measured close to the atom. In the case of z atomic shift, there is considerably less scatter in the calculated magnitudes of the IFCs than in the x atomic shift. This could be due to the fact that with the z atomic shift, much of the nanotube symmetry is preserved and this leads to the response of many atoms being equivalent. From physical intuition, we would expect the interatomic force constants to decrease monotonically with distance away from the shifted atom. It is important to note that the 3 unit cell calculation does not follow this trend for neighbors far away from the shifted atom. The increased IFCs in this range are actually an artifact of the supercell calculation which relies on periodic boundary conditions. Atoms in this region interact strongly with both the shifted atom and its mirror images in the repeated supercells. This effect is reduced in the 5 unit cell calculation, but there is still a noticeable increase in IFC magnitude for neighbors far from the shifted atom. The work of Mounet and Marzari also provides an excellent example of the effect of long-range interactions on the predicted phonon dispersion of diamond and graphene [82].

From the previous example, we see that although the interactions become very small beyond a certain number of nearest neighbors, they are still finite. Thus, for an extended system, the translation/rotation invariance conditions are only numerically satisfied if an extremely large number of neighbors are included. This is unpractical, and one needs to impose a cutoff at a few nearest neighbors in order to manage the calculation. Imposing this cutoff does not affect most of the vibrational spectrum noticeably, since the neglected interactions are very small. However, it can result in unphysical behavior, especially at low frequencies, which may invalidate the results of the calculation [17]. Therefore, it is essential to symmetrize the force constants so that invariance relations are satisfied. Here we follow the derivation given in [17] for the symmetrization approach, with some additional details.

The symmetry invariance relations have important consequences on the character of the lowest acoustic phonon dispersions. For example, in the case of a periodic nanowire parallel to the z axis, each of the three translations corresponds to the $k_z = 0$ eigenmode of a phonon branch: one longitudinal for the z translation, and

Fig. 5.2 The magnitude of interatomic force constants is shown as a function of distance from the atom shifted in the z (panel **a**) and the x (panel **b**) directions in a (8,0) boron nitride nanotubes. Results are given from supercell calculations containing 3 and 5 unit cells

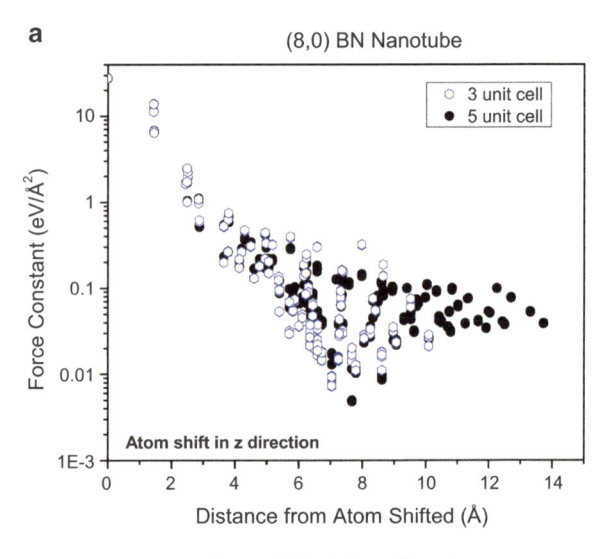

Atom shifted in z direction

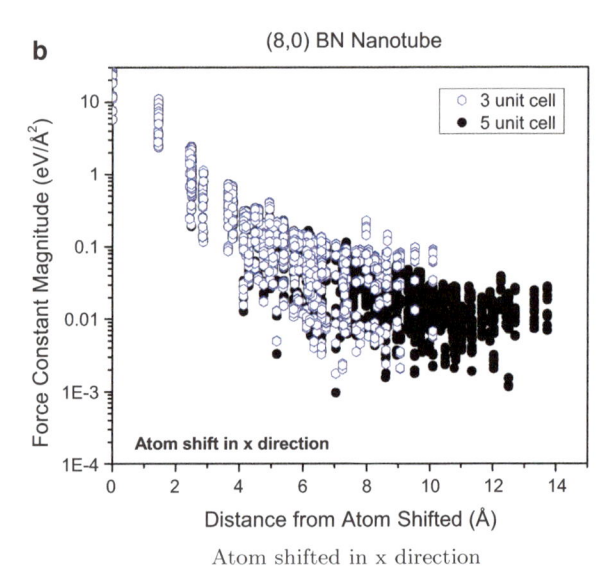

Atom shifted in x direction

two flexural for the x and y translations. Due to the invariance relations, Eq. (5.42), the frequencies of these three branches become 0 at $k_z = 0$. In addition, the rigid rotation around the z axis corresponds to the $k_z = 0$ eigenmode of the torsional branch, which also goes to zero at the origin, making a total of four acoustic branches for a nanowire.

The symmetrized force constants must be such that they differ from the original set by as little as possible. Let us denote the symmetrized constants by $\tilde{\Phi}$, so that

5 Ab Initio Thermal Transport

$$\tilde{\Phi}_{ij} \equiv \Phi_{ij} + D_{ij}, \tag{5.43}$$

where the D's denote the differences between the two sets. A measure of how different the two sets are is given by the scalar quantity f:

$$f \equiv \sum_{ij} D_{ij}^2 / \Phi_{ij}^2. \tag{5.44}$$

The symmetry conditions to be satisfied by the new set are

$$\phi_i^n \equiv \sum_j (D_{ij} + \Phi_{ij}) R_j^n = 0, \tag{5.45}$$

with $n = 1, \ldots, 6$. This can also be written as

$$\sum_j D_{ij} R_j^n = a_i^n, \tag{5.46}$$

with

$$a_i^n \equiv -\sum_j \Phi_{ij} R_j^n. \tag{5.47}$$

The quantities a_i^n are known. Now we need to determine the unknown quantities D_{ij} that satisfy Eq. (5.46), and which also yield the smallest possible f (5.44). Such a problem is solved by the method of Lagrange multipliers. The standard multiplier equations are [83]

$$\frac{\partial f}{\partial D_{ij}} + \frac{\partial f}{\partial D_{ji}} + \sum_n \lambda_i^n \frac{\partial \phi_i^n}{\partial D_{ij}} + \lambda_j^n \frac{\partial \phi_j^n}{\partial D_{ji}} = 0, \tag{5.48}$$

where the multipliers λ are not known, and we have used the fact that $D_{ij} = D_{ji}$. Generally, it is not necessary to explicitly determine the values of the multipliers. However, we found it convenient to solve for the λ's first, and evaluate the D's from them. The first term in Eq. (5.48) is

$$\frac{\partial f}{\partial D_{ij}} = 2 \frac{D_{ij}}{\Phi_{ij}^2}. \tag{5.49}$$

The second term is

$$\lambda_i^n \frac{\partial \phi_i^n}{\partial D_{ij}} + \lambda_j^n \frac{\partial \phi_j^n}{\partial D_{ji}} = \lambda_i^n R_j^n + \lambda_j^n R_i^n. \tag{5.50}$$

Substituting the two equations above into Eq. (5.48) yields the $D's$ in terms of the multipliers:

$$D_{ij} = \frac{\Phi_{ij}^2}{4} \sum_m \lambda_i^m R_j^m + \lambda_j^m R_i^m. \qquad (5.51)$$

Substituting this into the symmetry conditions, Eq. (5.46), yields

$$\sum_j \sum_m \frac{1}{4} \Phi_{ij}^2 \{ R_j^m R_j^n \lambda_i^m + R_i^m R_j^n \lambda_j^m \} = a_i^n. \qquad (5.52)$$

The equation can be rearranged in a more symmetric form as

$$\sum_j \sum_m B_{i,j}^{n,m} \lambda_j^m = a_i^n, \qquad (5.53)$$

$$B_{i,j}^{n,m} \equiv \frac{1}{4} \left(\sum_k \Phi_{ik}^2 R_k^m R_k^n \right) \delta_{ij} + \frac{1}{4} R_i^m R_j^n \Phi_{ij}^2. \qquad (5.54)$$

The λ's are obtained by solving Eq. (5.53), and the D's are directly obtained from them by Eq. (5.51). Then, the symmetrized force constants are immediately known via Eq. (5.43).

As already mentioned, the IFCs must satisfy various other symmetry conditions, for example involving also third order energy derivatives [80]. However, including those conditions make the formulation very complicated. In general, for bulk calculations, the most practical approach is to truncate harmonic and anharmonic force constants, and symmetrize the harmonic ones using only translational (but not rotational) invariance. For nano systems with four acoustic branches, such as nanotubes, thin nanowires, or nanoribbons, imposing rotational invariance around the longitudinal axis is also mandatory as discussed in [17].

5.2.2.3 Symmetrization of Anharmonic Force Constants in Periodic Crystals

It is essential that both the second and third order energy derivatives satisfy the proper space group symmetries of the system. We have found that the calculated thermal conductivity of a crystalline compound can be off by a non-negligible amount if the anharmonic force constants violate some of these symmetries.

Because of the space group symmetry of the system, IFCs are coupled. Considering a general space group symmetry operation $\sum_\alpha T^{\alpha'\alpha} R_i^\alpha + b^{\alpha'} = R_{Tb(i)}^{\alpha'}$, where T and **b** stand for the point group operator and translation operator, respectively, R_i^α is the α component of the equilibrium coordinate of the i-th atom, and Tb(i) specifies the atom to which the i-th atom is mapped under the corresponding operation. The third order IFCs tensor should satisfy the following relation,

$$\Phi_{Tb(i);Tb(j);Tb(k)}^{\alpha'\beta'\gamma'} = \sum_{\alpha\beta\gamma} T^{\alpha'\alpha} T^{\beta'\beta} T^{\gamma'\gamma} \Phi_{ijk}^{\alpha\beta\gamma}, \qquad (5.55)$$

5 Ab Initio Thermal Transport

from which we can find the independent nonzero IFCs, ϕ_i, where i ranges from 1 to the total number of independent nonzero elements. We just need to calculate the independent ϕ_i and obtain the remaining IFCs by using Eq. (5.55). An equivalent procedure can be used for any order IFCs.

In addition to the space group symmetry, the system energy does not change if the system as a whole is displaced by any amount, and we have sum rules for third order IFCs,

$$\sum_k \Phi_{ijk}^{\alpha\beta\gamma} = 0 \tag{5.56}$$

The equation is still valid if the summation is over i or j due to the permutation symmetry.

To enforce the sum rules, we add a compensation d_i to each independent nonzero IFC. To guarantee that the compensation is small, some additional constraints have to be considered. We will aim at minimizing the sum of the squares of the compensation for each independent nonzero element.

The sums in Eq. (5.56) are not completely independent, and the independent sums can be easily found numerically. If the constraints on these independent sums are fulfilled, all sum rules can be guaranteed. Since all the force constants can be deduced from the independent elements, the sums can be written in terms of the independent force constants elements as

$$\sum_j A_{ij} \phi_j = B_i. \tag{5.57}$$

Since the sum rules have to be satisfied, the constraints on the compensation are

$$g_i \equiv \sum_j A_{ij} d_j + B_i = 0. \tag{5.58}$$

The function to be minimized is

$$f = \frac{1}{2} \sum_j d_j^2. \tag{5.59}$$

After introducing Lagrange multiplier λ_i, the expression of d_j in terms of λ_i is obtained from

$$\frac{\partial (f + \sum_i \lambda_i g_i)}{\partial d_j} = 0, \tag{5.60}$$

from which it follows

$$d_j = -\sum_i \lambda_i A_{ij}. \tag{5.61}$$

Substituting this relation into Eq. (5.58), we have

$$\sum_i C_{ij} \lambda_j = B_i, \qquad (5.62)$$

with $C_{ij} = \sum_m A_{im} A_{jm}$. λ_j can be obtained by solving the linear equation arrays, and d_j is further obtained by using Eq. (5.61). Then the sum rules are completely satisfied by the force constants deduced from these independent elements after adding d_j to ϕ_j.

As an alternative of the presented method, a minimization of the sum of the square of the sums given in Eq. (5.56) in a high dimensional parameter space can also result in the enforcement the sum rule. However, the latter needs much more memory storage and much longer calculation time, and it does not always find the global minima in the predefined parameter ranges.

5.3 Examples

Here we review various materials that we have previously investigated. In the cases discussed, the *ab initio* approach has permitted us to achieve accurate thermal conductivity results to within 5–15% of experimental values for the isotopically enriched and naturally occurring crystals over a wide temperature range, and to within 30% error for the alloys. We also discuss the case of alloys with embedded nanoparticles, where the *ab initio* approach has also allowed us to confirm and refine the predictions of a previous simpler approach.

5.3.1 Bulk Si, Ge, and C

Previous work [14, 44, 84] has already documented the remarkable accuracy of *ab initio* plane-wave pseudopotential calculations within the LDA formalism at calculating lattice dynamical properties of semiconductors. To illustrate this, Fig. 5.3 shows the germanium phonon dispersion calculated from the DFPT generated harmonic IFCs [16, 21]. Excellent agreement with measured data [86] is evident.

We first present results for the thermal conductivities of Si and Ge. In our solution of the phonon BTE, hundreds of thousands of energy and momentum conserving three-phonon scattering events are needed throughout the Brillouin zone. We have applied the approach developed in [56] to calculate the necessary anharmonic IFCs out to seventh nearest neighbors. For the electronic states, an 8×8×8 Monkhorst–Pack grid [87] was used, and an energy cutoff of 24 Ry was taken for the plane wave expansion. The pseudopotentials were generated based on the approach of von Barth and Car [88].

Fig. 5.3 Phonon dispersions of germanium. Ab initio calculations—*solid line*; measured data—*squares* (from [85])

Fig. 5.4 Ab initio calculations of the lattice thermal conductivity of naturally occurring (*blue and red dashed curves*) and isotopically enriched (*blue and red solid curves*) Si and Ge as a function of temperature compared with the corresponding experimental values from [89,90]

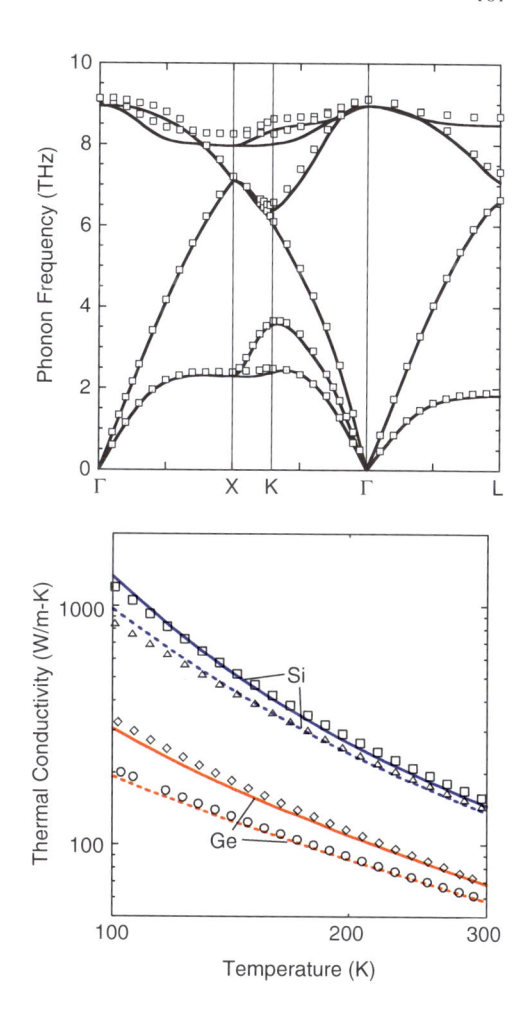

Figure 5.4 shows the calculated lattice thermal conductivity of naturally occurring (dashed curves) and isotopically enriched (solid curves) Si and Ge [16, 21] compared with the corresponding experimental values (symbols) [89,90]. Naturally occurring Si consists of 92.2% Si^{28}, 4.7% Si^{29}, and 3.1% Si^{30}, while naturally occurring Ge has 20.5% Ge^{70}, 27.4% Ge^{72}, 7.8% Ge^{73}, 36.5% Ge^{74}, and 7.8% Ge^{76}. The isotopically enriched samples have 99.983% Si^{28}, 0.014% Si^{29}, 0.003% Si^{30}, and 99.99% Ge^{70}, and 0.01% Ge^{73}. The corresponding volume fractions, (f in Eq. (5.20)) are: 2.33×10^{-6} (enriched Si), 2.01×10^{-4} (naturally occurring Si), and 1.84×10^{-7} (enriched Ge), 5.87×10^{-4} (naturally occurring Ge). All curves show a roughly $\sim 1/T$ behavior, which is characteristic of three-phonon limited thermal conductivity [2,3]. The agreement between theory and experiment is extremely good in all cases with differences of only 5–10%. This is particularly noteworthy given that no adjustable parameters have been used.

Fig. 5.5 Ab initio calculations of the lattice thermal conductivity of isotopically enriched (*red curve*) and naturally occurring type IIa (*blue curve*) diamond as a function of temperature compared to measured data (*symbols*) from [92–95]

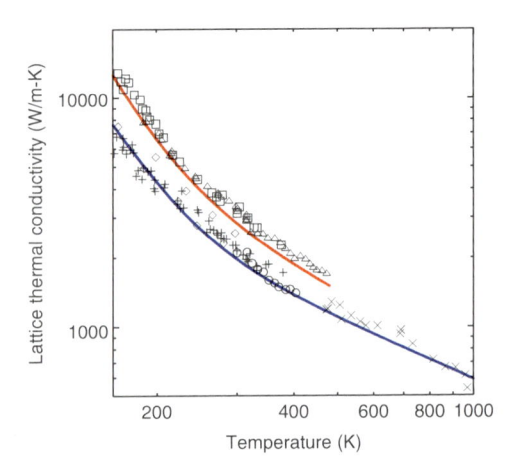

As a test of the robustness of the theory, we also consider diamond whose thermal conductivity is more than an order of magnitude higher than either Si or Ge. Here, we have used the BHS pseudopotential [91], for both harmonic and anharmonic IFCs. The harmonic IFCs were calculated using a 6×6×6 Monkhorst–Pack q-point mesh, while a 4×4×4 Monkhorst–Pack mesh was used to determine the anharmonic IFCs, out to seventh nearest neighbors. A much higher cutoff energy of 100 Ry was needed for diamond.

Figure 5.5 shows the calculated temperature dependence of the lattice thermal conductivity of isotopically enriched (99.93% C^{12}, 0.07% C^{13}) and type IIa naturally occurring (98.9% C^{12} and 1.1% C^{13}) diamond from our ab initio approach [21] compared with measured values. For this case, impurity fractions are 4.86×10^{-6} (isotopically enriched) and 7.54×10^{-5} (naturally occurring). The open squares [92] and triangles [93] show the measured values for isotopically enriched diamond with 99.93% C^{12}, 0.07% C^{13}, 99.9% C^{12}, and 0.1% C^{13}, respectively. The open diamonds [94], open circles [95], plusses [93], and crosses [92] are for naturally occurring type IIa diamond. The corresponding calculated results are given by the solid curves. It is again evident that the calculated curves for both the isotopically enriched and naturally occurring diamond are in very good agreement with the data, falling within 10–15% of the measured values over the wide temperature range considered.

The ab initio/BTE approach reveals important physical behavior that does not emerge from widely used relaxation time approaches. Since Normal and Umklapp processes are calculated explicitly, it is possible to directly assess the importance of each in a given material. It is well known that the momentum-conserving Normal processes cannot provide thermal resistance [2, 3] and if only they are present, the thermal conductivity will diverge. This is confirmed through the calculated divergence of κ in Eq. (5.26) with increasing iteration in the BTE solution when Umklapp processes are artificially turned off. In Si and Ge, around room temperature, the iteratively converged κ is only around 5% higher than that obtained

5 Ab Initio Thermal Transport

Fig. 5.6 Ab initio calculations of isotopically enriched lattice thermal conductivity of Si and diamond (*solid green and purple curves*) from the full calculations and for the case where the acoustic-optic phonon–phonon scattering channels have been omitted (*dashed green and purple curves*). Measured values for isotopically enriched diamond and Si are also included

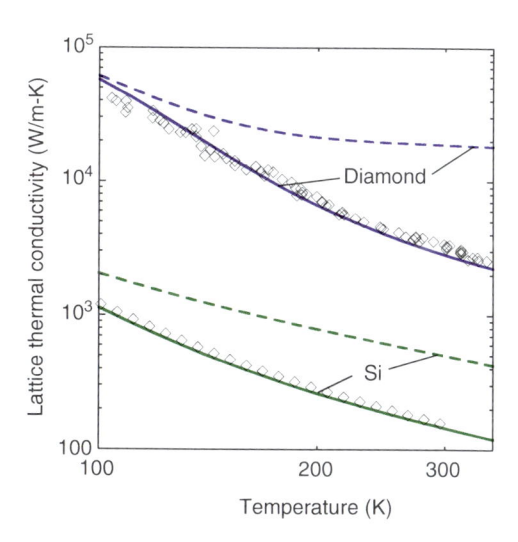

from the zeroth iteration. This demonstrates that the Umklapp processes are strong in these materials and has allowed us to accurately determine new functional forms for the relaxation times for these materials [96].

In contrast, for diamond, the converged room temperature κ is about 50% higher than the zeroth order value indicating the dominance of Normal processes. This is in part a result of the high frequency scale for diamond resulting from the light carbon atoms and strong covalent bonding. Another important factor is that the calculated phase space for three-phonon scattering is unusually small [21, 97].

Calculations of the lattice thermal conductivity frequently ignore the optic phonons because they have small group velocities and typically reside at higher frequencies than the three acoustic phonon branches. Our calculations confirm that the acoustic phonons carry most of the heat, representing well over 90% of the total thermal current in Si, Ge and diamond at room temperature. However, we find that the optic phonons provide essential scattering channels for the acoustic phonons [16, 21], as has also been found previously [29, 30]. This is illustrated in Fig. 5.6, which shows the calculated lattice thermal conductivities of isotopically enriched diamond and Si as a function of temperature. The solid curves give results from the full calculations already presented in Figs. 5.3 and 5.4, plotted with the corresponding measured data. The dashed curves give the results with the scattering between acoustic and optic phonons removed. At room temperature, the κ for Si increases by more than a factor of three, while for diamond, an even more striking sixfold increase is found. The dramatic increase in diamond is a consequence of the extremely small three-phonon scattering phase space and the fact that about 80% of the scattering processes involve acoustic and optic phonons. Overall, our ab initio/BTE approach demonstrates the important role that optical branches play in generating thermal resistance and this work also confirms the general predictions made in previous works based on fitted scattering times for Normal and Umklapp processes between acoustic and optical branches [98, 99].

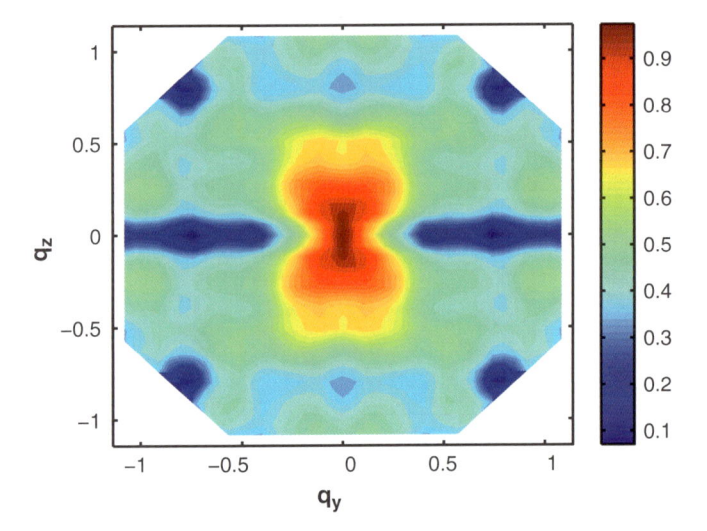

Fig. 5.7 Contour plot of the contribution to the room temperature intrinsic lattice thermal conductivity, κ, of Si for a two-dimensional slice through the Brillouin zone with $q_x = 0.002695$ nm^{-1}. The thermal gradient is taken to be along the z direction

The contributions to the intrinsic lattice thermal conductivity from different wave vectors can be obtained by summing the integrand in Eq. (5.26) over all phonon branches. These contributions are sharply peaked near the center of the Brillouin zone. Figure 5.7 illustrates a representative two-dimensional slice of the contribution to κ for Si at room temperature and for a small q_x ($q_x = 0.002695$ nm^{-1}). For the calculation, the thermal gradient is taken to be along the z direction. The magnitude of the contribution on this contour is indicated by the color scheme and is presented on a log scale. Although it is not shown in the current figure, plotting an isosurface where contributions to κ are large (≥ 0.8) generates a dumbbell shaped region centered at the Γ point and oriented along the z axis. The figure highlights a very sharp peak near the zone center that drops rapidly with increasing wave vector magnitude, and which provides a significant contribution to κ. There are two primary factors that can lead to the peak in thermal conductivity contributions near the center of the Brillouin zone. As noted in Eq. (5.26), the thermal conductivity depends on the square of the group velocity in the direction of the thermal gradient, $(v_\lambda^z)^2$. The phonon dispersion of silicon shows that the group velocities of the TA branches along the Γ-X symmetry line become very small near the edge of the peaked region in Fig. 5.7. The general shape of the peaked region is also affected by the phase space for three phonon scattering events which tends to zero as $\mathbf{q} \to 0$. This causes the scattering times, $\tau_{j\alpha}(\mathbf{q})$, to be large near the zone center, thereby giving large contributions to κ. We note that the vast majority of Umklapp processes can only occur for phonons away from $\mathbf{q} = 0$. Their increasing occurrence, as $\mathbf{q}$ becomes large also leads to a reduction in the contributions to κ away from the Γ point which is observed in Fig. 5.7. It is important to note that Fig. 5.7 comes from the integrand of Eq. (5.26), which does not take account of the compensating effect of the phonon density of states, which increases with frequency. Plotting instead the integrand of Eq. (5.27) shows the contribution of each frequency to the thermal conductivity is more evenly distributed over the frequency range of the acoustic branches, with large contributions coming away from the zone center.

5.3.2 SiGe Alloys, and the Effect of Embedded Nanoparticles

The case of alloy scattering is more complicated than that of small isotopic impurity concentrations. As we discussed in Sect. 5.2.1.3, at elevated impurity concentrations, it becomes unclear which species is the matrix and which one is the impurity. So some sort of mean field model is required if one is to continue using a picture based on propagating waves scattered by localized impurities. The simplest of these models was proposed by Abeles to understand SiGe alloys [100], and it has later been successfully applied to interpret the thermal conductivity of a large range of semiconducting compounds [101].

The thermal conductivity of SiGe provides one of the most important case studies for thermal transport in alloys or solid solutions. Historically, the experimental discovery that SiGe had much lower thermal conductivities than pure Si or Ge was a surprise [102]. Its correct theoretical interpretation was provided a bit later [100]. This, combined with the fact that electron mobilities were not much reduced by alloying, prompted a high degree of activity to develop better thermoelectric materials based on SiGe solid solutions, and this material still remains one of the best thermoelectrics at high temperature [103]. Research in this area is far from closed, for it has been shown that nanostructuring the alloys in various ways can result in even better thermoelectric figures of merit [104] (see also next subsection).

In the case of alloy scattering, using approximated phonon dispersions and the Rayleigh scattering approximation can lead to large errors in the calculated thermal conductivity. For example, using the Tersoff potential in combination with Rayleigh's formula yields a thermal conductivity of about 20 W/m-K for the $Si_{0.5}Ge_{0.5}$ alloy, more than 2 times larger than the experimental one [105]. Thus, in approximated calculations, adjustable parameters are needed. In order to accurately predict the thermal conductivity of an alloy or solid solution, it is very important to have an accurate description of the phonon dispersion, and also a reliable atomistic calculation of the phonon scattering rates. This illustrates the importance of performing a rigorous *ab initio* calculation versus the use of simpler parametrized models.

We have calculated values of thermal conductivity versus Ge concentration for SiGe solid solutions. The calculation includes mass disorder scattering through Tamura's formula, Eq. (5.20), in addition to the ab initio calculated anharmonic scattering rates. The values obtained are close to the experimental ones, but still about 30% higher for the intermediate concentrations. This reportedly is an effect introduced by the use of the effective medium approximation. Calculations using

Fig. 5.8 Calculated thermal conductivity of $Si_{0.5}Ge_{0.5}$ with embedded Ge nanoparticles, at $T = 800$ K, for three different values of the nanoparticle volume fraction, as a function of nanoparticle diameter. *Lines* correspond to the Born approximation, and *symbols* to the exact T-matrix solution

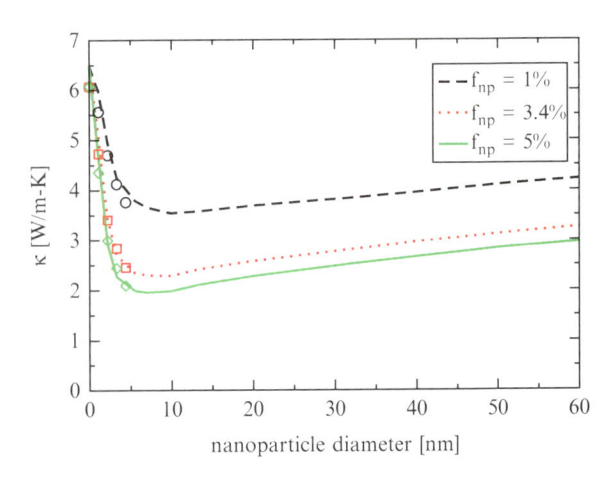

averaging over a supercell have shown that mass disorder actually enhances three phonon scattering, resulting in a lower thermal conductivity than if three phonon terms were computed for the effective medium directly [70].

Embedding nanoparticles in an alloy is an effective way to reduce its thermal conductivity. A direct experimental demonstration of this was provided in [106], whereby embedding ErAs nanoparticles in an InGaAs alloy reduced its thermal conductivity by a factor of two. The reason for this strong effect was theoretically explained in [107] in terms of the larger Rayleigh cross section of nanoparticles at low frequency, where alloy scattering is not very effective. Subsequently, it was shown that there is an optimal nanoparticle size that minimizes the alloy thermal conductivity [108]. The role of the alloy matrix was also highlighted, showing that nanoparticles affect the thermal conductivity of alloys much more strongly than that of non alloys.

Figure 5.8 shows the *ab initio* calculated results for the hypothetical inclusion of Ge nanoparticles into a SiGe matrix. In the black curve, the Tamura formula, Eq. (5.20), has been employed for the low frequency scattering. The symbols correspond to a calculation using the exact nanoparticle scattering rates calculated using Green's functions. The calculation in [108] had employed a simple parameterized model of phonon dispersions and alloy scattering. The *ab initio* results are qualitatively similar to those in [108] and confirm the validity of the conclusions in that reference.

5.3.3 Nanowires

The effects of boundary scattering on thermal conductivity are very important, especially at lower temperatures. Traditionally this effect has simply been included as an added constant scattering rate proportional to the system size. In the case of not too thin wires, defined as those where confinement effects are not important, it is possible to include the size effect in a more exact way, via the numerical solution of the space-dependent Boltzmann Equation [23]. A brief summary of the method is included here.

Now the distribution function $f_{\mathbf{r},\lambda} = f_0(\omega_\lambda) + g_{\mathbf{r},\lambda}$ is space dependent. Expressing $g_{\mathbf{r},\lambda}$ in terms of phonon lifetime $\tau_{\mathbf{r},\lambda}$ defined as $g_{\mathbf{r},\lambda} = -\frac{dT}{dz} v_\lambda^z \frac{df_0}{dT} \tau_{\mathbf{r},\lambda}$, where the temperature gradient and the axes of NWs are taken along z direction, and v_λ^z is z component of the group velocity, the BTE can be written as

$$1 = (\tau_\lambda^0)^{-1} \tau_{\mathbf{r},\lambda} - \Delta_{\mathbf{r},\lambda} + \mathbf{v}_\lambda \cdot \nabla \tau_{\mathbf{r},\lambda}. \tag{5.63}$$

It is not an easy task to solve for $\tau_{\mathbf{r},\lambda}$ numerically since the nanowire surface boundary condition is hard to implement. Approximating $\Delta_{\mathbf{r},\lambda}$ by its average value $\bar{\Delta}_\lambda$ over the cross section, which is evaluated using $\bar{\tau}$, the average of $\tau_{\mathbf{r}}$, we have [2]

$$\tau_{\mathbf{r},\lambda} = \tau_\lambda^0 (1 + \bar{\Delta}_\lambda) \left\{ 1 - e^{-\left|\frac{\mathbf{r}-\mathbf{r}_b}{\tau^0 v_\lambda}\right|} G_{\mathbf{r},\lambda} \right\}, \tag{5.64}$$

where $\mathbf{r}_b$ is the point on the surface where the phonon of mode λ can reach moving backward from $\mathbf{r}$, and $G_{\mathbf{r},\lambda}$ is determined by the boundary conditions. For completely diffusive boundary conditions, as considered here, $G_{\mathbf{r},\lambda} = 1$. $\bar{\tau}_\lambda$ can thus be obtained as

$$\bar{\tau}_\lambda = \tau_\lambda^0 (1 + \bar{\Delta}_\lambda) \left(\frac{1}{S_c} \int_{S_c} \left\{ 1 - e^{-\left|\frac{\mathbf{r}-\mathbf{r}_b}{\tau^0 v_\lambda}\right|} \right\} d\mathbf{s} \right), \tag{5.65}$$

with S_c being the nanowire cross section. Equation (5.65) can be solved iteratively starting with the zeroth order solution

$$\bar{\tau}_\lambda^{(0)} = \tau_\lambda^0 \left(\frac{1}{S_c} \int_{S_c} \left\{ 1 - e^{-\left|\frac{\mathbf{r}-\mathbf{r}_b}{\tau^0 v_\lambda}\right|} \right\} d\mathbf{s} \right). \tag{5.66}$$

$\bar{\tau}_\lambda^{(0)}$ is equivalent to the RTA. The iteration is repeated until the convergence is achieved.

The method is completely general and it can be applied to any compound. Figure 5.9 illustrates the case of diamond nanowires. The graph shows how isotope scattering effects become relatively less important as the size of the wire gets smaller. This calculation made use of the iterative scheme, so all Normal and Umklapp processes are fully included. Furthermore, boundary effects were not approximately added as a separate scattering rate, but occur naturally when solving the space-dependent BTE described above. More details can be found in [23].

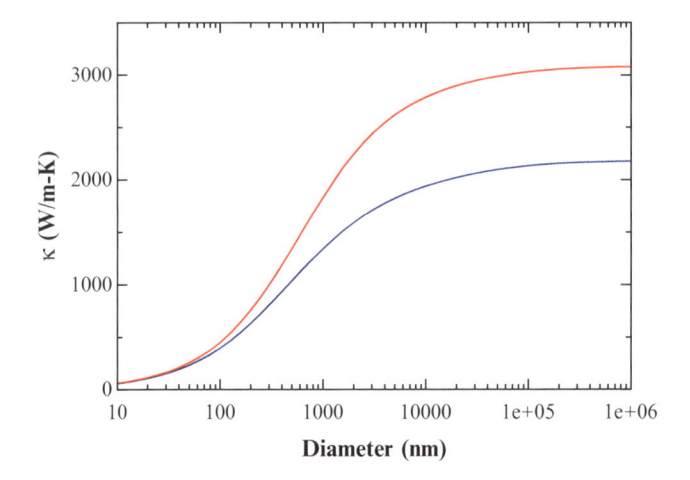

Fig. 5.9 Calculated thermal conductivity of diamond wires at room temperature, as a function of diameter. The *red curve* is for isotopically pure diamond and the *blue* one is for naturally occurring diamond (See [23])

5.4 Other Ab Initio Techniques

It is important to note recent work in the geosciences community on predicting the thermal conductivity of MgO (periclase) from first principles molecular dynamics and lattice dynamics calculations. MgO is the second most abundant phase in the lower mantle of the Earth [109] and an accurate description of its thermal conductivity at high temperatures and pressures is crucial for geologic thermal circulation models [110, 111]. Since empirical potentials typically used in molecular dynamic approaches rely on fits to experimental data, it is difficult to develop accurate parameters for these high temperature and pressure regimes where available data is often sparse or non-existent. This makes first principle thermal transport approaches ideal for geologic applications in extreme physical environments.

Molecular dynamics thermal conductivity calculations typically use either a Green–Kubo approach [112, 113] (general fluctuation-dissipation theorem) or direct non-equilibrium molecular dynamics simulations where energy is artificially exchanged between separate hot and cold regions in the supercell. In a recent investigation of thermal transport in half-Heusler compounds [114], the authors used a combination of potentials derived from first principle calculations and classical equilibrium molecular dynamics simulations to calculate the thermal conductivity with the Green–Kubo approach.

Stackhouse et al. [109, 115] calculated the thermal conductivity of MgO using ab initio non-equilibrium molecular dynamic simulations based on a plane wave density functional approach. All simulations were run for 34 picoseconds (1 femtosecond time step) and the supercell size was varied from 2×2×6 conventional unit cells (192 atoms) up to 2×2×16 (512 atoms). Finite size effects due to the small supercells will clearly affect the calculated thermal conductivity. These effects can be estimated in terms of an additional phonon mean free path that depends on phonon-boundary scattering and the supercell length (L). By calculating the thermal conductivity for a series of different supercell sizes, the bulk thermal conductivity can be extrapolated in the limit as L goes to infinity. The calculated values are in good agreement for temperatures (300–2,000 K) at atmospheric pressure (0 GPa) and reasonable agreement for different pressures at a temperature of 2,000 K [109]. It should be emphasized that this extrapolation works well for materials with low thermal conductivity. However, for high thermal conductivity materials with long phonon mean free paths, the error due to finite supercell size effects can be quite large [115] and the computational cost of going to sufficiently large supercells could be prohibitive.

First principle molecular dynamics has also been used to determine MgO thermal conductivity based on the anharmonic phonon dispersion and relaxation time of individual phonon modes [116]. Within the relaxation time approximation, the phonon spectral density can be determined by taking the Fourier transform of the velocity autocorrelation function and the relaxation times can be related to the width of the spectral peaks. It should be noted that the exact wavevectors q that the velocity autocorrelation function can be evaluated at are restricted due to the finite size of the supercell. Although phonon frequencies can shift due to anharmonicity, this effect is typically very small. Therefore, it is possible to calculate the harmonic phonon dispersions from a material using density functional perturbation theory or finite displacement approach and couple this with relaxation times extracted from ab initio molecular dynamics runs to determine the thermal conductivity. This was done by de Koker for MgO and the calculated values are in reasonable agreement with available experimental and theoretical results for different pressures [116, 117]. However, as we demonstrated earlier in the chapter, the use of the relaxation time approximation to the full Boltzmann transport equation can lead to 50% underestimate for the thermal conductivity in certain materials (e.g., diamond). Tang and Dong also recently calculated the thermal conductivity of MgO at various geologic conditions [118] using a Boltzmann transport equation with the relaxation time approximation where the relaxation times are based on anharmonic force constants calculated in real space with supercell density functional calculations [69].

5.5 Concluding Remarks

The field of ab initio thermal transport is just starting, but quick advances in computing power will likely enable its increasingly rapid development. The examples discussed in this chapter, for elementary group IV crystals, show that it is possible to predict materials' thermal conductivity with remarkable accuracy, in the absence of any experimental input. In principle, a myriad of compounds lie now ahead awaiting to be put under the test, to determine whether DFT and BTE suffice for a reliable prediction of their thermal conduction properties. Caution must be exercised however, since every new system comes with potential new issues. For example, considerable testing will be required to assess whether polar crystals, such as III-V compounds, can be described as accurately as the group IV systems. The validity of the BTE may also break down in some cases. Using a 2D model, it has been shown that at sufficiently high temperature Peierls' BTE should have important higher order corrections, which unfortunately are far too complex to treat explicitly [119]. Deficiencies have also been pointed out in a lot of the molecular dynamics work because they typically ignore vertex corrections [120]. Although nobody knows to what extent these issues may limit the reach of ab initio thermal transport, it is likely that a large number of simple and not so simple compounds will be well described by the theory presented here.

Indeed, first principles approaches based on DFPT have recently been used to study phonon–phonon interactions, phonon lifetimes, and phonon thermal transport in graphene [121–123], half Heuslers [114], and lead telluride [76, 124–126].

A main challenge now is to undertake a systematic approach to study as many possible material types as possible, towards a more complete understanding of the reach and limitations of this predictive approach. Integrating and standardizing techniques such as those reviewed here within established ab initio packages would allow for faster progress in this area.

Acknowledgments We thank A. Ward, I. Savic, S. Wang, G. Deinzer, M. Malorny, K. Esfarjani, A. Kundu, and N. A. Katcho, for their contribution to the works cited or summarized in this chapter. We are grateful to A. Shakouri, L. Shi, F. Mauri, M. Lazzeri, and N. Vast for helpful discussions. We acknowledge support from the National Science Foundation under grant Nos. 1066634 and 1066406, the EU, Agence Nationale de la Recherche, CEA, and Fondation Nanosciences. L.L. acknowledges support from DARPA and from the NRC/NRL Research Associateship Program. A portion of the calculations discussed in this chapter were calculated using the Intel Cluster at the Cornell Nanoscale Facility, part of the National Nanotechnology Infrastructure Network funded by the NSF.

References

1. Cahill DG, Ford WK, Goodson KE, Mahan GD, Majumdar A, Maris HJ, Merlin R, Phillpot SR (2003) J Appl Phys 93:793
2. Ziman JM (1960) Electrons and phonons. Oxford University Press London
3. Callaway J (1991) Quantum theory of the solid state. Academic Press, New York
4. Srivastava GP (1990) The physics of phonons. Adam Hilger, Bristol
5. Asen-Palmer M, Bartkowski K, Gmelin E, Cardona M, Zhernov AP, Inyushkin AV, Taldenkov A, Ozhogin VI, Itoh KM, Haller E (1997) Phys Rev B 56:9431
6. Han YJ, Klemens PG (1993) Phys Rev B 48:6033
7. Tamura SI, Tanaka Y, Maris HJ (1999) Phys Rev B 60:2627
8. Khitun A, Wang KL (2001) Appl Phys Lett 79:851
9. Berber S, Kwon YK, Tomanek D (2000) Phys Rev Lett 84:4613
10. Ponomareva I, Srivastava D, Menon M (2007) Nano Lett 7:1155
11. Lukes J, Zhong H (2007) J Heat Trans 129:705
12. Volz S, Chen G (1999) Appl Phys Lett 75:2056

13. Che J, Cagin T, Deng W, Goddard III WA (2000) J Chem Phys 113:6888
14. Baroni S, de Gironcoli S, Corso AD, Giannozzi P (2001) Rev Mod Phys 73:515
15. Mingo N, Yang L (2003) Phys Rev B 68:245406
16. Broido DA, Malorny M, Birner G, Mingo N, Stewart DA (2007) Appl Phys Lett 91:231922
17. Mingo N, Stewart DA, Broido DA, Srivastava D (2008) Phys Rev B 77:033418
18. Savic I, Mingo N, Stewart DA (2008) Phys Rev Lett 101:165502
19. Stewart DA, Savic I, Mingo N (2009) Nano Lett 9:81
20. Savic I, Stewart DA, Mingo N (2008) Phys Rev B 78:235434
21. Ward A, Broido DA, Stewart DA, Deinzer G (2009) Phys Rev B 80:125203
22. Kundu A, Mingo N, Broido DA, Stewart DA (2011) Phys Rev B 84:125426. DOI 10.1103/PhysRevB.84.125426. URL http://link.aps.org/doi/10.1103/PhysRevB.84.125426
23. Li W, Mingo N, Lindsay L, Broido DA, Stewart DA, Katcho NA (2012) Phys Rev B 85:195436. DOI 10.1103/PhysRevB.85.195436. URL http://link.aps.org/doi/10.1103/PhysRevB.85.195436
24. Peierls RE (1929) Ann Phys (Liepzig) 3:1055
25. Peierls RE (1955) Quantum theory of solids. Clarendon, Oxford
26. Omini M, Sparavigna A (1995) Phys B 212:101
27. Omini M, Sparavigna A (1996) Phys Rev B 53:9064
28. Omini M, Sparavigna A (1997) Nuovo Cimento D 19:1537
29. Sparavigna A (2002) Phys Rev B 65:064305
30. Sparavigna A (2002) Phys Rev B 66:174301
31. Broido DA, Ward A, Mingo N (2005) Phys Rev B 72:014308
32. Pascual-Gutierrez JA, Murthy JY, Viskanta R (2009) J Appl Phys 106:063532
33. Mingo N, Esfarjani K, Broido DA, Stewart DA (2010) Phys Rev B 81:045408
34. Tamura S (1983) Phys Rev B 27:858
35. Gilat G, Raubenheimer LJ (1966) Phys Rev B 144:390
36. Lambin P, Vigneron JP (1984) Phys Rev B 29:3430
37. Yates JR, Wang X, Vanderbilt D, Souza I (2007) Phys Rev B 75:195121. DOI 10.1103/PhysRevB.75.195121. URL http://link.aps.org/doi/10.1103/PhysRevB.75.195121
38. Hohenberg P, Kohn W (1964) Phys Rev 136:B864
39. Kohn W, Sham LJ (1965) Phys Rev 140:A1133
40. Martin RM (2004) Electronic structure: basic theory and practical applications. Cambridge University Press, Cambridge
41. Mattsson AE, Schultz PA, Desjarlais MP, Mattsson TR, Leung K (2005) Model Simulat Mater Sci Eng 13:R1
42. Hellmann H (1937) Einfuhrung in die quantenchemie. Deuticke, Leipzig
43. Feynman RP (1939) Phys Rev 56:340
44. Giannozzi P, de Gironcoli S, Pavone P, Baroni S (1991) Phys Rev B 43:7231
45. Zein EN (1984) Sov Phys Solid State 26:1825
46. Baroni S, Giannozzi P, Testa A (1987) Phys Rev Lett 59:2662
47. Mounet N, Marzari N (2005) Phys Rev B 71:205214
48. Giannozzi P, et al (2009) J Phys Conds Matter 21:395502
49. Gonze X, et al (2009) Comput Phys Commun 180:2582
50. Clark SJ, Segall MD, Pickard CJ, Hasnip PJ, Probert MJ, Refson K, Payne MC (2005) Zeitschrift fur Kristallographie 220:567
51. Gonze X, Vigneron JP (1989) Phys Rev B 39:13120
52. Gonze X (1995) Phys Rev A 52:1096
53. Debernardi A, Baroni S, Molinari E (1995) Phys Rev Lett 75:1819
54. Lang G, Karch K, Schmitt M, Pavone P, Mayer AP, Wehner RK, Strauch D (1999) Phys Rev B 59:6182
55. Debernardi A (1998) Phys Rev B 57:12847
56. Deinzer G, Birner G, Strauch D (2003) Phys Rev B 67:144304
57. Coldwell-Horsfall RA (1963) Phys Rev 129:22
58. Soler JM, Artacho E, Gale JD, Garcia A (2002) J Phys Condens Matter 14:2745

59. Ozaki T (2003) Phys Rev B 67:155108
60. Bowler DR, Miyazaki T (2010) J Phys Condens Matter 22:074207
61. Skylaris CK, Haynes PD, Mostofi AA, Payne MC (2005) J Chem Phys 122:084119
62. Kresse G, Furthmuller J, Hafner J (1995) Europhys Lett 32:729
63. Frank W, Elsasser C, Fahnle M (1995) Phys Rev Lett 74:1791
64. Parlinski K, Li ZQ, Kawazoe Y (1997) Phys Rev Lett 78:4063
65. Artacho E, Anglada E, Diéguez O, Gale JD, García A, Junquera J, Martin RM, Ordejón P, Pruneda JM, Sánchez-Portal D, Soler JM (2008) J Phys Condens Matter 20:064208
66. Alfe D (2009) Comp Phys Comm 180:2622
67. Togo A, Oba F, Tanaka I (2008) Phys Rev B 78:134106
68. Esfarjani K, Stokes HT (2008) Phys Rev B 77:144112
69. Tang X, Dong J (2009) Phys Earth Planet Inter 174:33
70. Garg J, Bonini N, Kozinsky B, Marzari N (2011) Phys Rev Lett 106(4), 045901. DOI 10.1103/PhysRevLett.106.045901
71. Chaput L, Togo A, Tanaka I, Hug G (2011) Phys Rev B 84:094302
72. Lyddane RH, Sachs RG, Teller E (1941) Phys Rev 59:673
73. Cochran W, Cowley RA (1962) J Phys Chem Solid 23:447
74. Detraux F, Ghosez P, Gonze X (1998) Phys Rev Lett 81:3297
75. Parlinksi K, Li ZQ, Kawazoe Y (1998) Phys Rev Lett 81:3298
76. An J, Subedi A, Singh DJ (2008) Solid State Comm 148:417
77. Kilian O, Allan G, Wirtz L (2009) Phys Rev B 80:245208
78. Deinzer G, Schmitt M, Mayer AP, Strauch D (2004) Phys Rev B 69:014304
79. Leibfried G, Ludwig W (1961) Solid State Phys 12:275
80. Maradudin AA, Horton GK (1974) Dynamical properties of solids. North-Holland, Amsterdam
81. Mahan GD, Jeon GS (2004) Phys Rev B 70:075405
82. Mounet N, Marzari N (2005) Phys Rev B 71:205214
83. Arfken GB (1985) Mathematical methods for physicists. Academic Press, New York
84. Yin MT, Cohen ML (1982) Phys Rev B 26:3259
85. Nilsson G, Nelin G (1972) Phys Rev B 6:3777
86. Warren JL, Yarnell JL, Dolling G, Cowley RA (1967) Phys Rev 158:805
87. Monkhorst HJ, Pack JD (1976) Phys Rev B 13:5188
88. von Barth U, Car R (1993) (Unpublished) for a brief description of this method, see Corso AD, Baroni S, Resta R, de Gironcoli S, Phys Rev B 47:3588
89. Inyushkin AV, Taldenkov AN, Gibin AM, Gusev AV, Pohl HJ (2004) Phys Stat Solid C 1:2995
90. Ozhogin VI, Inyushkin AV, Taldenkov AN, Tikhomirov AV, Popov GE (1996) JETP Lett 63:490
91. Bachelet GB, Hamann DR, Schluter M (1982) Phys Rev B 26:4199
92. Olson JR, Pohl RO, Vandersande JW, Zoltan A, Anthony TR, Banholzer WF (1993) Phys Rev B 47:14850
93. Wei L, Kuo PK, Thomas RL, Anthony TR, Banholzer WF (1993) Phys Rev Lett 70:3764
94. Berman R, Hudson PRW, Martinez M (1975) J Phys C Solid State Phys 8:L430
95. Onn DG, Witek A, Qiu YZ, Anthony TR, Banholzer WF (1992) Phys Rev Lett 68:2806
96. Ward A, Broido DA (2010) Phys Rev B 81:085205
97. Lindsay L, Broido DA (2008) J Phys Condens Matt 20:165209
98. Srivastava GP (1980) In: Maris HJ (ed) Phonon scattering in solids. Plenum Press, New York, p 149
99. Srivastava GP (1980) J Phys Chem Solids 41:357
100. Abeles B (1963) Phys Rev 131:1906. DOI 10.1103/PhysRev.131.1906
101. Adachi S (2007) J Appl Phys 102(6):063502. DOI 10.1063/1.2779259
102. Abeles B, Beers DS, Cody GD, Dismukes JP (1962) Phys Rev 125:44. DOI 10.1103/PhysRev.125.44
103. Snyder GJ, Toberer ES (2008) Nat Mater 7:105. DOI 10.1038/nmat2090
104. Lan Y, Minnich AJ, Chen G, Ren Z (2010) Adv Funct Mater 20:357

105. Savic I (2009) Private communication
106. Kim W, Zide J, Gossard A, Klenov D, Stemmer S, Shakouri A, Majumdar A (2006) Phys Rev Lett 96(4):045901. DOI 10.1103/PhysRevLett.96.045901
107. Kim W, Majumdar A (2006) J Appl Phys 99(8):084306. DOI 10.1063/1.2188251
108. Mingo N, Hauser D, Kobayashi NP, Plissonnier M, Shakouri A (2009) Nano Lett 9:711
109. Stackhouse S, Stixrude L, Karki BB (2010) Phys Rev Lett 104:208501
110. Manthilake GM, de Koker N, Frost DJ, McCammon CA (2011) Proc Natl Acad Sci USA 108:17901
111. Stamenković V, Breuer D, Spohn T (2011) Icarus 216:572
112. Green MS (1954) J Chem Phys 22:398
113. Kubo R (1957) J Phys Soc Jpn 12:570
114. Shiomi J, Esfarjani K, Chen G (2011) Phys Rev B 84:104302. DOI 10.1103/PhysRevB.84.104302. URL http://link.aps.org/doi/10.1103/PhysRevB.84.104302
115. Stackhouse S, Stixrude L (2010) Rev Mineral Geochem 71:253
116. de Koker N (2009) Phys Rev Lett 103:125902
117. de Koker N (2010) Earth Planet Sci Lett 292:392
118. Tang X, Dong J (2010) Proc Natl Acad Sci 107:4539
119. Sun T, Allen PB (2010) Phys Rev B 82(22):224305. DOI 10.1103/PhysRevB.82.224305
120. Sun T, Shen X, Allen PB (2010) Phys Rev B 82:224304
121. Bonini N, Lazzeri M, Marzari N, Mauri F (2007) Phys Rev Lett 99:176802
122. Bonini N, Rao R, Rao AM, Marzari N, Menéndez J (2008) Phys Stat Sol (b) 245:2149
123. Bonini N, Garg J, Marzari N (2012) Nano Lett 12:2673
124. Delaire O, Ma J, Marty K, May AF, McGuire MA, Du MH, Singh DJ, Podlesnyak A, Ehlers G, Lumsden MD, Sales B (2011) Nat Mater 10:614
125. Shiga T, Shiomi J, Ma J, Delaire O, Radzynski T, Lusakowski A, Esfarjani K, Chen G (2012) Phys Rev B 85:155203
126. Tian Z, Garg J, Esfarjani K, Shiga T, Shiomi J, Chen G (2012) Phys Rev B 85:184303

Chapter 6
Interaction of Thermal Phonons with Interfaces

David Hurley, Subhash L. Shindé, and Edward S. Piekos

Abstract In this chapter we will first explore the connection between interface scattering and thermal transport using the Boltzmann transport equation (BTE). It will be shown that Boltzmann transport provides a convenient method for considering boundary scattering in nanochannel structures. For internal interfaces such as grain boundaries found in polycrystals, it is more natural to consider transmission and reflection across a single boundary. In this regard we will discuss theories related to interface thermal resistance. Our qualitative discussion of the theories of phonon transport will be followed by a discussion of experimental techniques for measuring thermal transport. We end this chapter by giving a detailed description of two complementary experimental techniques for measuring the influence of interfaces on thermal phonon transport.

6.1 Introduction

Manipulating phonon-mediated thermal transport using interfaces is an exciting field of research with many technological applications. For example, thermal management strategies in micro-electronic and micro-electromechanical devices are becoming increasingly important with decreasing device size. Ultra-nanocrystalline diamond films [1] provide an attractive and cost-effective solution to channel heat away from sensitive devices. However, these films have grain dimensions on the order of the phonon mean free path and, as a result, the grain boundary thermal

---

D. Hurley (✉)
Idaho National Laboratory, PO Box 1625, Idaho Falls, ID 83415-2209, USA
e-mail: david.hurley@inl.gov

S.L. Shindé (✉) • E.S. Piekos
Sandia National Laboratories, PO Box 5800, Albuquerque, NM 87185, USA
e-mail: slshind@sandia.gov

resistance plays a dominate role in limiting thermal transport [2]. Thermoelectric energy conversion for both power generation and solid state cooling provides another example. A key challenge for improving the efficiency of thermoelectric materials is balancing the competing requirements of high electrical conductivity and Seebeck coefficient with low thermal conductivity. To date, the highest thermoelectric figure of merit comes from materials that are excellent electron conductors and poor phonon conductors. Great strides towards achieving this goal have been made by tailoring phonon transport properties using specially designed nano-sized materials (e.g., quantum dots, nanorods, nanowires, superlattices) [3–7]. These examples illustrate that establishing a clear understanding of the fundamental role interfaces play in controlling phonon transport is central to the development of new phononic structures.

Phonon interaction with interfaces falls into two broad categories defined by two distinct geometries. Interfaces that form the boundary of nanometer size channels (e.g., nanowires and thin films) define the first geometry. In this case, vibrational energy can't be transmitted across the interface and, as a result, the interface serves to guide phonons. If we neglect the influence of confinement on group velocity, it can be shown that, in the limit of specular reflection, the boundary does not limit thermal transport. This behavior is in stark contrast to the influence of internal interfaces (e.g., grain boundaries, superlattice interfaces). In this case, thermal transport requires that vibrational energy be transmitted across an internal boundary. Naturally, some vibrational energy is also reflected by the interface. Consequently, an internal interface will always serve as an impediment to thermal transport regardless of the model used to describe the interaction in this chapter.

6.2 From Phonons to Thermal Conductivity

Thermal transport can be understood from both a macroscopic and an atomistic viewpoint. On the macroscopic scale one considers thermal transport within a continuum framework to obtain the temperature distribution. On an atomistic scale one considers thermal transport within a quantum framework to obtain an expression of the thermal conductivity. At the quantum level thermal energy is transported (carried) by electrons and quantized lattice vibrations called "phonons." In insulators and semiconductors, due to the scarcity of free electrons, thermal conductivity is determined almost entirely by high frequency incoherent phonons. These thermal phonons do not travel far before interacting (scattering). Phonon scattering is the physical process that ultimately limits thermal conductivity in insulators and semiconductors. At room temperature in nearly perfect single crystal materials, phonon scattering is dominated by scattering with other phonons. Interface scattering dominates at lower temperatures in single crystals or at room temperature in boundary-rich materials (e.g., nanocrystalline and nanoporous materials as well as superlattice structures).

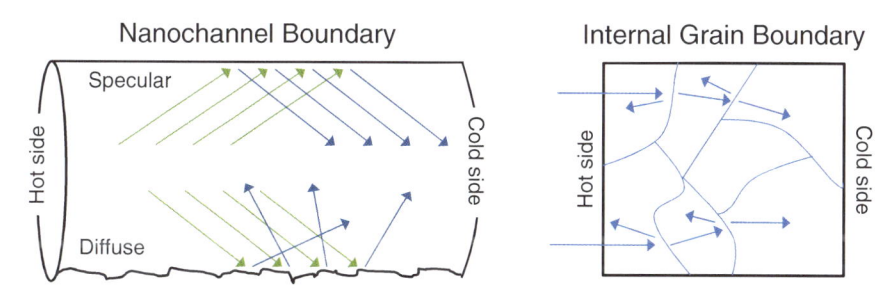

6 Interaction of Thermal Phonons with Interfaces

Fig. 6.1 Interfaces fall into two broad categories defined by two distinct geometries

Fig. 6.2 The dispersion relation for a linear chain of N identical atoms of mass M, held together by linear forces acting only between adjacent atoms

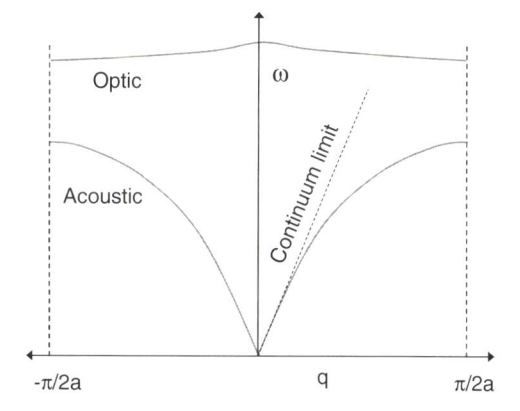

The emphasis of this section will be on the development of the Boltzmann transport equation (BTE) as applied to phonons with an emphasis on the role of interfaces. The essential ideas for understanding the properties of phonons (normal vibrational modes) can be obtained by considering a linear chain of N identical atoms of mass M, held together by linear springs acting only between adjacent atoms. For such a system the dispersion equation relating angular frequency, ω, and wave number, q, is given by

$$\omega = 2\sqrt{\frac{k}{M}} \left|\sin\left(\frac{qa}{2}\right)\right| \qquad (6.1)$$

where k is the linear spring constant and a is the equilibrium distance between adjacent atoms. This dispersion relationship is shown in Fig. 6.2. The group velocity, defined as $d\omega/dq$, at $q = \pm\pi/a$ is zero so these modes correspond to standing waves. The mode corresponding to $q = 0$ represents rigid body motion. Within the wavenumber zone shown in Fig. 6.1 there are N values of q representing different relative motions of the atoms. Any range of q which extends over $2\pi/a$ contains all possible modes. It is for this reason that we have only shown the portion of the dispersion curve between $-\pi/a$ and π/a (known as the first Brillouin zone). For small wavenumber (long wave length) phonons, the dispersion relation is almost linear. This is referred to as the continuum limit because the slope of the dispersion curve in this region is related to the speed of sound of an elastic continuum. An additional high frequency mode exists for a linear chain composed of atoms having different masses. The high frequency mode is called the "optic" branch and the low frequency mode is called the "acoustic" branch. In most semiconducting solids optical modes exhibit rather weak wave vector dispersions and thus do not play a direct role in thermal transport [8]. The picture is more complicated in three dimensions, however; some general observations can be made. There will be three acoustic modes for a primitive cell of the lattice containing P atoms. In the small wavenumber limit, two of these modes will be purely transverse and one purely longitudinal in character along high symmetry directions. There will also be $3P-3$ optical modes.

To further explore phonon-mediated thermal transport we must consider the quantization of vibrational energy. From quantum mechanics, a vibrational mode of frequency ω has energy given by $(n + 1/2)\,\hbar\omega$, where n is any positive integer and $\hbar$ is Planck's constant divided by 2π. The term $\hbar\omega/2$ is the energy of a mode at absolute zero. In real crystals, energy is exchanged between modes and the average occupation number, n, of a mode with energy $\hbar\omega$ in thermal equilibrium is given by Bose–Einstein relation:

$$n^0 = \frac{1}{\exp(\hbar\omega/k_{\rm B}T) - 1} \tag{6.2}$$

From this relation, it may be seen that the equilibrium population in a particular mode increases with increasing temperature. Another way of stating this is that phonons are created by increasing the temperature and destroyed by decreasing the temperature, i.e., the number of phonons is not conserved. This is one reason why the thermal conductivity has a strong dependence on temperature.

To consider the mechanisms that govern thermal transport, consider a system pushed out of equilibrium by, for example, a thermal gradient imposed by an external heat source. In this case, the phonon distribution, $N(q)$, is no longer described by Eq. (6.2). The thermal conductivity depends on the extent from which the phonon distribution can deviate from equilibrium for a specified thermal gradient. The process whereby the system attains a new equilibrium is governed by the time evolution of n (i.e., phonon transport). The change with time in phonon distribution within a particular region is governed by two processes: drift and scattering.

Drift can change the local phonon distribution in the presence of a temperature gradient because phonons leaving the region of interest are replaced by phonons from regions with differing distributions. For a thermal gradient along the z-axis, this term is given by:

$$\left(\frac{\partial N(\omega)}{\partial t}\right)\bigg|_{\rm drift} = -V_{\rm G}\frac{\partial N(\omega)}{\partial T}\frac{\partial T}{\partial z} \tag{6.3}$$

where $V_{\rm G}$ is the phonon group velocity.

Scattering can change the distribution via the replacement of existing phonons with new phonons having different wavenumbers and/or polarizations.

Thermal equilibrium is established when the distribution function becomes independent of time:

$$\frac{\partial N(\omega)}{\partial t} = \left(\frac{\partial N(\omega)}{\partial t}\right)\bigg|_{\text{drift}} + \left(\frac{\partial N(\omega)}{\partial t}\right)\bigg|_{\text{scatter}} \quad (6.4)$$

This equation is referred to as the BTE for phonons. This equation is in general very difficult to solve and one is faced with treating the scattering term using approximate methods. For a detailed evaluation of computational techniques for solving the BTE, please see a recent article by Chernatynskiy and Phillpot [9].

A common means of simplifying the BTE is called the relaxation time approximation. Using this approximation, the scattering term is given as:

$$\left(\frac{\partial N(\omega)}{\partial t}\right)\bigg|_{\text{scatter}} = \frac{n^0 - n}{\tau} \quad (6.5)$$

where τ is the relaxation time. With some further linearizing assumptions [10], the thermal conductivity can be expressed in terms of the relaxation time as follows:

$$k = 13 \int_0^{\omega_{\max}} \hbar \omega V_G^2 \tau f(\omega) \frac{\partial N^0}{\partial T} d\omega \quad (6.6)$$

where $f(\omega)$ density of phonon modes.

In the 1950s and 1960s, many analytical solutions to this equation were presented [11–13]. An excellent review of these efforts and many others can be found in Berman [14]. In order to obtain a closed form solution, much of the early work replaced the actual dispersion relationship with the linear dispersion relationship corresponding to that of a continuum. As shown in Fig. 6.2, linear dispersion is only appropriate for small wavenumber phonons. While these solutions offered insight into the physical mechanisms that govern heat transport, the simplifying assumptions led to discrepancies with experimental observations. For instance, neglecting dispersion resulted in significant errors in the calculation of the mean free path for phonons in silicon at room temperature [15, 16]. The reason for this is that the group velocity of large wavenumber phonons, which are fully excited at room temperature, is overestimated using the continuum limit. More recently, solution techniques have been employed that compute scattering rates directly without resorting to average or mode-dependent Grüneisen parameters and with no underlying assumptions of crystal isotropy [8, 17, 18]. This new computationally intensive approach requires explicit consideration of the interatomic force constants and can distinguish between normal and Umklapp scattering and provides information regarding the contribution to heat transfer carried by each phonon mode.

In addition to properly treating dispersion, solution methods must account for the relative importance of various phonon scattering mechanisms. In the expression above we have used the relaxation time approximation, where every phonon mode is assigned a relaxation time corresponding to different scattering mechanisms. The combined relaxation time, τ_c, is obtained by adding the individual relaxation times according to Matthiessen's rule:

$$\frac{1}{\tau_C} = \frac{1}{\tau_1} + \frac{1}{\tau_2} + \frac{1}{\tau_3} + \cdots \tag{6.7}$$

As will be noted in the next section, a predominant scattering mechanism is due to three phonon interactions. The relaxation time approach treats the magnitude and, often, the frequency dependence of the three-phonon relaxation time as adjustable parameters. While this method is not suitable for the analysis of materials whose thermal properties are not already known, it does provide valuable physical insight into the relative importance of various scattering mechanism. It is in this context that we examine phonon scattering in the next section.

6.3 Phonon Scattering Within Relaxation Time Framework

In general, phonons scatter due to distortions of the crystalline lattice. These can be momentary distortions due to the vibrational motion of phonons or can be permanent distortions due to lattice defects. The phonon–phonon scattering mechanism is better understood by considering the interatomic bonding energy curve (shown in Fig. 6.3). The force acting on an atom is obtained by taking the spatial derivative of the potential energy curve ($F = \partial E/\partial x$). The potential energy profile is determined by both long-range attractive forces and short-range repulsive forces. This potential can be conveniently represented using a Taylor series expansion (graphically shown in Fig. 6.2). The first term in the series is a quadratic term (linear force term) and it is this term that primarily defines the elastic constants and the phonon velocity. Because this term describes a linear spring, superposition applies and thermal phonons would propagate without interaction if it were the sole contributor to the

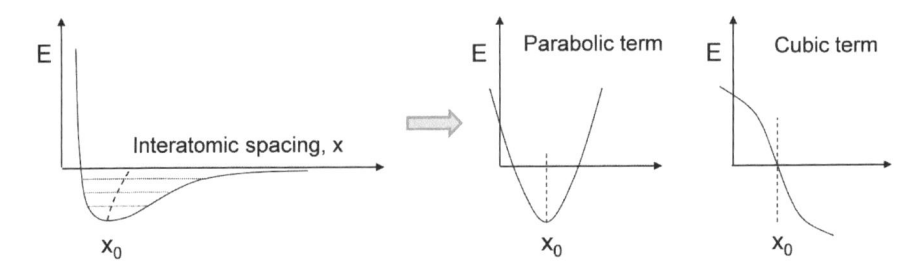

Fig. 6.3 Interatomic potential. *Left*: Thermal expansion is due to asymmetric energy well. *Right*: The parabolic and cubic term from a Taylor expansion of the potential

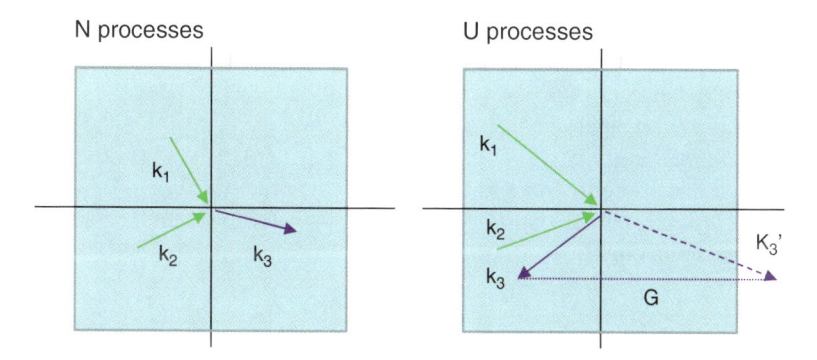

Fig. 6.4 Graphical illustration of normal (N) and Umklapp (U) scattering processes

interatomic potential energy curve. Thus, in a perfect crystal there would be nothing driving an arbitrary phonon distribution back to equilibrium. It is the second term in the expansion (the cubic term) that governs three phonon scattering processes and is primarily responsible for limiting thermal conductivity in perfect crystals. It should also be noted that the cubic term is responsible for thermal expansion. At $T = 0$ K the equilibrium interatomic distance is x_0. However, heating adds kinetic energy as shown in Fig. 6.3. Because the energy well is asymmetric, due primarily to the cubic term, the mean interatomic spacing increases.

Three phonon scattering is divided into normal (N) processes that conserve momentum and Umklapp (U) processes that do not conserve momentum. These two processes are illustrated in Fig. 6.4. The incoming phonons involved in N process have small magnitude wave vectors and, as a result, the wave vector of the outgoing phonon remains within the first Brillouin zone (shaded region in Fig. 6.4). Conversely, U processes involve incoming phonons with large magnitude wave vectors and the outgoing phonon would be scattered outside the first Brillouin zone. The outgoing wave vector is projected back into the first zone by the addition of a reciprocal lattice vector G. The addition of a backward-pointing vector destroys phonon momentum and thus plays an influential role in limiting thermal conductivity. At higher temperatures, there are more phonons with large magnitude wave vectors and the chance of Umklapp processes therefore increases. The temperature dependence for Umklapp scattering is reflected in the relaxation time expression first given by Klemens [11] as

$$\frac{1}{\tau_U} = 2\gamma^2 \frac{k_B T}{\mu V_0} \frac{\omega^2}{\omega_D} \qquad (6.8)$$

where γ is the Grüneisen anharmonicity parameter, μ is the shear modulus, V_0 is the volume per atom, and ω_D is the Debye frequency. In this expression, the scattering time increases linearly with temperature and, as a result, Umklapp scattering is the dominant process for limiting conductivity at high temperatures. When investigating other scattering mechanisms it is often advantageous to freeze out this mechanism by conducting experiments at low temperatures.

Now consider how we can include the influence of boundaries. Combining resistive scattering processes that happen uniformly throughout the crystal, such as Umklapp scattering, with processes that happen only at boundaries is greatly simplified if the boundary character is spatially homogeneous. This approach is a good approximation for phonon transport guided by nanostructures. In these cases, one can define a scattering time in a sample of characteristic dimension D that is a function of a single parameter, p, known as the "specularity parameter" and related to the boundary character:

$$\frac{1}{\tau_B} = \frac{V_G}{D}(1-p) \qquad (6.9)$$

If the influence of confinement on group velocity can be neglected, then for purely specular interfaces ($p=1$) boundary scattering does not contribute to the thermal resistivity. For purely diffuse scattering ($p=0$), Eq. (6.8) leads to Casimir's result of the mean free path in a tube being equal to its diameter in the absence of other scattering mechanisms [19].

The situation is different for a polycrystal where it is unrealistic to consider that each boundary is identical in character. In this case grain boundary scattering depends on the crystal misorientation across the boundary, as well as the boundary orientation relative to the heat current. These defining characteristics change from boundary to boundary. While the BTE has been applied to polycrystals by adding a boundary scattering relaxation time [20] into the total relaxation time, the thermal conductivity can be calculated as a function of average grain size only after including a number of adjustable parameters. It is therefore more natural to consider an effective medium approach that addresses more directly the fundamental role of the boundary. This approach was first developed by Nan and Birringer [21]. For polycrystals with isotropic spherical crystallites, the thermal conductivity can be written in terms of the bulk conductivity, k_{bulk}, as follows:

$$k = \frac{k_{bulk}}{1 + \frac{k_{bulk} R_k}{d}} \qquad (6.10)$$

where d is the average grain diameter and R_k is the Kapitza resistance. The Kapitza resistance defines the impediment to thermal transport created by a single interface and is the subject of the next section.

6.4 Theories of Interface Resistance of Isolated Boundaries

A discontinuity in the temperature field exists at an interface between two different materials through which heat flows. This phenomenon, first observed by Kapitza [22] between liquid helium and a solid, has been the subject of many modeling efforts. At low temperatures, where the Debye models serve as a good approximation for the density of states, the acoustic mismatch (AM) model and the diffusive mismatch (DM) have been developed to explain the observed behavior. The AM model [23] relates the acoustic impedance mismatch across an interface to phonon transmission. The AM model provides a reference for phonon transmission across perfect interfaces. To account for phonon scattering due to interface imperfections DM model was developed [24]. Unfortunately neither model provides good predictive capability across a wide range of materials. The problem is that interface scattering processes depend sensitively on atomic structure and are difficult to treat in analytic models. Recently, molecular dynamic (MD) models [25, 26] have been developed that investigate the influence of atomic structure on phonon transport across interfaces. In the following we will present a qualitative discussion of analytical and numerical models of phonon transport across interfaces.

6.4.1 Analytical Models

The AM and DM models do not explicitly consider the atomic structure of an interface; they both compute a transmission probability from the bulk properties of the materials on either side of it. To qualitatively understand these models, first consider heat flux across a boundary between material 1 and material 2:

$$Q = \frac{1}{4}\sum_j \int_0^{\omega_{max}} N_{1,j}(\omega, T_2)\hbar\omega V_{1,j} w_{1\to 2,j}(\omega) d\omega$$
$$- \frac{1}{4}\sum_j \int_0^{\omega_{max}} N_{1,j}(\omega, T_1)\hbar\omega V_{1,j} w_{1\to 2,j}(\omega) d\omega \qquad (6.11)$$

where N corresponds to the phonon density of states times the Bose Einstein occupation factor (Eq. 6.2) and $w_{1\to 2,j}$ is the fraction of energy transmitted to material 2 due to acoustic phonons of mode j and velocity $V_{1,j}$ in material 1. For simplicity we have neglected the dependence on the angle of incidence. The Kapitza resistance is then defined as:

$$R_{th} = \frac{T_2 - T_1}{Q} \qquad (6.12)$$

The difference between the AM and DM models lies in the determination of the transmission coefficient $w_{1\to 2,j}$. The AM model treats the interface as an elastic discontinuity. By assuming that no scattering takes place at the interface (specular interface) and by imposing boundary conditions based on continuum elasticity theory, the AM model gives the transmission coefficient for normal incidence phonons as:

$$w_{12} = \frac{4Z_1 Z_2}{(Z_1 + Z_2)^2} \qquad (6.13)$$

$$Z_i = \rho_i V_i$$

In the DM model the specular interface is replaced with the opposite extreme. In this case all phonons are diffusively scattered by the interface. In the DM framework, the transmission coefficient w is typically referred to as a transmission probability and is proportional to the density of phonon states on either side of the interface.

To better understand the difference between these two models, consider two limiting cases. The first case involves a Kapitza boundary between metal sample and liquid helium. In the DM model, the probability that a phonon will forward scatter is very high when the phonon is going from the metal to liquid helium because of the vastly higher density of states in the helium. Conversely in the AM model, because there is a large acoustic mismatch across the interface, the phonon will likely be reflected. The second limiting case involves an interface between two acoustically identical materials. The AM model will give transmission coefficient of unity while the DM model gives a transmission probability of 50 %.

6.4.2 Atomic-Scale Simulation

In contrast to analytical methods discussed above, molecular dynamics (MD) simulation can explicitly consider the atomic structure of the interface. This simulation approach involves application of Newton's equations of motion to atoms in a solid that interact with each other through a known interatomic potential. It should be noted that MD simulation uses a classical distribution function and thus is not applicable for studying thermal transport at temperatures well below the Debye temperature. Accurate prediction of phonon transport using MD simulation requires realistic interatomic potentials. Typically potentials are obtained empirically. Molecular dynamics simulations can be used to estimate the thermal conductivity of bulk crystals and nanoscale structures (like thin films or nanowires) via the direct heat flux and the Green–Kubo methods. Calculation of the Kapitza resistance of a grain boundary or bicrystal interface is often done using the direct heat flux and/or the lattice dynamics methods.

The direct heat flux approach, which mimics experiment, involves the application of a heat current followed by computation of the temperature gradient. For a given system length, the thermal conductivity can be deduced from the temperature gradient using Fourier's law of heat conduction. The thermal conductivity of a bulk crystal in a given crystallographic direction at a given temperature is determined by extrapolation over an entire sample length. The Kapitza conductance is computed by determining the temperature jump at the interface. This approach does have a potential issue in that it is necessary to apply a large temperature gradient in order to make it discernible from the statistical noise, which can lead to a nonlinear response.

In the lattice dynamics technique [27] localized wave packets are created through a superposition of normal modes of a bulk perfect crystal. The wave packets are propagated using MD simulation. After interacting with an interface, the energy transmission coefficient is determined for each polarization and wave vector as the fraction of incident energy. The Kapitza conductance can be determined from the energy transmission coefficient, the phonon group velocity and the dispersion relation. The lattice dynamics method is often used to obtain a better description of interfacial phonon scattering for a range of temperatures while the direct heat flux method is used to obtain a value for the thermal conductivity or Kapitza conductance at a given temperature.

Using a molecular dynamic approach (MD), Schelling et al. [26] recently determined the Kapitza resistance for both high and low energy twist grain boundaries in Si. This work, however, gave values of boundary conductance that were high in comparison with experimental studies of Cahill [28] and Lee [29]. One reason for the discrepancy is that the simulated grain boundaries were defect free and atomically flat, thus presumably minimizing the thermal resistance.

For multiple interfaces such as superlattices, in addition to the nature of the interface, it is essential to consider interference of phonons reflected from multiple interfaces. Furthermore, the phonon dispersion relation is modified and zone folding occurs, resulting in multiple phonon band gaps [30]. Several effects can result from modification of phonon dispersion. The phonon group velocities are reduced significantly, especially for higher energy acoustic phonons. Additionally, because ω–k relations are modified, there are more possibilities for conservation of crystal momentum and energy involved in normal and Umklapp scattering. Hence the scattering rate increases. Comparison with experimental data, however, reveals that modification of phonon dispersion alone cannot explain the magnitude of the thermal conductivity reduction [31, 32].

6.5 Experimental Measurement of Thermal Transport

There are a vast number of experimental methods for measuring thermal transport properties. The intent of this section is to give a brief overview of commonly used methods. Emphasis will be placed on the application of these methods to measure thermal transport across and along interfaces. Measurement of phonon transport can be divided into spectroscopic approaches and thermal gradient approaches. Spectroscopic approaches typically measure phonon lifetime and phonon dispersion and can be characterized as providing input to the BTE. Conversely thermal gradient approaches, which measure either thermal conductivity directly or a combination of conductivity and specific heat, can be characterized as providing validation of the output of the BTE.

6.5.1 Spectroscopic Techniques

Techniques that employ neutron scattering [33, 34] provide phonon dispersion and lifetime over the entire Brillouin zone with unparalleled accuracy. These studies require the use of either a research reactor or spallation neutron source and experiments must be performed on large single crystal samples. Similar information, albeit with reduced phonon energy resolution, is provided using inelastic X-ray scattering (IXS) [35]. A related approach, termed X-ray thermal diffuse scattering (TDS) [36, 37] is used to measure phonon dispersion with submicron spatial resolution. Both TDS and IXS require synchrotron X-ray sources. Optical spectroscopic techniques can provide complimentary phonon information. Raman scattering has long been applied to measuring the lifetime of optical phonons [38]. However, in most semiconducting solids optical phonons play a limited role in determining thermal transport properties. Brillouin scattering provides information about acoustic phonons with frequency below 100 GHz. At room temperature, however, the majority of thermal energy is carried by phonons at frequencies in excess of 1 THz.

6.5.2 Thermal Gradient

This approach involves localized heating and spatially or temporally resolved temperature measurement. The measurement approach is based on Fourier's law of heat conduction, and is typically the method of choice for measuring the influence of isolated interfaces. These methods can be divided into steady state and transient heating regimes. Steady state heating directly measures the thermal conductivity while transient heating experiments measure either the thermal diffusivity or thermal effusivity. Both approaches take a different tack for investigating the influence of isolated boundaries.

The steady state approach relies on micro- and nano-fabrication techniques to either produce nanoscale thermal transport channels or make nanoscale probes for temperature sensing. In the case of nanochannels, the large surface to volume ratio associated with these nanoscale structures serves to emphasize the role of phonon scattering at the sample surface. This measurement approach has become the accepted technique for the measurement of the thermal properties of nanowires and nanotubes [39, 40]. At the opposite extreme, an approach termed scanning thermal microscopy (SThM) [41] involves bringing a sharp temperature-sensing tip in close proximity to the sample surface. The temperature sensor can consist of either a thermocouple junction or a thin film resistor placed at the probe tip. Most platforms for SThM utilize cantilever-based probes such as atomic force microscopes [42]. Recent advances have made possible the thermal imaging of nanostructures with ~50 nm spatial resolution [43]. A primary issue associated with SThM involves development of clear understanding of the tip-sample heat transfer [44].

Transient methods require either a pulsed heating source or a sinusoidally modulated heating source and the respective spatial resolution is determined by either the diffusion length or the thermal wavelength. A large number of methods are used for transient heating and temperature sensing; in this section, we limit our discussion to methods that use electrical resistance or lasers.

The 3-omega method, widely used for characterizing thin films, is based on using a heating wire for heat pulse excitation and temperature sensing. An alternating current at frequency ω is passed through a long heating wire deposited onto the sample surface. Joule heating causes the electrical resistance to oscillate at 2ω. The resulting voltage drop across the heating wire at 3ω is detected by a four-terminal measurement using a lock-in amplifier. The 3-omega method has been applied to measurement of Kapitza resistance between a thin film and substrate [29]. However, this approach introduces a second interface between the heater wire and film making it difficult to isolate the influence of the Kapitza resistance between the film and substrate [45, 46].

The physics of phonon transport has been greatly advanced as a result of the "heat pulse" technique [47–49]. This approach records the phonon flux reaching a detector some distance away from a thin metallic film heater source. A short pulse of current heats the metallic film and launches phonons into the sample material substrate. The detector is typically composed of a thin film superconducting bolometer deposited on the opposite face of the sample crystal. Elegant experiments by Wyatt and Page [50] showed that in addition to the specular refraction of phonons into the helium, there also existed a large fraction of diffuse scattering occurring at all angles. More recently, a generation technique using the optical absorption of a pulsed laser beam has been exploited to control the generation process and image phonon transport in both the ballistic and diffusive regimes [51]. Related techniques that use superconducting tunnel junctions do allow some degree of tunability; however, full spectrum selective detection like that provided by neutron scattering is still lacking [52].

Purely laser-based methods have emerged as a leading candidate for making precise thermal transport measurements due to their non-contact nature and well-defined optical coupling conditions. Laser excitation can be implemented in the frequency domain using amplitude modulated continuous wave (CW) laser heating [53–55] or in the time domain using pulsed laser heating [56, 57]. The probe beam either reflects off the material surface, sampling the heated region (through thermoreflectance), or propagates along the surface, sampling the heated gas just above the surface (through the mirage effect) [58]. Time domain measurements offer the highest spatial resolution but typically consider only 1D heat flow in the depth direction of the sample and thus are limited to investigating buried interfaces parallel to the sample surface [59–61]. Frequency domain measurements are often used to look at thermal transport in the lateral direction. The ability to look in the lateral direction greatly facilitates investigation of individual interfaces found in bulk structures that intersect the sample surface [62, 63].

6.6 Measurement of the Kapitza Resistance Across a Bicrystal Interface

In this section time resolved thermal wave microscopy (TRTWM) [64, 65], an approach that incorporates elements of both time domain and frequency domain methods, is used to image lateral thermal transport across a bicrystal interface [66].

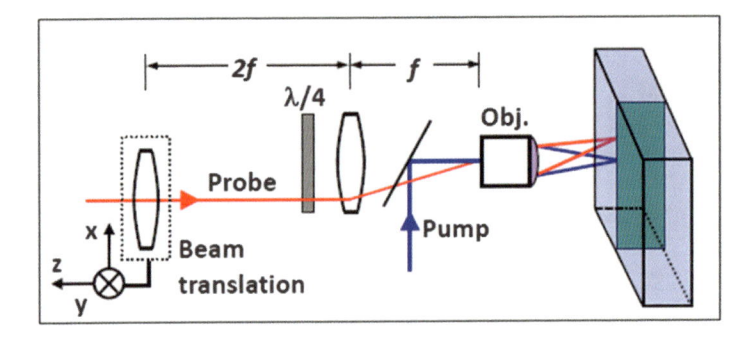

Fig. 6.5 Experimental setup. The pump and probe are focused onto the same side of the sample using a single objective

The structure of the interface is characterized using high resolution transmission electron microscopy and electron energy loss spectroscopy. The Kapitza resistance of the interface is estimated by comparing experimental results with a continuum thermal transport model that included the influence of the oxide layer.

6.6.1 Time Resolved Thermal Wave Microscopy

The basis for TRTWM as applied to semiconductors is to eliminate the obscuring influence of the diffusing electron-hole plasma by promoting fast carrier recombination through mechanical polishing of the surface. For delay times that are large compared to the carrier recombination time, the probe only senses the thermal wave component of the signal. Thus a primary advantage of this approach is that it can be applied to uncoated semiconductor surfaces [67].

The thermal wave imaging system involves translation of a microscopic probe beam relative to a fixed pump beam (Fig. 6.5). The pump and probe beams, with wavelengths of 400 and 800 nm and incident pulse energies of 0.26 and 0.026 nJ, respectively, are derived from a Ti:sapphire laser with a pulse duration of ∼200 fs. The pump and probe beam have approximately the same spot diameters at the sample surface of 1.9 μm at full width at half maximum intensity[1]. The pump beam is chopped at a frequency in the range $f_0 = 10$–50 kHz to enable measurements of thermal waves at these frequencies. A high carrier injection density (∼10^{20} cm^{-3}) is used to obtain a large signal-to-noise ratio for accurate signal analysis[2].

---

[1]The cross-correlation of the pump and probe spot sizes was found by fitting the experimentally measured thermal wave phase profile to theory.

[2]The maximum signal-to-noise ratio for the amplitude signal is ∼3,000. For a typical scan of 20 μm, the phase signal varies by ∼50° while the maximum deviation between experiment and theory is ∼0.5°.

The maximum transient temperature rise caused by a single pump pulse is ~90 K [68], and the steady state temperature rise is 5 K [54]. Further details on the experimental approach as applied to bulk Si can be found in a previous paper [67].

6.6.2 Sample Geometry, Preparation and Characterization

An artificial interface made by fabricating a thin film on a substrate is straightforward to examine using transient heating methods. Lateral positioning of pump and probe beams is not an issue because the interface lies in the plane of the sample surface. Thus to effectively probe the interface all that is required is to tune the thermal diffusion length to be of the same order as the film thickness. On the other hand, interfaces in bulk samples that intersect the free surface (top left pane of Fig. 6.6) offer additional challenges. Namely, specialized samples must be fabricated that can be characterized using high precision laterally resolved experimental techniques. One approach that offers the potential to systematically study interfaces is to look across a single bicrystal interface. In the following we describe the fabrication and characterization of a Si bicrystal that is used to examine the influence of interface scattering on thermal transport.

There are a host of special boundaries that have been well characterized, and a very detailed understanding of the atomic arrangements at the boundaries has been developed based on coincident site lattice (CSL) models [69]. For this study, we selected a highly disordered (Σ29 boundary) structure. As illustrated in the top right pane of Fig. 6.6 the Σ29 structure is a disordered boundary created by twisting one (1 0 0) surface with respect to the other by 43.6°. This is very convenient since silicon single-crystal wafers can be purchased with (1 0 0), (1 1 0), or (1 1 1) crystallographic surface orientations. The wafers are mirror polished and generally have a very thin oxide (SiO_2) layer. The wafer-to-wafer bonding therefore requires techniques for (1) precise rotation of one wafer with respect to the other, (2) surface cleaning and activation procedures to remove all the surface oxide, followed by hydrogen (H) termination of the surfaces to be bonded, and (3) bonding at low temperatures (<200°C) followed by high-temperature annealing in a reducing ambient to drive off any remaining surface oxide layer at the bonded interface. This work therefore involves careful fixturing, developing various surface-activation techniques and special processing, and optimization of the annealing. Acoustic imaging of the interface is used to locate the voids. Specimens are diced from void-free regions and mirror polished edge-on to prepare the surface for measurements. After fabrication electron backscatter diffraction measurements are used to verify the misorientation of the bicrystal interface. An inverse pole diagram showing the crystal orientation in false color for the Σ29 boundary is shown in the middle left pane of Fig. 6.6. It should be noted that the bicrystal boundary is not visible in the optical image used for alignment of the pump and probe beams. However the defects denoted by the dashed circles in the SEM image and the EBSD image are clearly visible in the optical image and are used as fiducial marks for locating the interface.

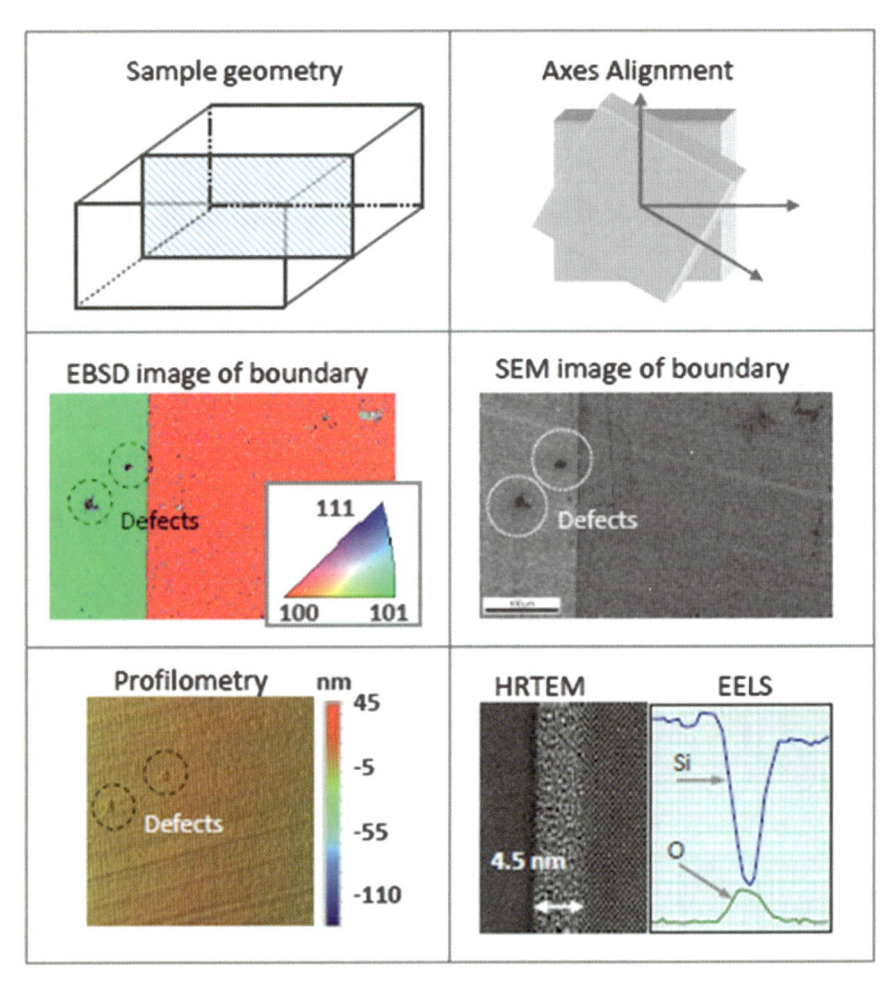

Fig. 6.6 Fabrication requires precise crystal alignment. An electron backscatter diffraction (EBSD) micrograph clearly shows the bicrystal interface. The defects denoted by the *dashed circles* are used as fiducial marks for locating the interface. Optical profilometry reveals minimal topography variations near the boundary region. The HRTEM and EELS images show a thin native oxide layer at the interface

In addition to characterizing the crystal misorientation across the bicrystal boundary it is important to visualize changes in topography of the sample surface near the boundary region. In a previous study it was shown that these changes in topography associated with polishing can greatly influence laser-based thermal diffusivity measurements. The surface is polished using diamond paste to avoid preferential etching near the boundary region. Scanning tunneling microscopy and optical profilometry are used to determine if there was a step created at the Σ29 boundary interface. The bottom left pane of Fig. 6.6 shows an optical profilometry

Fig. 6.7 Experimental thermal wave phase profile. The bicrystal interface is located at the origin. The right-hand side is fitted using the background solution to obtain the bulk diffusivity. Interface location represented by *dashed line*

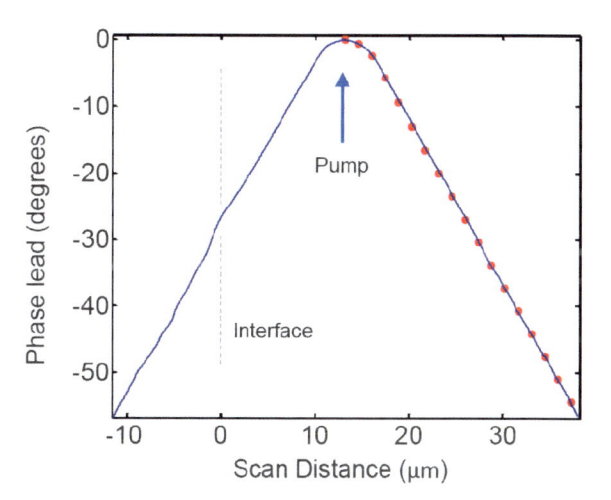

micrograph of the sample surface across the boundary region. Using electron backscatter data and prominent surface features such as scratches as fiducial marks, the faint line across the center of this image is identified as the bicrystal interface. This image confirms that the polishing method does not preferentially etch the boundary region, eliminating any significant changes in topography.

Additionally, the interface is characterized using cross-sectional HRTEM and EELS. The HRTEM image in the bottom right pane of Fig. 6.6 reveals a second phase at this interface. The extent of this region is determined to be about 4.5 nm. The EELS peak positively identifies this phase as SiOx (silicon oxide). Its stoichiometry cannot be fully determined using EELS; however, because the specimens were annealed at 1,000°C, we believe that it is SiO_2.

6.6.3 Results and Discussion

Typically the phase of the thermal wave temperature field is used for analysis because the phase is less susceptible to spatial variations in the sample's unperturbed optical reflectivity. A typical experimental thermal wave phase profile versus scan distance, y, is shown in Fig. 6.7. The small perturbation due to the interface located on the left-hand side of the pump is clearly visible near $y = 0$. A best fit model of the thermal wave background [62, 66] that does not include the influence of the interface is plotted for the data on the right-hand side of the pump (dotted). The thermal wave source used for the model is represented by a Gaussian profile in the lateral direction having a spot size, $a_c = (a_1^2 + a_2^2)^{1/2}$, determined by the cross-correlation of the pump, a_1, and probe, a_2, spot sizes. The close comparison between model and experiment confirms that the data far removed from the interface is negligibly influenced by the interface.

Fig. 6.8 Comparison between experiment (*solid*) and theory (*dotted*). The perturbation in the experimental data tracks with the location of the bicrystal boundary. The boundary in the top graph is located at the center of the pump (denoted by the *blue arrow*)

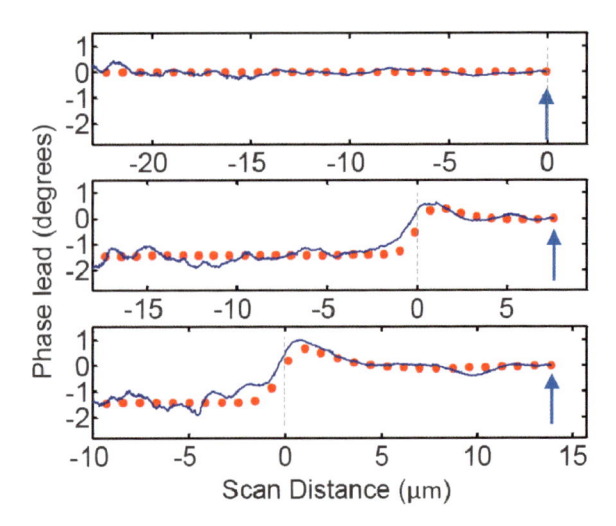

Experimental, background free phase profiles for three different pump/boundary separation distances are shown in Fig. 6.8 (solid line). In the bottom pane, the bump in the phase profile at a distance of approximately 13 μm from the pump pulse (indicated by dashed line) corresponds to the perturbation caused by the interface. In the middle pane the Kapitza boundary has been moved closer to the pump illustrating that the perturbation tracks with boundary location. In the top pane the boundary is placed in the center of the pump. In this case there is no thermal gradient and, as a consequence, the perturbation vanishes. It should be noted that if the perturbation were due to beam deflection caused by slight variations in topography near the boundary, the perturbation would not vanish.

In order to extract the Kapitza resistance we adapted a continuum model first developed by McDonald et al. [62] to include the influence of the thin SiO_2 layer. Because the SiO_2 layer thickness, 4.5 nm, is much larger than the mean free path of the phonons responsible for thermal transport [70], our model contained three distinct regions and two SiO_2/Si interfaces. The best fit model results are also present in Fig. 6.8 (dotted). For this analysis the SiO_2 conductivity was fixed at 1 W/m²K. The best fit value of the Kapita resistance was found by minimizing the sum of the square of the residuals between experiment and theory. The weighted average value for the Kapitza resistance for the three different measurement locations considered is 2.3×10^{-9} m² K/W.

There are few studies in the literature involving measurement or modeling of the Kapitza resistance of a Si/SiO_2 interface. Lee and Cahill have investigated the Kapitza resistance between Si substrates and SiO_2 layers of varying thickness using the 3-omega method. By attributing the apparent decrease in SiO_2 layer conductivity with decreasing layer thickness entirely to a composite interface resistance, they obtain an upper limit on the interface resistance at room temperature equal to

2×10^{-8} m² K/W. However, their approach introduced a second interface between the heater film and SiO$_2$ layer making it difficult to isolate the influence of the Kapitza resistance between the SiO$_2$ layer and Si Substrate. Using a heater film and a probe laser for temperature sensing, Kato and Hatta [45] considered the Kapitza resistance between a Si substrate and a thermally oxidized SiO$_2$ thin film. Their approach also involved a composite measurement of the interface resistance between the heater film and the SiO$_2$ layer as well as the interface resistance between the SiO$_2$ layer and the Si substrate. Due to large scatter in their measurements they were not able to make firm conclusions regarding the magnitude of the Kapitza resistance. Hu et al. [71] have calculated the Kaptiza resistance between Si and SiO$_2$ utilizing two analytical models. They predicated a value of 2.4×10^{-9} m²K/W using the AM model and a value of 3.5×10^{-9} m²K/W using the DM model. While our measurement does appear to more closely agree with the AM model, it is noted that these models make many simplifying assumptions and a quantitative comparison with experiment is only appropriate at low temperatures.

6.7 Measurement of the Thermal Resistance Along a Boundary

Thermal resistance in the in-plane direction is controlled by the specularity parameter, p. As discussed in a previous section, specularly reflecting surfaces ($p = 1$) will provide no resistance to thermal transport along their length because tangential "momentum" is conserved in phonon interactions with the surface. In contrast, diffusely reflecting surfaces ($p = 0$) destroy phonon momentum and therefore have a resistive effect on thermal transport. Determining the specularity parameter of bounding surfaces is therefore of utmost importance to predicting the thermal transport along thin films and within nanowires.

Despite its importance in determining thermal transport, attempts to quantify the specularity parameter are largely absent in the literature. The attempts that have been made are typically based on matching an analytical BTE approximation that includes a boundary relaxation time (Eq. 6.9) to thin film or nanowire measurement results [72–74]. While this is a seemingly reasonable approach, the results have ranged from nearly specular to diffuse in a manner that generally defies physical explanation. This situation may be traceable to shortcomings in the models, run-to-run variation in the preparation of the samples of varying thickness, as well as difficulties in performing absolute measurements of thermal conductivity in thin films. In this section, a test structure is described that employs a differential approach to infer the specularity parameter. The differential approach can greatly reduce the effect of sample variation, avoid absolute measurement of thermal conductivity, and provide a powerful means for assessing the effects of surface treatments that can be applied after the device is fabricated.

Fig. 6.9 Schematic of straight and bent structures. Internal divisions are present only to illustrate construction from identical elements

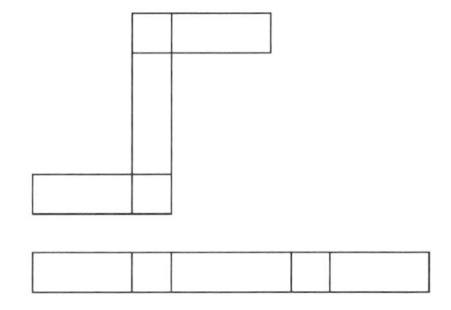

6.7.1 Theory of Operation

To create a differential measurement for the specularity parameter, a geometry that tends to induce a large number of phonon–surface interactions is coupled to one that tends to induce a lesser number of interactions. A simple means of effecting this difference is by creating geometries, one straight and one not, such that the probability of suffering a surface collision traversing the straight structure is lower than that traversing the nonstraight structure. To eliminate differences due to interphonon and impurity scattering, the two structures should have identical lengths. To amplify the effect of boundary scattering compared to these competing effects, the structures should be small compared to the phonon mean free path.

A geometry that satisfies these conditions is shown in Fig. 6.9. Here, the straight and stair-step structures have identical path lengths along the centerline (and either edge). In other words, the two geometries are composed of identical amounts of material. This can be illustrated by breaking the bent structure into squares and rectangles and placing them in the straight structure, as illustrated by the internal divisions in the figure. To minimize the equalizing effect of the walls in the plane of the page, which are parallel in both the bent and straight geometries, the structures should be very tall in the out-of-page direction compared to their cross-channel width.

6.7.2 Noncontinuum Simulation

Because the differential structures are designed to be small compared to the phonon mean free path, continuum (Fourier's law-based) tools cannot be used to predict their behavior. At the same time, in order to ease manufacturability and to avoid increased interphonon scattering due to confinement effects, they are likely to be too large to be amenable to molecular dynamics simulation. Occupying a spatial scale gap between these techniques, Monte Carlo simulation of the Boltzmann equation provides a means for analyzing these structures. By applying the Boltzmann equation, we are operating under the so-called phonon gas assumption, which presumes that the length scales are sufficiently large compared to the phonon coherence length such that wave effects, such as interference, can be neglected [75].

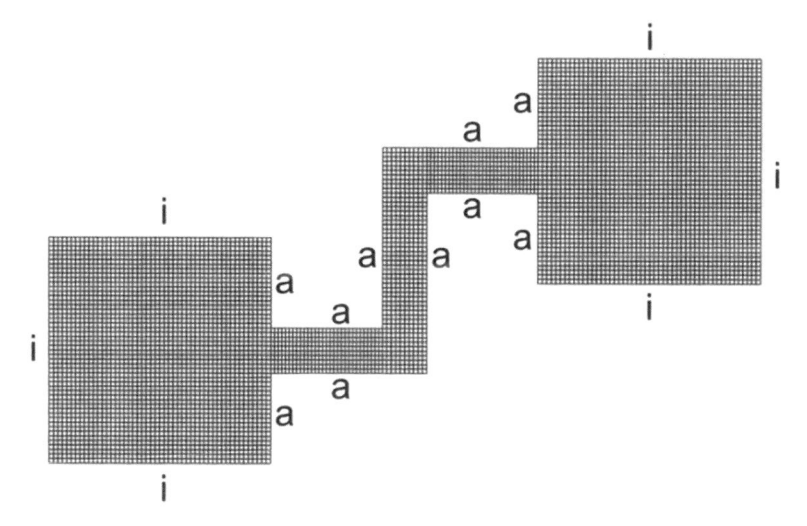

Fig. 6.10 Computational domain showing a bent ligament and portions of the platforms on either side. The grid used for all simulations is also shown. Boundary conditions on each face are marked with an "i" for isothermal and an "a" for adiabatic

The solution scheme employed in this section most closely resembles the method known as direct simulation Monte Carlo (DSMC). This technique was developed by Bird [76] for simulating gas behavior surrounding reentry vehicles at high altitude, where the continuum assumption breaks down. Later analysis showed rigorously the relationship between DSMC and the Boltzmann equation [77], and it has been employed in flows ranging from free molecular to fully continuum [78, 79].

Mazumder and Majumdar [80] published the first application of a similar technique to phonons in 2001. The simulation tool employed in the current work, described in detail previously [81], draws from Mazumder and Majumdar as well as Bird. The relaxation time-based interphonon and phonon-impurity scattering model is a modified version of Holland's model [13] with new coefficients fit to the thermal conductivity of both natural and isotopically enriched silicon. Boundary scattering is handled separately when computational particles reach walls. Only the direction is changed during these interactions, performing either a reversal of the normal velocity (specular reflection) or a randomization of the outgoing direction as if the particle were emitted from the wall (diffuse reflection). Intermediate specularity parameters are achieved by choosing diffuse reflection over specular reflection for a particular interaction even with probability equal to the specularity parameter.

6.7.2.1 Simulation Parameters

The simulation domain consists of either a bent or a straight ligament and a thermal reservoir on either side of it, as shown in Fig. 6.10. The ligaments are 100 nm in width and 1 µm in total length along their centerline. All simulations are performed in two dimensions. An isothermal boundary condition is enforced on the three reservoir edges furthest from the ligament. An adiabatic boundary condition, with a varying specularity parameter, is enforced on reservoir edges adjacent to the ligament, as well as along the edges of the ligament.

The simulation grid of uniform, 100 nm square cells is also shown in Fig. 6.10. While the numerical technique allows the cell size to be enlarged as the temperature decreases due to the increasing phonon mean free path, the grid shown is employed for all simulations presented herein. This grid size was set conservatively at the highest temperature, so it is very conservative at the lowest temperature. The target particle count is set at two million for all simulations. This target provides approximately 300 particles per cell, which is also conservative.

6.7.2.2 Results

A temperature distribution for the complete domain of a bent ligament at 200 K with specular reflection is shown in Fig. 6.11. A key feature of note in these results is that the temperature varies very little across the narrow dimension of the ligament. This fact is exploited herein by averaging the observed temperature in cells across the ligament when computing a temperature distribution. This increases the effective number of samples embodied in the reported temperature by a factor of 10.

An in-plane flux distribution for the complete domain of a bent ligament with diffuse reflection at 200 K is also shown in Fig. 6.11. The in-plane flux is determined by computing a vector magnitude from the average energy times speed reported by the simulation in the two spatial directions shown in Fig. 6.10. In this figure, the parabolic profile expected in pressure-driven fluid flows in channels (Poiseuille flow) is visible. Also in parallel with viscous fluids, an "entrance region" is visible at the high-temperature end of the ligament, where the velocity distribution transitions from uniform to parabolic across the channel as the boundary layers from each wall grow and coalesce.

A comparison of flux distributions for the straight and bent ligaments at 200 K with either fully specular or fully diffuse surface interaction is shown in Fig. 6.12. In both cases, the straight ligament is shown to have a greater net flux. Confirming the operating principle of the measurement device, the difference in net flux between the ligaments is greater for the specular case, where the straight channel walls have no effect on the tangential momentum of traversing phonons.

In order to explore the operation of this device over a range of specularity parameters and temperatures, a more convenient means of comparison is desirable. Toward this end, it is asserted that an effective thermal conductivity can be defined for the ligaments, despite the fact that they operate in a noncontinuum regime and thermal conductivity is an inherently continuum parameter.

The effective thermal conductivity for each case is computed by dividing the net flux through the ligament, measured at its isothermal surfaces, by a temperature gradient measured in its interior. This temperature gradient is found by fitting a straight line to the cross-ligament averaged temperature over a region centered on

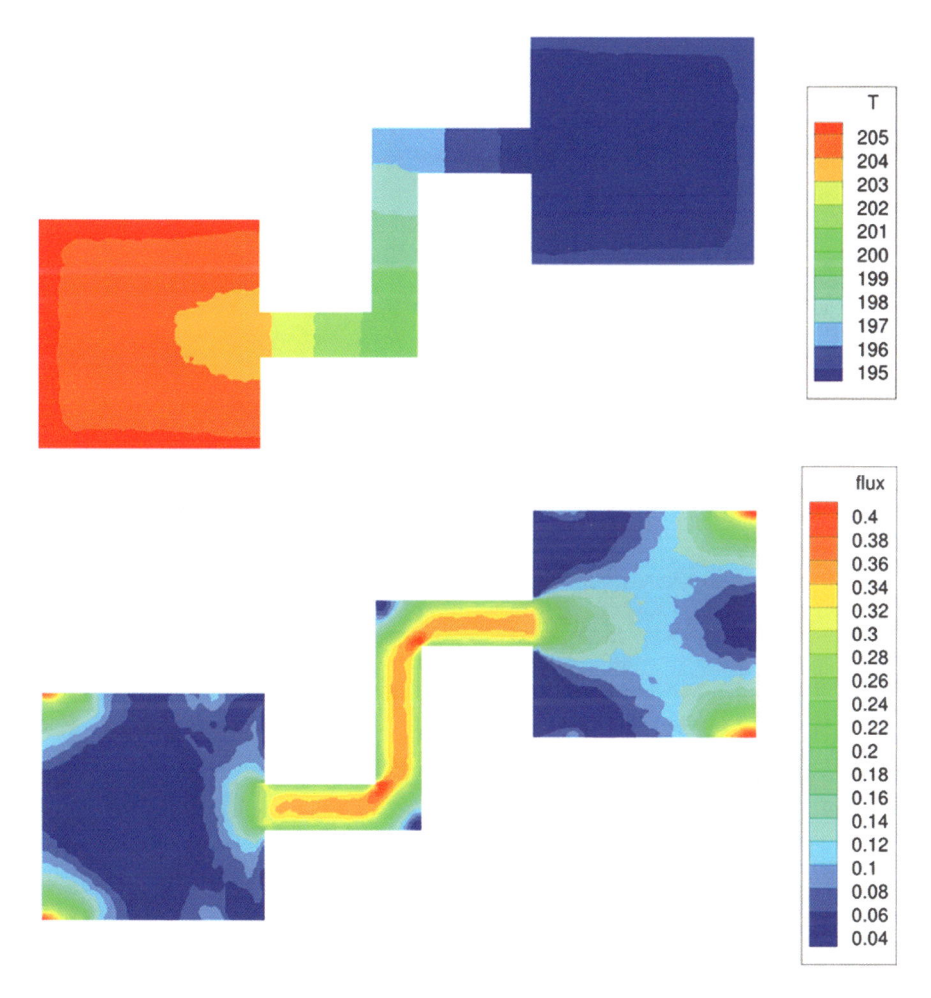

Fig. 6.11 Temperature and flux distributions for the bent ligament at 200 K with fully diffuse reflection on adiabatic walls. Temperature units are K and flux units are GW/m^2

the ligament midpoint. To maintain parity between the straight and bent ligaments, the size of this region is set at 300 nm. This choice corresponds to the portion of the vertical segment of the bent ligament with adiabatic boundaries on both sides.

It is expected that the most challenging case for determining an effective thermal conductivity will be the straight ligament with fully specular reflections at the lowest temperature. In this case, phonon scattering by surfaces and other phonons is minimized, making it difficult to reach a local equilibrium. The temperature distribution for the complete ligament, along with the fit computed from data in the 300 nm center region is presented in Fig. 6.13. While significant nonlinearity is indeed visible at the ends, the center third is well approximated by a linear fit.

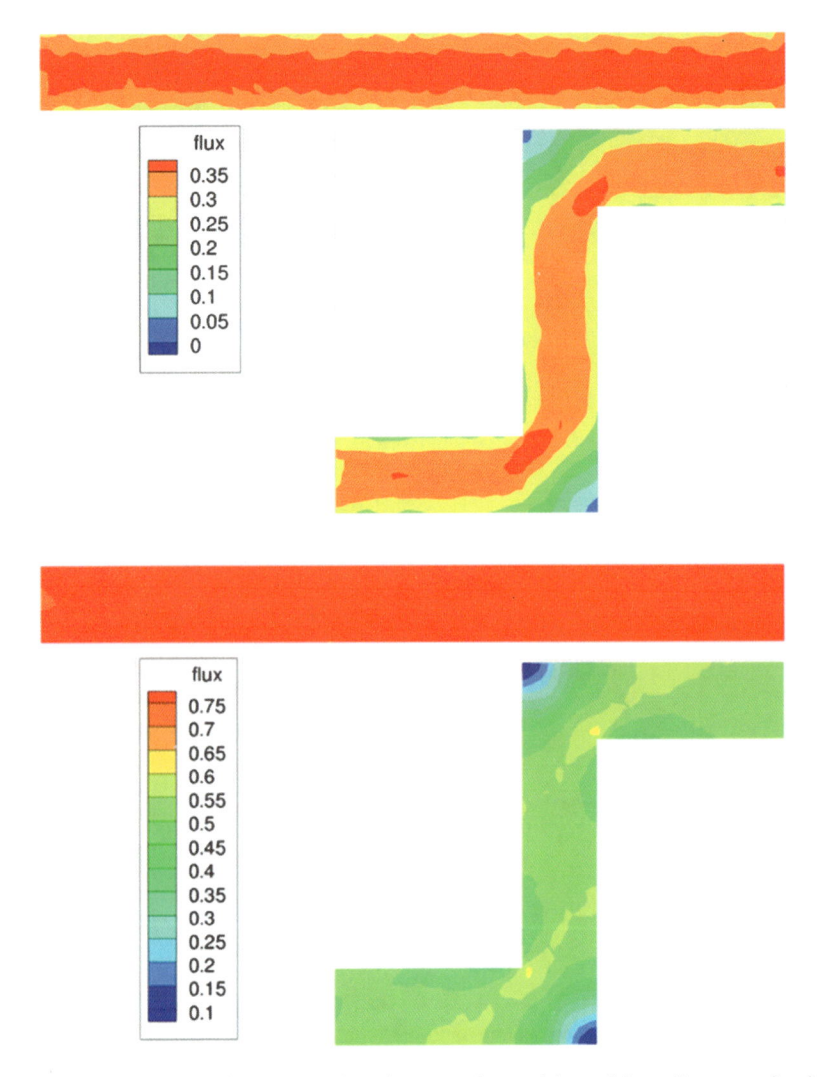

Fig. 6.12 Comparison of flux distributions between the straight and bent ligaments for fully diffuse and fully specular reflections at 200 K

A check on the validity of the "effective thermal conductivity" formulation may be made by repeating the process with a different temperature gradient to ensure that the result is independent of this parameter, as required by Fourier's law. For a 10 and 5 K temperature difference across the ligament, the computed thermal conductivities were found to differ by only 2 %, so this requirement appears to be satisfied.

The difference in effective thermal conductivity between the straight and bent ligaments is shown in Fig. 6.14. Several key features of the differential arrangement are visible in this figure. First, the results are presented as the effective thermal conductivity of the bent ligament subtracted from that of the straight ligament.

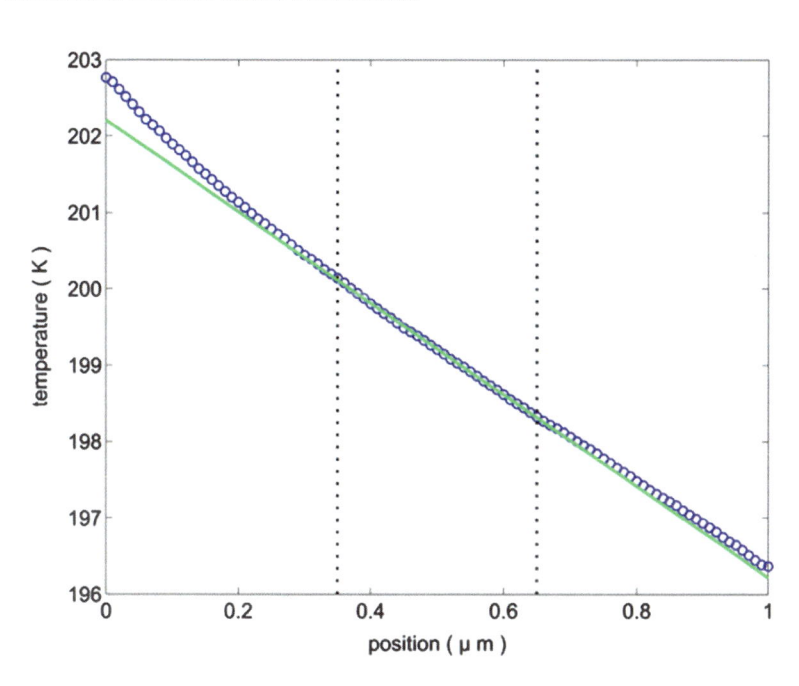

6 Interaction of Thermal Phonons with Interfaces

Fig. 6.13 Temperature distribution and linear fit for the straight ligament at 200 K. The *dotted lines* bound the region included in the fit

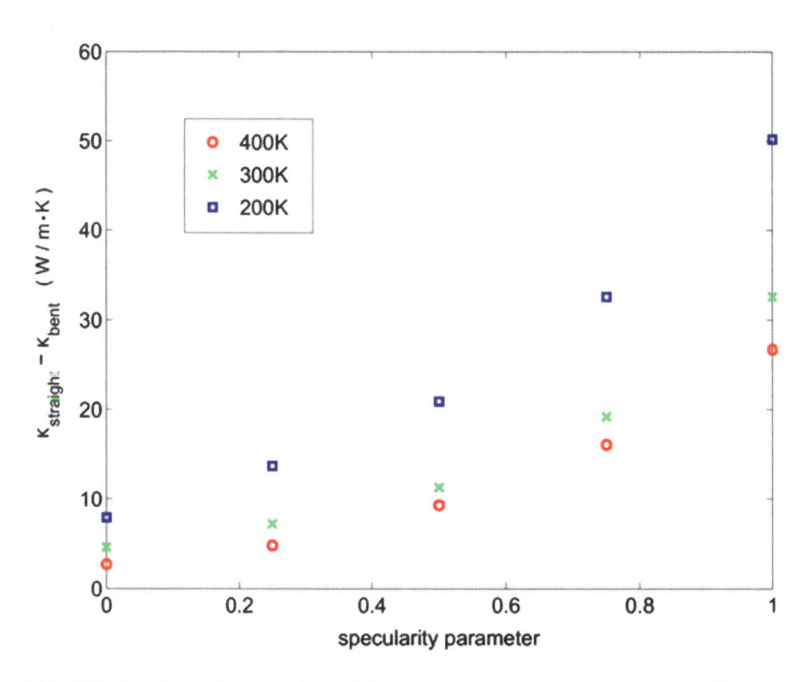

Fig. 6.14 Effective thermal conductivity difference between the straight and bent ligaments as a function of specularity parameter and temperature

Because the straight ligament provides an opportunity for phonon travel without surface interactions that is absent in the bent ligament, this difference is expected to be always positive. The data does not contradict this expectation. Second, for all temperatures, the difference between the straight and bent ligaments increases with the specularity parameter. This behavior may be explained by noting that specular reflections do not destroy tangential momentum, thus they provide no resistance to flow parallel to a surface. Under purely specular reflection, therefore, a surface can only provide resistance by assuming an orientation with a component perpendicular to the flow direction, so the sensitivity to geometry is maximized. Third, for similar reasons, the difference in thermal conductivity between the straight and bent ligaments increases as the temperature decreases. In this case, increasing sensitivity to geometry is caused by weakening interphonon interactions, which are also destroyers of tangential momentum.

6.7.3 Continuum Simulation

Computing an effective thermal conductivity enables the use of continuum simulation techniques in the design of the remaining components required to construct a differential surface interaction measurement structure. A schematic of a complete device, designed by John Sullivan of the Center for Integrated Nanotechnologies (http://cint.lanl.gov/) and adapted for their "Nanomechanics and Thermal Transport Discovery Platform," is shown in Fig. 6.15. This device is constructed from a silicon-on-insulator (SOI) wafer using microfabrication techniques described in detail in [82]. In this device, the center platform is warmed via an electrical heater and the resulting temperature on the platforms on either side, connected to the center platform by a bent or a straight ligament, is detected via electrical resistance thermometry. The test ligaments are formed monolithically with the platforms to eliminate uncertainties due to contact resistance.

Electrothermal simulations of the structure shown in Fig. 6.15 were performed using the Sandia National Laboratories "Sierra" finite element-based coupled physics simulation suite. A representative result of one such calculation is presented in Fig. 6.16. In this figure, it can be seen that the design intent of the structure is realized. In particular, the "legs" attaching the center platform to the frame are sufficiently long to force a significant amount of heat through the test ligaments and into the sensor platforms. Similarly, the leg length is sufficient to maintain a measurable temperature rise in the sensor platforms. Finally, due to the thermal conductivity difference between the straight and the bent ligaments, the temperature of the sensor platform connected to the bent ligament is lower than that of the platform connected to the straight ligament.

Performing these simulations for a range of specularity parameters provides a prediction of device behavior. The results of these calculations are shown in Fig. 6.17 for the same heater current at two ambient temperatures. From this figure, it may be seen that the response to specularity parameter is stronger at higher temperature than at lower temperature, which is somewhat counterintuitive because

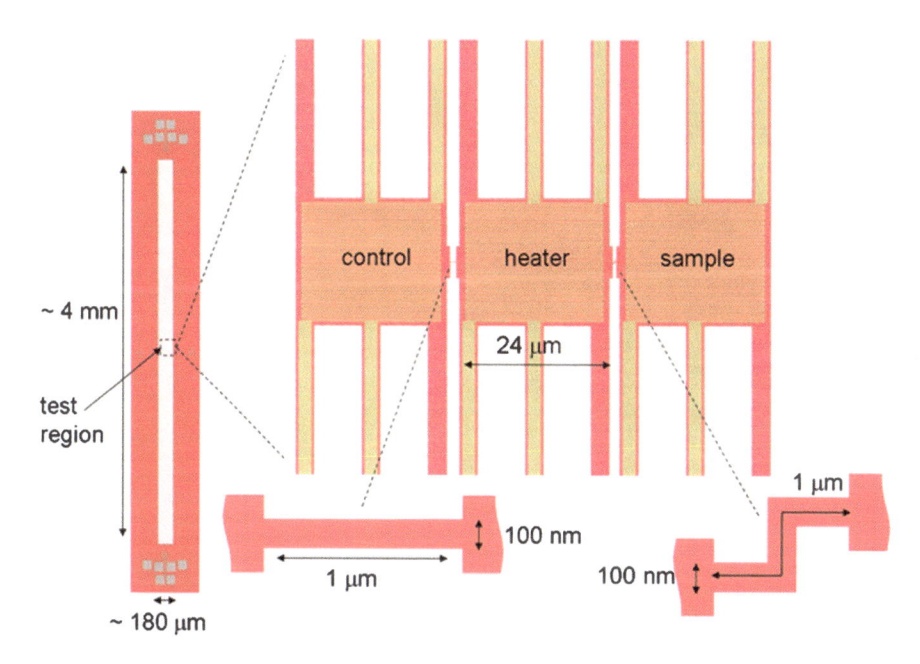

6 Interaction of Thermal Phonons with Interfaces 201

Fig. 6.15 Schematic of a complete differential surface interaction measurement device. To minimize end-wall effects in the ligaments, the device is realized on an SOI wafer with a 2 μm device layer thickness (out of the page), giving the ligaments a bladelike aspect ratio. The beige components are meandering platinum heater/sensor lines and their associated electrical connections

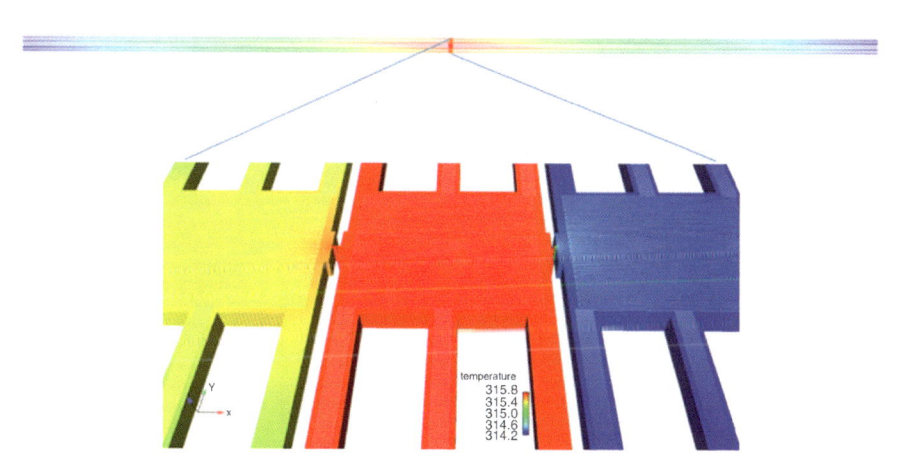

Fig. 6.16 Temperature distribution of a differential structure with 0.1 mA heater current at 300 K ambient temperature. The thermal conductivity of each ligament was derived from Monte Carlo simulations with fully specular reflection. The lower limit of the temperature scale was increased from 300 to 314.2 K in the figure inset to highlight the temperature difference between platforms

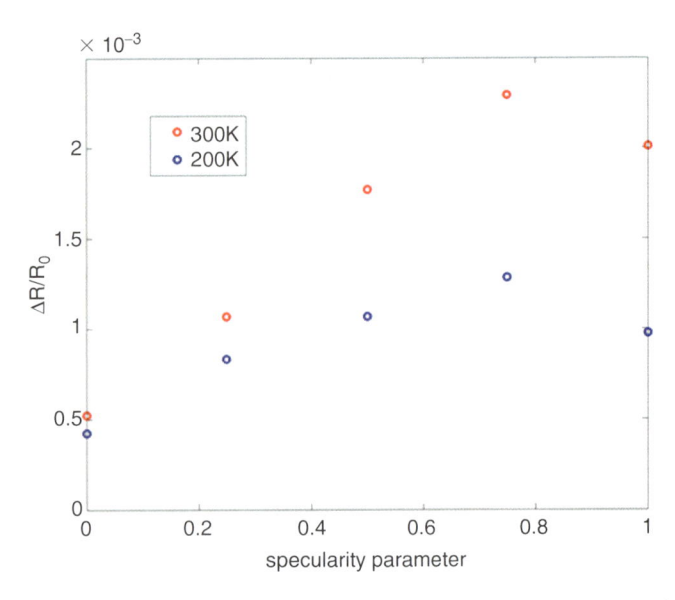

Fig. 6.17 Predicted resistance difference between sensor platforms as a function of specularity parameter at two temperatures with 0.1 mA heater current and platinum heaters and sensors

interphonon scattering becomes stronger as the temperature increases, which should mask the effects of the boundaries. This behavior is a result of the constant heater current; at higher temperatures, the thermal conductivity of the suspension legs decreases, as does the electrical conductivity of the heater. The center platform therefore experiences a greater temperature increase over ambient and is thus able to force more heat across the test ligaments, increasing sensitivity. A second notable feature of the data is that the response is nearly linear below $p = 0.75$. Increasing the specularity parameter further to $p = 1$, however, the observed resistance difference decreases. This is somewhat surprising behavior, given the monotonic increase in thermal conductivity difference between the ligaments with increasing specularity parameter shown in Fig. 6.14. Fortunately, this is unlikely to be a problem in operation, because it is very difficult to realize specularity parameters in this range at reasonable temperatures. This structure, with appropriate computational support, could therefore be used to determine the specularity parameter of phonon transport along the surfaces of its test ligaments.

6.8 Summary

In this chapter we have developed a framework for understanding phonon interactions with interfaces. Phonon interactions with interfaces fall into two broad categories defined by two distinct geometries. Interfaces that form the boundary of nanometer size channels (e.g., nanowires and thin films) define the first geometry and internal interfaces (e.g., grain boundaries, superlattice interfaces) define the second geometry. It was shown that the BTE provides a convenient model for considering boundary scattering in nanochannel structures. For internal interfaces, such as the grain boundaries found in polycrystals, it is more natural to consider transmission and reflection across a single boundary.

The important aspect of correlation between theory and experiments has also been considered. The experimental techniques available for measuring phonon transport applicable to nanoscale system have been reviewed, and our own experimental results on time resolved thermal wave microscopy on specimens with grain boundaries having known atomic structure have been presented. In addition, we have developed the concept of differential nanostructures with varying surface specularity parameter as a means for quantifying phonon transport along boundaries in nanoscale systems. A Monte Carlo simulation methodology for such structures has been presented, and a process for inputting results from these simulations into continuum electrothermal calculations to compare to experimental results has been described.

Overall, our intention has been to create a tight coupling between theoretical and experimental approaches to develop detailed understanding of phonon transport in nanoscale systems with multiple interfaces. This field is still evolving, and with sustained effort, it is likely to provide new solutions to problems in such diverse fields as thermal management, energy harvesting, and directed energy applications.

References

1. Gruen DM (1999) Annu Rev Mater Sci 29:211
2. Angadi MA, Watanabe T, Bodapati A, Xiao XC, Auciello O, Carlisle JA, Eastman JA, Keblinski P, Schelling PK, Phillpot SR (2006) J Appl Phys 99:114301
3. Dresselhaus MS, Chen G, Tang MY, Yang R, Lee H, Wang D, Ren Z, Fleurial JP, Gogna P (2007) New directions for Low-dimensional thermoelectric materials. Adv Mater 19:1043
4. Majumdar A (2004) Thermoelectricity in semiconductor nanostructures. Science 303:777
5. Rowe M (ed) (2006) Thermoelectrics handbook, macro to nano. Taylor & Francis, Boca Raton
6. Hochbaum AI, Chen R, Delgado RD, Liang W, Garnett EC, Najarian M, Majumdar A, Yang P (2008) Enhanced thermoelectric performance of rough silicon nanowires. Nature 451:163
7. Boukai AI, Bunimovich Y, Tahir-Kheli J, Yu J-K, Goddard WA III, Heath JR (2008) Silicon nanowires as efficient thermoelectric materials. Nature 451:161
8. Pascual-Gutierrez JA, Murthy JY, Viskanta R (2009) J Appl Phys 106:063532
9. Chernatynskiy A, Phillpot SR (2010) Phys Rev B 82:134301
10. Majumdar A (1993) J Heat Transf 115:7
11. Klemens PG (1958) In: Seitz F, Turnbull D (eds) Solid state physics, vol 7. Academic, New York, p 1
12. Callaway J (1958) Phys Rev 113:1046
13. Holland MC (1963) Phys Rev 134:A471
14. Berman R (1976) Thermal conduction in solids. Clarendon, Oxford
15. Sood KC, Roy MK (1993) J Phys Condens Matter 5:301
16. Ju YS, Goodson KE (1999) Appl Phys Lett 74:3005

17. Sparavigna A (2003) Phys Rev B 67:144305
18. Broido DA, Malorny M, Birner G, Mingo N, Stewart DA (2007) Appl Phys Lett 91:231922
19. Casimir HBG (1938) Physica (Amsterdam) 5:495
20. McConnell A, Uma S, Goodson KE (2001) Thermal conductivity of doped polysilicon layers. J Microelectromech Syst 10:360
21. Nan C-W, Birringer R (1997) Phys Rev B 57(14):8264
22. Kapitza PL (1941) J Phys (USSR) 4:181
23. Khalatnikov IM (1952) Eksp Teor Fiz 22:687
24. Swartz ET, Pohl RO (1989) Thermal boundary resistance. Rev Mod Phys 61:605
25. Pickett WE, Feldman JL, Deppe J (1996) Model Simul Mater Sci Eng 4:409
26. Schelling PK, Phillpot SR, Keblinski P (2002) Comparison of atomic-level simulation methods for computing thermal conductivity. Phys Rev B 65:144306
27. Aubry S, Kimmer C, Skye A, Schelling P (2008) Comparison of theoretical and simulation-based predictions of grain-boundary Kapitza conductance in silicon. Phys Rev B 78:064112–064120
28. Cahill DG, Bullen A, Lee S-M (2000) High Temp High Press 32:125
29. Lee S-M, Cahill DG (1997) Heat transport in thin dielectric films. J Appl Phys 81:2590
30. Simkin MV, Mahan GD (2000) Minimum thermal conductivity of superlattices. Phys Rev Lett 84:927
31. Tamura S, Tanaka Y, Maris HJ (1999) Thermal-conductivity measurements of GaAs/AlAs superlattices using a picosecond optical pump-and-probe technique. Phys Rev B 60:2627
32. Yang B, Chen G (2001) Microscale Thermophys Eng 5:107
33. Nilsson G, Nelin G (1971) Phys Rev B 3:364
34. Cowley RA, Woods ADB, Dolling G (1966) Phys Rev B 150:487
35. Wong J, Krisch M, Farber DL, Occelli F, Schwartz AJ, Chiang TC, Wall M, Boro C, Xu R (2003) Science 301:1078
36. Wong J, Wall M, Schwartz AJ, Xu R, Holt M, Hong H, Zschack P, Chiang TC (2004) Appl Phys Lett 84:3747
37. Xu R, Chiang TC (2005) Z Kristallogr 220:1009
38. Smith GO, Juhasz T, Bron WE, Levinson YG (1992) Phys Rev Lett 68:2366
39. Kim P, Shi L, Majumdar A, McEuen PL (2001) Thermal transport measurements of individual multiwalled nanotubes. Phys Rev Lett 87:215502
40. Shi L, Li D, Yu C, Jang W, Kim D, Yao Z, Kim P, Majumdar A (2003) Measuring thermal and thermoelectric properties of one-dimensional nanostructures using a microfabricated device. J Heat Transf 125:881
41. Williams CC, Wickramasinghe HK (1986) Appl Phys Lett 49:1587
42. Binnig G, Quate CF, Gerber C (1986) Phys Rev Lett 56:930
43. Shi L, Plyasunov S, Bachtold A, McEuen PL, Majumdar A (2000) Appl Phys Lett 77:4295
44. Cahill DC, Ford WK, Goodson KE, Mahan GD, Majumdar A, Maris HJ, Merlin R, Phillpot SR (2003) J Appl Phys 93:793
45. Kato R, Hatta I (2008) Int J Thermophys 29:2062
46. Tong T, Majumdar A (2006) Rev Sci Instrum 77:104902
47. von Gutfeld RJ, Nethercot AH Jr (1964) Heat pulses in quartz and sapphire at Low temperatures. Phys Rev Lett 12:641
48. Levinson YB (1986) Phonon propagation with frequency down-conversion. In: Eisenmenger W, Kaplyanski AA (eds) Nonequilibrium phonons in nonmetallic crystals. Elsevier, New York
49. Bron WE (1986) Phonon generation, transport and detection through electronic states in solids. In: Eisenmenger W, Kaplyanski AA (eds) Nonequilibrium phonons in nonmetallic crystals. Elsevier, New York
50. Wyatt AFG, Page GJ (1978) The transmission of phonons from liquid He to crystalline NaF. J Phys C 11:4927
51. Wolfe JP (1998) Imaging phonons: acoustic wave propagation in solids. Cambridge University Press, Cambridge
52. Bron WE (1980) Spectroscopy of high-frequency phonons. Rep Prog Phys 43:301

53. Jackson WB, Amer NM, Boccara C, Fournier D (1981) Appl Opt 20:1333
54. Opsal J, Rosencwaig A, Willenborg DL (1983) Appl Opt 22:3169
55. Hartmann J, Voigt P, Reichling M (1997) J Appl Phys 81:2966
56. Paddock CA, Eesley GL (1986) J Appl Phys 60:285
57. Koh YK, Singer SL, Kim W, Zide JMO, Lu H, Cahill DG, Majumdar A, Gossard AC (2009) J Appl Phys 105:54303
58. Boccara AC, Fournier D, Badoz J (1980) Appl Phys Lett 36:130
59. Stoner RJ, Maris HJ (1993) Phys Rev B 48:16373
60. Capinski WS, Maris HJ, Ruf T, Cardona M, Ploog K, Katzer DS (1999) Phys Rev B 59:8105
61. Cahill DG (2004) Rev Sci Instrum 75:5119
62. McDonald FA, Wetsel GC Jr, Jamieson GE (1986) Can J Phys 64:1265
63. Mansanares AM, Velivov T, Bozoki Z, Fournier D, Boccara AC (1993) J Appl Phys 75:3344
64. Bienville T, Belliard L, Siry P, Perrin B (2004) Superlatt Microstruct 35:363
65. Hurley DH, Telschow KL (2005) Phys Rev B 71:241410
66. Hurley DH, Khafizov M, Shindé S (2011) J Appl Phys 109:83504
67. Hurley DH, Wright OB, Matsuda O, Shindé SL (2010) J Appl Phys 107:023521
68. Thomsen C, Grahn HT, Maris HJ, Tauc J (1986) Phys Rev B 34:4129
69. Wolf D, Yip S (eds) (1992) Materials interfaces: atomic-level structure and properties. Chapman and Hall, London
70. Cahill DG, Pohl RO (1987) Phys Rev B 35:4067
71. Hu C, Kiene M, Ho PS (2001) Appl Phys Lett 79:4121
72. Klitsner T, VanCleve JE, Fischer HE, Pohl RO (1988) Phonon radiative heat transfer and surface scattering. Phys Rev B 38:7576
73. Tighe TS, Worlock JM, Roukes ML (1997) Direct thermal conductance measurements on suspended monocrystalline nanostructures. Appl Phys Lett 70:2687
74. Liang LH, Li B (2006) Size-dependent thermal conductivity of nanoscale semiconducting systems. Phys Rev B 73(15):153303
75. Chen G (2000) Particularities of heat conduction in nanostructures. J Nanoparticle Res 2:199
76. Bird GA (1994) Molecular gas dynamics and the direct simulation of gas flows. Oxford Engineering Science, Oxford University Press, New York
77. Wagner W (1992) A convergence proof of Bird's direct simulation Monte Carlo method for the Boltzmann equation. J Stat Phys 66(3–4):1011
78. Oran ES, Oh CK, Cybyk BZ (1998) Direct simulation Monte Carlo: recent advances and applications. Annu Rev Fluid Mech 30:403
79. Reese JM, Gallis MA, Lockerby DA (2003) New directions in fluid dynamics: non-equilibrium aerodynamic and microsystem flows. Philos Trans R Soc A 361(1813):2967
80. Mazumder S, Majumdar A (2001) Monte Carlo study of phonon transport in solid thin films including dispersion and polarization. J Heat Transf 123(4):749
81. Piekos ES, Graham S, Wong CC (2004) Multiscale thermal transport. Technical Report SAND2004-0531, Sandia National Laboratories, Albuquerque
82. Shindé SL, Piekos ES, Sullivan JP, Friedmann TA, Hurley DH, Aubry S, Peebles DE, Emerson JA (2010) Phonon engineering for nanostructures. Technical Report SAND2010-0326, Sandia National Laboratories, Albuquerque

Chapter 7
Time-Resolved Phonon Spectroscopy and Phonon Transport in Nanoscale Systems

Masashi Yamaguchi

Abstract Length scale-dependent phonon interaction is a key concept for the fundamental understanding of thermal transport in nanoscale materials. Thermally distributed phonons with various wavelengths belong to various transport regimes in nanoscale materials depending on the relative size of wavelength, mean-free-path vs. characteristic sizes of nanoscale materials. In this chapter, first a brief review is given on the phonon dispersion measurements using conventional scattering experiments and their limitations. Then a recently developed acoustic transport experiment is described. The method uses tunable acoustic source in GHz–THz frequency range which is excited by using ultrafast pulse shaping technique. Frequency-dependent mean-free-path and group velocity directly at the frequency range where phonon wavelength becomes comparable to the size of the nanoscale materials.

7.1 Introduction

Phonon transport in nanoscale materials plays a crucial role in various aspects of technology [1]. For example, thermal management in nanoscale electronic devices is an important part of the device design today [2]. The transistor gate size is well deep into nanoscale, and the reduction of thermal transport due to the spatially confined structure is the major issue in this case. Further miniaturization of these devices is limited by physical properties of nanoscale materials rather than technological challenges, and further understanding of phonon transport properties in nanoscale materials is a crucial step for the next generation of nanoscale devices. Another example is the need for the efficient thermoelectronic devices for various

---
M. Yamaguchi (✉)
Department of Physics, Applied Physics, and Astronomy, Rensselaer Polytechnic Institute, 110 eighth street, Troy, NY 12180, USA
e-mail: yamagm@rpi.edu

applications [3–7]. Contrary to the previous example, reducing thermal transport while keeping high electrical conduction is required for further improvement of thermoelectronic devices with higher figure-of-merit [5]. Phonon transport is a dominant contribution to the thermal transport in insulating and lightly doped semiconductors. In a confined structure, the phonon wavelength and mean-free-path become comparable to the size of the sample, and this has been a subject of intense research efforts [1]. Particularly, reduced thermal conductivity in nanoscale materials such as nanowires, nanocomposites, and superlattices due to the surface and interface scatterings has been studied extensively [8–11]. Spatial confinement generally affects phonon properties such as phonon–phonon, phonon–carrier, and surface scattering rates, phonon density of states and group velocity, electron–phonon coupling, and resultant phonon transport properties [8, 12, 13]; however, the exact mechanisms are yet to be explored.

Ultrafast laser spectroscopy has been utilized as a powerful tool for the observation of phonon dynamics of solid materials in picosecond and subpicosecond time scales [14–18]. The pump-probe scheme is often used to excite and observe phonons. Intense short laser pulse—shorter than the oscillation period of phonons—is used to excite phonons impulsively, and the time evolution of the excited phonon response can be monitored with variably delayed weak probe pulses as a change of the optical properties. There are practical advantages to using time domain approach for the study of nanoscale transport, although time domain spectroscopy shares many of the common features with frequency domain spectroscopy [19, 20]. For example, the frequency range of the generated phonon in picosecond acoustics is up to hundreds of GHz through the photothermal process [21], and up to THz frequency range from the interface of polar semiconductor interfaces [22]. These are generally higher than the frequency range of Brillouin light scattering with commonly used visible light sources. In particular, acoustic transport properties, time of flight, and mean-free-path of acoustic pulse can be measured directly by using picosecond acoustics, unlike scattering-based experiments.

Phonon transport in nanoscale materials shows distinct features depending on the characteristic length scales of phonons, such as the wavelength and mean-free-path [8–10, 23]. Such dispersive acoustic natures are expected to happen across the frequency where the characteristic length scales of phonons and feature sizes of nanoscale materials become comparable, and acoustic transport in different regime has been studied using different theoretical models [8, 12, 24–27]. For the fundamental understanding of thermal transport and properties in nanoscale materials, investigations on the acoustic transport in a wide frequency range are necessary.

In the following sections, a brief review of thermal transport in nanoscale materials will be given first, and then relevant time-resolved phonon spectroscopy using ultrafast laser techniques for acoustic transport experiments will be discussed. The understanding of the dispersive acoustic transport properties is the key to the complete understanding of the nanoscale thermal transport phenomena, and tunable acoustic spectroscopy in GHz–THz frequency range using pulse shaping technique and its application to the study of acoustic transport in glass materials will be described in detail. Finally, future perspective of phonon transport in nanoscale materials using ultrafast laser technique will be given.

7.2 Phonon Interactions and Frequency-Dependent Thermal Transport

In this section, phonon interaction mechanism and its relation to the frequency-dependent thermal transport is briefly reviewed. Thermal conductivity, κ, is defined by Fourier's law as $\mathbf{Q} = -\kappa \nabla T$, where $\mathbf{Q}$ is a rate of heat energy flow per unit area and T is the local temperature. Microscopic mechanism of thermal transport depends on the material, and thermal conductivity of common solid materials has a wide distribution from 10^{-2} W/m/K in aerogels to 10^4 W/m/K in diamond at low temperature. Heat energy in solid materials can be carried by phonons or electrons/holes. Optical phonon contributions to the thermal conductivity are typically smaller than acoustic branches in dielectric crystals because of the slow group velocity due to relatively flat dispersion relation and lower occupation in low temperature due to the higher eigen frequencies and the short relaxation time in bulk materials compared to acoustics phonons [28–30]. For example, optical phonons contribution to thermal conductivity is 5% in bulk silicon [28], and it has been reported that the contribution becomes up to 20% larger in nanoscale silicon. In this review, we limit the discussion to the major thermal conduction by acoustic phonons. Thermal conductivity in crystals can be expressed under relaxation time approximation for an isotropic material as follows [13]:

$$\kappa(T) = \frac{1}{3} \sum_i \int_0^{\omega_D} C_i(\omega) v_i(\omega) \ell_i(\omega) d\omega \qquad (7.1)$$

The index, i, of the summation is for three acoustic phonon branches, one longitudinal and two transverse. $C_i(\omega)$ is the contribution to the lattice specific heat due to phonons of i th branch at frequency ω, $v_i(\omega)$ and $\ell_i(\omega)$ are sound velocity and mean-free-path of phonons of i-th branch, respectively. Intuitively, $C_i(\omega)$ determines the number of heat carriers, i.e. acoustic phonons in this case. $v_i(\omega)$ and $\ell_i(\omega)$ determine how fast and how far phonons can carry heat without collisions. Thermal conductivity is expressed as a product of these quantities. $C_i(\omega)$ and $v_i(\omega)$ are closely related to the dispersion relation of phonons, and $\ell_i(\omega)$ is governed by phonon scattering processes. Frequency-dependent mean-free-path is rarely measured directly in high frequency regions, where the wavelength and feature size of nanomaterials become comparable.

Scattering of phonons is a result of both intrinsic causes such as phonon–phonon scattering and phonon–electron scattering, and extrinsic causes such as imperfections of crystals and surface scatterings. Total contributions from different scattering mechanisms to the mean-free-path are accounted for in the relaxation approximation as a summation of the rates of the scattering events, or inverse of the scattering time using Matthiessen's rule. Mean-free-path and the relaxation time are related as $\ell_i(\omega) = v_i(\omega)\tau_{\text{total}}$, and the total scattering time is given by a sum of the inverse of the relaxation times of various mechanisms as follows:

$$\frac{1}{\tau_{\text{Total}}} = \frac{1}{\tau_{\text{phonon-phonon}}} + \frac{1}{\tau_{\text{impurity}}} + \frac{1}{\tau_{\text{surface}}} + \cdots \qquad (7.2)$$

The first contribution, phonon–phonon scattering, is caused by a harmonicity of the lattice vibrations and has the largest contribution at room temperature in pure dielectric crystals [13]. In the scattering process, total phonon momenta before and after the scattering event have the same direction and magnitude. Hence, the scattering event does not cause thermal resistance. On the other hand, in Umklapp scattering process, total momenta before and after the scattering event are different by $\hbar \mathbf{G}$, where $\mathbf{G}$ is a reciprocal lattice vector. Umklapp scattering time at high temperature is proportional to the square of the phonon frequency and is given by Klemens [31] as

$$\frac{1}{\tau_{\text{phonon-phonon}}} = 2\gamma^2 \frac{k_B T}{\mu V_0} \frac{\omega^2}{\omega_D} \qquad (7.3)$$

where γ is Grüneisen parameter, μ is the shear modulus, V_0 is the volume per atom, and ω_D is the Debye frequency. In nanoscale materials, both phonon density of states and dispersion depend on the characteristic size of the sample, and the above equation for the bulk should be modified depending on the geometry of the sample. The second contribution in (7.2) is the phonon scattering time by mass disorder, i.e. isotopes, which has stronger frequency dependence than acoustic phonon–phonon scattering and is given by

$$\frac{1}{\tau_{\text{impurity}}} = \frac{V_0 \omega^4}{2\pi v^3} \Gamma \qquad (7.4)$$

where v is the group velocity of phonon, $v = \left|\frac{\partial \omega}{\partial q}\right|$, Γ is a constant determined by the ratio of the mass of isotope and regular atoms, and a fraction of content of isotopes [10]. The rate of the scattering is proportional to ω^4. Surface scattering becomes significant in nanoscale materials, and it has been studied extensively for various sample geometries [13, 32–34]. Models of surface scattering often phenomenologically separate specular and diffusive contributions. At the limit of complete diffusive scattering (known as Casimir limit), surface scattering time is given by following equation:

$$\frac{1}{\tau_{\text{surface}}} = \frac{v}{W} \qquad (7.5)$$

Fig. 7.1 Acoustic phonon dispersion relation for five lowest confined branches in a free-standing silicon cylindrical nanowires with diameter of 20 nm. Reprinted from [10] with permission

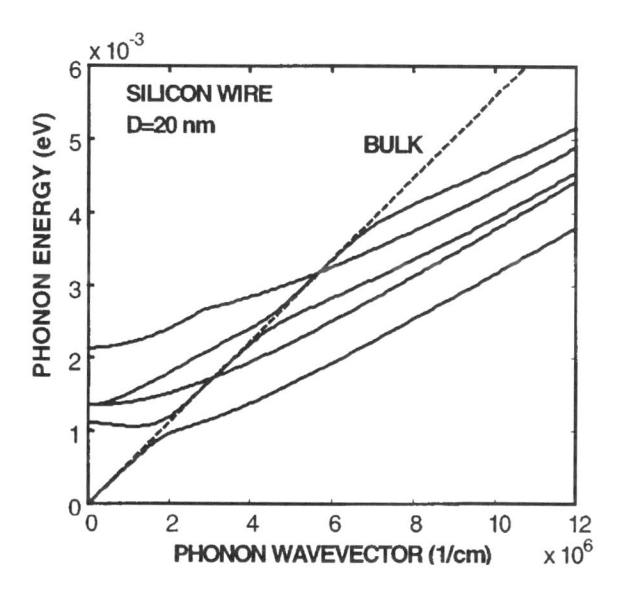

where W is characteristic thickness of the sample and does not have frequency dependence. However, in reality, the surface scattering generally should depend on the surface roughness relative to the phonon wavelength, in addition to the characteristic size of nanostructures. Each of the phonon scattering mechanisms shows distinct frequency dependence of the relaxation time or mean-free-path. However, frequency-dependent mean-free-path is rarely measured directly in the high frequency regime due to the lack of the experimental means.

The effect of spatial confinement on phonon dispersion directly affects the frequency-dependent specific heat and group velocity in the Eq. (7.1). Phonon dispersions in various nanostructure geometries have been studied extensively [8, 10]. As an example, Zou and Balandin [10] studied longitudinal phonon dispersion in silicon nanowire using continuum cylindrical model (Fig. 7.1). Debye-like dispersion relation in bulk material is split into multiple dispersion curves in nanowire depending on the transverse confinement. The spatial confinement effects significantly modify the slope of the dispersion curve, i.e. group velocity, and the frequency dependence of specific heat.

7.3 Phonon Dispersion and Inelastic Scattering Experiments

Phonon dispersion and frequency-dependent mean-free-path are the key elements for the fundamental understanding of thermal transport in nanoscale materials, as described in the previous section. Experimentally, neutron inelastic scattering and X-ray Brillouin scattering can measure phonon dispersion relation in a wide frequency range [35] while these experiments generally require a rather large amount

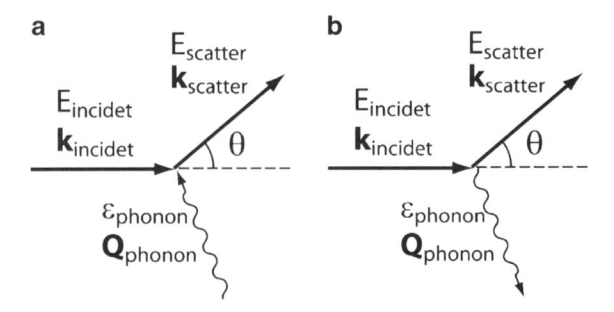

Fig. 7.2 Diagrams of (**a**) anti-stokes (phonon absorption) and (**b**) stokes scattering (phonon emission). *Straight arrows* represent incident and scattered photons, and *wavy arrow* represents phonon. θ is the scattering angle. Energy and momentum are required to be conserved in the scattering events

of samples, which is a practical limitation for many nanoscale samples. On the other hand, inelastic light scattering has an advantage of higher frequency resolution and higher sensitivity. In particular, Sandercock-type, multi-pass tandem Fabry–Perot allows Brillouin light scattering measurements in a wide frequency range with high resolution [36]. However, observable phonon momentum transfer with light scattering is limited to the phonons near the Brillouin zone center. In this section, the condition of the scattering event is briefly reviewed and a possibility of observing phonons without the limitation of scattering condition is discussed. Figure 7.2 shows diagrams of stokes and anti-stokes scattering processes by a phonon. In each scattering event, both total energy and momentum of photons and phonons are conserved. Hence, $E_{scatter} = E_{incident} \pm \varepsilon_{phonon}$ and $\mathbf{k}_{scatter} = \mathbf{k}_{incident} \pm \mathbf{Q}_{phonon}$, where $E_{scatter}$, $E_{incident}$, ε_{phonon}, $\mathbf{k}_{scatter}$, $\mathbf{k}_{incident}$, and $\mathbf{Q}_{phonon}$ are energies and wave vectors of scattered, incident photon and phonon, respectively, and the plus and minus sign correspond to the stokes and anti-stokes scattering processes. A phonon is absorbed upon the scattering event and the energy and momentum of scattered photon is increased by the amount of the energy and momentum of the absorbed phonon (anti-stokes scattering), while a phonon is emitted and the scattered photon has smaller energy and momentum in the case of stokes scattering. In optical frequency range, the condition $|\mathbf{k}_{scatter}| \approx |\mathbf{k}_{incident}|$ is satisfied due to much larger energy of photons compared to phonons, and the momentum conservation condition can be replaced with $|\mathbf{Q}_{phonon}| = 2|\mathbf{k}_{incident}|\sin\frac{\theta}{2}$, where θ is the angle between the momenta of incident and scattered photons [37]. The wave vector of the observable phonon depends on the scattering angle. In other words, by changing the angle between the incident and scattered photons, scattering by phonons with a particular wave vector can be selected. The phonons with the largest wave vector can be observed in a backscattering configuration where the incident and scattered photon are 180° away. The range of wave vector of observable phonons is limited to a narrow range near the gamma point in Brillouin zone. For example, if the incident photon has the wavelength $\lambda = 532$ nm in vacuum, the observable maximum phonon wave

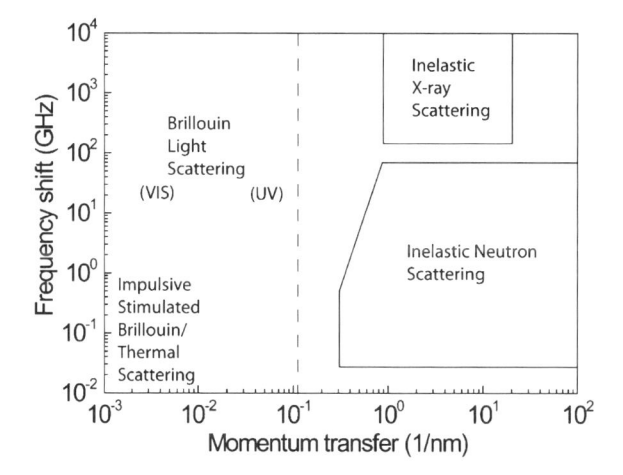

Fig. 7.3 Accessible regions in frequency-momentum space with conventional experimental methods for acoustic phonons. Inelastic light scattering experiments have advantages in low frequency-low momentum region. Use of UV light extends the accessible momentum transfer up to 0.14 (1/nm) [74] compared to conventional Brillouin scattering with visible light. However, the accessible momentum transfer is limited to the value near the Brillouin zone center. X-ray and neutron inelastic scattering can cover wider frequency-momentum space, while the access to the low frequency range is limited

vector is $Q = 11.9 \times 10^6$ (1/m) in back scattering geometry for sapphire, where the refractive index of 1.77 was used for the calculation. The lattice constant of sapphire in a-axis is 4.785 A, hence the zone boundary wave vector at X-point is 6.6×10^9 (1/m). The accessible wave vector range by inelastic light scattering using visible light source is limited up to about 1/1,000 of the phonon wave vector at the zone boundary. The determination of the full phonon dispersion relation is not possible with inelastic light scattering using phonons in the visible wavelength range. The same limitation applies to time domain equivalent to inelastic scattering, such as impulsive stimulated Brillouin scattering and impulsive stimulated thermal scattering [19, 20]. On the other hand, inelastic neutron scattering and X-ray Brillouin scattering using synchrotron radiation can gain access to the phonons with larger wave vectors due to the larger wave vector of incident neutron and X-ray photons. Various instrumentations have been developed to observe a phonon dispersion relation using inelastic scattering event (Fig. 7.3). The limitation of the accessibility to the phonons with high wave vector comes from the momentum conservation law of inelastic scattering event. Tabletop equipment, which can reach a wider wavelength range in Brillouin zone with optical sensitivity, is a desirable tool for the study of phonon transport in nanoscale materials. To expand the accessible wave vector range of phonons with optical experiment, it is necessary to use an experiment which is not based on the inelastic scattering of photons. Novel narrowband tunable acoustic spectroscopy based on tunable optical pulse sequence is described in a later section.

7.4 Time-Resolved Phonon Spectroscopy Using Ultrafast Laser

Ultrafast laser techniques have been extensively utilized as a powerful tool to study phonon properties in various condensed matter systems. Various physical mechanisms have been used for the generation and detection of phonons. A brief review of phonon spectroscopy based on ultrafast lasers is given in this section. The emphasis is on the generation and detection mechanism of acoustic phonons, which is relevant to acoustic phonon transport experiments.

7.4.1 Generation of Acoustic Waves

In opaque materials, absorption of optical pulses causes the local heating and resultant thermal stress near the surface of the sample. The local stress launches a propagating elastic strain pulse into the sample [14, 17, 38–41]. Acoustic waves generated by the irradiation of thin metal films have been extensively studied and are understood fairly well [14, 17, 38, 39]. Spectral bandwidth of the generated acoustic pulse depends on the penetration depth of optical pump, sound velocity, electron and thermal diffusions [42], and electron–phonon coupling [43]. The central frequency and bandwidth of about 100 GHz in metallic film excited with femtosecond laser pulses has been reported, and 440 GHz in metallic film thinner than penetration depth [21]. Here, we shall briefly review the acoustic generation process by photothermal mechanism based on a simple one-dimensional model, and a detailed description of the acoustic generation can be found elsewhere [17].

When a metallic film is irradiated by a short laser pulse of energy Q, the resulting temperature raise is given as follows [14]:

$$\Delta T(z) = \frac{(1-R)Q}{CA\varsigma} \exp\left(-\frac{z}{\varsigma}\right) \quad (7.6)$$

where R is the optical reflectivity of the film, C is the heat capacity, A is the area of irradiation, ς is the penetration depth of the optical pulse, and z is the distance from the surface of the sample. Here we assumed the diameter of the laser spot is much larger than the penetration depth. In that case, generated acoustic pulse contains main contributions from longitudinal acoustic phonons with the wavevector perpendicular to the sample surface. The relation above assumes the instantaneous temperature rise of the sample due to the laser pump pulses, and the spatial profile along the depth direction is determined by the penetration depth of the pump pulse. The local temperature rise sets up the isotropic thermal stress, $\sigma_{th} = -\gamma C_l \Delta T(z)$, where γ and C_l are the Grüneisen parameter and lattice heat capacity per volume. This impulsive stress launches longitudinal acoustic phonons traveling both away from and toward the surface. The partial wave, initially traveling toward the surface, will be reflected at the sample surface and changes the direction of the propagation. In the case of free surface, the polarity of the acoustic wave, which was initially propagating towards the surface, will be reversed upon the reflection. The reflected portion and the other portion, which were originally propagating into the sample, form bipolar strain pulse as follows:

$$\eta(z,t) = -\frac{(1-R)Q\gamma C_1}{2AC_\varsigma \rho v_s^2} \exp[-(z-v_s t)/\varsigma] \operatorname{sgn}[z-v_s t] \qquad (7.7)$$

where v_s is the sound velocity. In a real-life situation, several factors influence the spatial temperature profile and resultant strain pulse profile. The electron diffusion tends to broaden the spatial distribution of excited electrons until the excited electrons give up the energy to the lattice. For example, the penetration depth of Al in 400 nm is 10 nm; however, the electron diffusion length is about 50 nm. The effect of the electron diffusion to the picosecond acoustic generation was originally discussed by Tas and Maris [42] in metals and Wright and Kawashima [44] in semiconductors. Contrary to the Eq. (7.6) where the instantaneous temperature rise is assumed, the two-temperature model provides the time evolution of both electron and lattice temperatures in the time scale greater than the internal thermalization time of electrons [45]. In this model, the system is divided into electron and lattice subsystems, and each of them has own quasi-equilibrium temperatures, T_e and T_l, respectively.

$$C_e(T_e)\frac{\partial T_e}{\partial t} = \frac{\partial}{\partial t}\left(\kappa_e(T_e)\frac{\partial T_e}{\partial z}\right) - g(T_e - T_l) + S(z,t) \qquad (7.8)$$

$$C_l \frac{\partial T_l}{\partial t} = -g(T_e - T_l) \qquad (7.9)$$

where $C_e(T_e)$, C_l, and $\kappa_e(T_e)$ are the specific heats of electron and lattice, and thermal conductivity of electrons, and g is the electron phonon coupling constant which determines the energy transfer from the electron subsystem to the lattice subsystem. $S(z, t)$ is the source term which describes the power density deposited on the electron subsystem by the laser pulse. Temperature dependence of C_l is negligible due to the much larger heat capacity of the lattice system compared to the electron heat capacity, and the resultant temperature rise of the lattice is typically small under conventional experimental conditions. For example, in Aluminum, $C_l = 2.443$ Jcm^{-3}K^{-1} and $C_e = 0.028$ Jcm^{-3}K^{-1} (300 K) [42]. In a two-temperature model, the time evolution of spatial temperature profile is determined by both electron–phonon coupling and electron diffusion. The spectral bandwidth of the acoustic pulse generated in photothermal mechanism is limited by the spatial profile of the acoustic pulse width. The effect of diffusion becomes important when $D \geq \varsigma v_s$, where D is the electron diffusion constant. The electron diffusion tends to broaden the spatial distribution of excited electrons, hence the bandwidth is narrower. Shorter penetration depth with larger electron–phonon coupling is preferable for a wider bandwidth of generated acoustic pulses. The effects of the electron diffusion on the spatiotemporal profile of generated strain pulse have been discussed in detail for the metals with stronger electron–phonon coupling such as Cr and Ni [25], and weaker coupling such as noble metals [38, 46].

Acoustic phonon generation using interaction mechanisms other than photothermal effect has also been reported. Cummings and Elezzabi generated high frequency longitudinal acoustic phonons at L_1 and X_1 points in Brillouin zone through displacive mechanism in InSb [47]. Excitation of acoustic phonons in this mechanism requires the excitation of carriers with very high density and is evidenced with cosine-like oscillation behavior. Ultrafast screening of piezoelectric field by photon excited carriers at GaN/GaInN [22] and AlGaN/GaN [48] multiple quantum well has been used to generate longitudinal acoustic phonons. Excited phonon causes a large modulation in optical properties through Franz–Keldysh effect due to induced piezoelectric field [22].

Transverse acoustic phonons have been generated using the anisotropy of a crystal structure in semiconducting materials. Matsuda et al. [49] generated shear acoustic waves through photoelastic mechanism in Zinc. They used a Zinc crystal grown in a direction where c-axis is oblique to the SiO_2 substrate surface where zinc has a hexagonal close packed structure. They also generated a shear wave in the cubic GaAs crystal on lower symmetry (4 1 1) plane. From the pump power dependence of the generated waves, they suggested that the generation mechanism is the ultrafast screening of piezoelectric field [49].

7.4.2 Detection of Acoustic Waves

In optically opaque materials, acoustic phonons are optically excited near the surface, and then propagate into the sample. The phonons can be detected at the other side of the sample surface or same side of the surface when the phonons reflect back. Optical detection of acoustic phonons is based on the induced elastic strain or surface displacement by the phonons through various modulation mechanisms to the optical properties. Elastic strain induces the changes of dielectric tensor through elasto-optical coefficients as follows:

$$\Delta\varepsilon_{ij} = P_{ijkl} s_{kl} \qquad (7.10)$$

where $\Delta\varepsilon_{ij}$, P_{ijkl}, and s_{kl} are the components of induced change in dielectric tensor, Pockels elasto-optical tensor and elastic strain elements, respectively. The acoustic phonons can be detected as a modulation in the optical reflectivity at the surface of the sample due to the change of the dielectric tensor. Hurley and Wright [50] employed the Sagnac interferometer to detect the induced reflectivity change by acoustic waves. Heterodyne nature of the interferometer allows detecting both the reflective index and the phase change. This detection method was further applied to the imaging of surface phonons [15].

Surface displacement is also used to detect phonons optically. Wright and Kawashima [44] reported the phonon detection scheme by surface velocity. The deflection angle of probe pulse due to the surface displacement was measured using a bicell detector. Time-dependent surface velocity was obtained by the time derivative of the deflection angle, $\frac{d\theta}{dt} \propto v(t)$, where θ and $v(t)$ are the deflection angle and surface velocity. The latter is proportional to the elastic strain, and the temporal profile of elastic strain can be determined. Choi et al. [51] used a common path interferometry to detect the change of the optical path of probe pulse due to the surface displacement by acoustic pulse. They used a transmission grating to generate two parallel beams, and the first beam was reflected from the location on the sample surface where acoustic pulse arrives. The other was reflected from the surface where no acoustic pulse arrives. The change of the optical path was detected as the change of the phase of the interference of these two beams. This method has been applied to the detection of narrowband acoustic waves, as described further in the later section.

7.5 Phonon Transport Experiments Using Time-Resolved Phonon Spectroscopy

Phonon spectroscopy based on picosecond acoustic is principally a phonon transport experiment. Generated acoustic phonons propagate into and interact with the materials. Typical signal of picosecond acoustics from a single metallic layer consists of a train of acoustic echoes due to the multiple internal reflections of the acoustic pulse within the sample. A comparison of the different echo peaks allows determining the mean-free-path of acoustic pulses, group velocity, and the reflectivity at the interfaces, in contrast to scattering experiments. High temporal (<1 ps) and longitudinal spatial (<10 nm) resolution of picosecond acoustic spectroscopy has advantages for multilayered nanoscale materials, and the method has been applied to the study extensively [52–54]. In addition, upper limit of the accessible phonon frequency in picosecond acoustic spectroscopy is higher than in Brillouin scattering using visible light since it is not limited by the scattering event. In the next sections, time-resolved acoustic spectroscopy and tunable acoustic spectrometer using pulse shaping techniques are described.

7.5.1 Tunable Acoustic Spectroscopy and Phonon Localization in Glasses

Phonon transport properties strongly depend on the length scale due to the dispersion of interaction with environment, and there has been a considerable interest to study such properties in amorphous [55–59] and nanoscale materials [1, 10, 60] where the length scale becomes comparable to the size of the characteristic length of the material system. Conventional acoustic echo experiments use piezoelectric thin film transducers as an acoustic source, and recent developments have led to the devices working at a frequency as high as 20 GHz [61]. Efforts to measure frequency-dependent phonon transport properties have been pursued using ultrafast laser-based techniques by taking the advantage of the accessible high frequency limit. With the use of femtosecond lasers, higher frequency phonons have been generated. Broadband acoustic wavepackets with frequency components up to 440 GHz have been generated in thin metallic transducer by a subpicosecond laser pulses, and Fourier transform of the broadband pulse has been used to study frequency-dependent acoustic transport properties in silica glasses [21]. Narrowband acoustic pulses have been generated by using the mechanical resonance of metallic thin films [62], and zone folded phonons in semiconductor superlattice [16, 63]. The frequencies of the generated narrowband phonons are determined by the thickness of the transducer, and the period of the super lattice in these cases. Tunable narrow band acoustic phonons in GHz frequency region have been excited using pulse trains in metallic transducer [51, 64], and semiconductor interfaces [65].

Choi et al. [51] and Klieber et al. [64] generated narrowband acoustic pulses based on the excitation by optical pulse trains using a simple optical pulse shaper (Death star Pulse shaper). Absorption of a femtosecond laser pulse train leads to a periodic thermal expansion through photothermal expansion in metallic transducer and generates an acoustic pulse train. The center frequency of the excited phonons depends on the time separation of optical excitation pulses. The generated acoustic pulses propagate into the sample. The use of optical pulse shaping technique made it possible to tune the frequency of the generated acoustic phonons. Propagated acoustic phonon through the sample layer is detected at the other metallic transducer attached to the other side of the sample, hence the direct measurement of frequency-dependent phonon acoustic transport is possible.

7.5.1.1 Tunable Acoustic Spectroscopy Using Amplitude Pulse Shaping Technique

Figure 7.4 shows the experimental setup of a tunable acoustic spectrometer. Amplified Ti-sapphire laser pulses at 250 kHz were used as a light source. A single laser pulse is introduced into the pulse shaper system, which consists of three reflectors and a partial reflector. The introduced pulse completes the path shown in the figure. Each time the pulse reaches the partial reflector, a part of the pulse is transmitted through it. Fixed delay paths compensate the differences in the travel time of individual pulses for each of pulses. A pair of parallel mirrors with the distance equal to the half of the distance of the adjacent pulse when the delay line was set at its minimum delay was used for the compensation so that all seven pulses overlap temporally and spatially at the sample position [64]. Tuning of the center frequency is achieved by changing the position of the delay line in Fig. 7.4, which changes the temporal separations between adjacent pulses evenly.

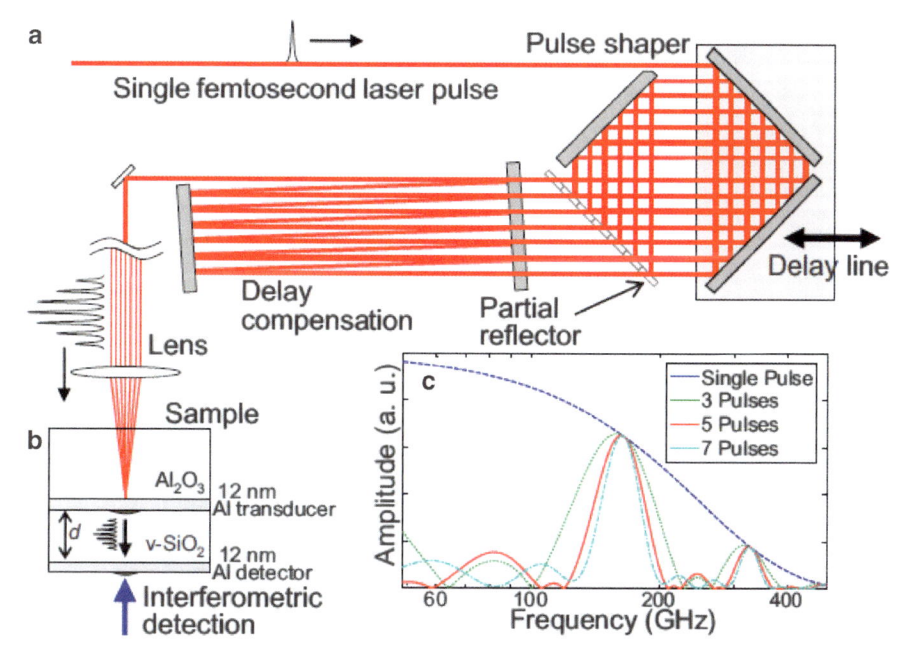

Fig. 7.4 Experimental setup of the "Death star" tunable acoustic spectrometer. (**a**) The setup generates evenly spaced pulse sequences with adjustment of a signal delay line to change the separation. (**b**) The pulse train is focused on the metallic transducer and generates acoustic pulse train. The acoustic train propagates into the sample adjacent to the transducer and reach to the metallic film for the detection. (**c**) Calculated power spectrum of photo-acoustic response in polycrystalline Al. Reprinted from [64] with permission

The lowest tunable frequency is limited by the length of the translation stage and is roughly 2 GHz for this particular setting. The highest tunable frequency of the pulse shaper is constrained by the excitation pulse duration, and it is roughly 2 THz with the excitation pulse used. It should be noted that the excited acoustic bandwidth is constrained by the metallic transducer and limited to roughly 400 GHz as described below. The partial reflector is custom designed and the reflector has different reflectivity in different regions so that each time an excitation pulse hit the portion of the reflector. The reflectivity of each portion was designed so that the envelope of transmitted pulse train has a Gaussian profile. Figure 7.5 shows the optical autocorrelation signal of the output from the pulse shaper tuned at various frequencies between 250 GHz and 1 THz.

For phonon transport experiments, the sample has a sandwich structure with metallic transducer layers on both sides of the sample. Acoustic phonons are excited on the excitation transducer. In the current system, thin Aluminum film with the thickness of 15 nm has been used. Narrowband acoustic pulse excited with an optical pulse train propagates through the sample and reaches the other side of the sample.

Fig. 7.5 Autocorrelation traces of the optical excitation pulses for acoustic generation. These optical pulses were used for the excitation of narrowband acoustic pulses at various frequencies

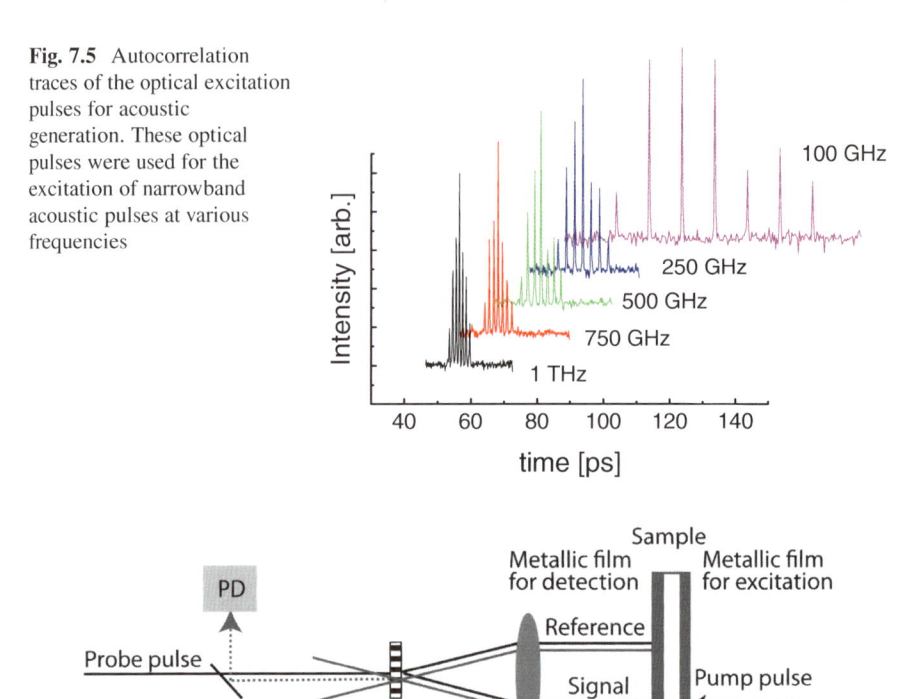

Fig. 7.6 Grating interferometer for the acoustic detection. Probe and reference arms are generated and recombined at same phase mask. Beams reflected from sample are indicated *dark* and *gray lines*

Acoustic displacement at the opposite side of the sample from the excitation transducer was detected with a variably delayed probe pulses using a phase mask-based interferometer [66]. A schematic of the surface displacement detection is shown in Fig. 7.6. A binary phase mask splits a probe beam, and the depth of the binary pattern is chosen to maximize ±1 order of diffraction at the wavelength of the probe beam. The two arms are the reference and probe arms. Pulses in both arms were sent to the sample surface, and the probe pulse was reflected from the excited region by the acoustic pulse, while the reference pulse was reflected from a portion of the sample surface without acoustic excitation. Both of the reflected pulses are diffracted by the same phase mask into the direction of incident beam. The interferometric signal is proportional to the phase difference caused by the acoustically induced displacement at the excited area due to the arrival of acoustic pulse. The resolution of the interferometric detection is $\sim\lambda/1{,}000$, where λ is wavelength of the probe pulse [66]. The use of the common path interferometer significantly reduces noise due to the air current and mechanical vibrations from the optical table.

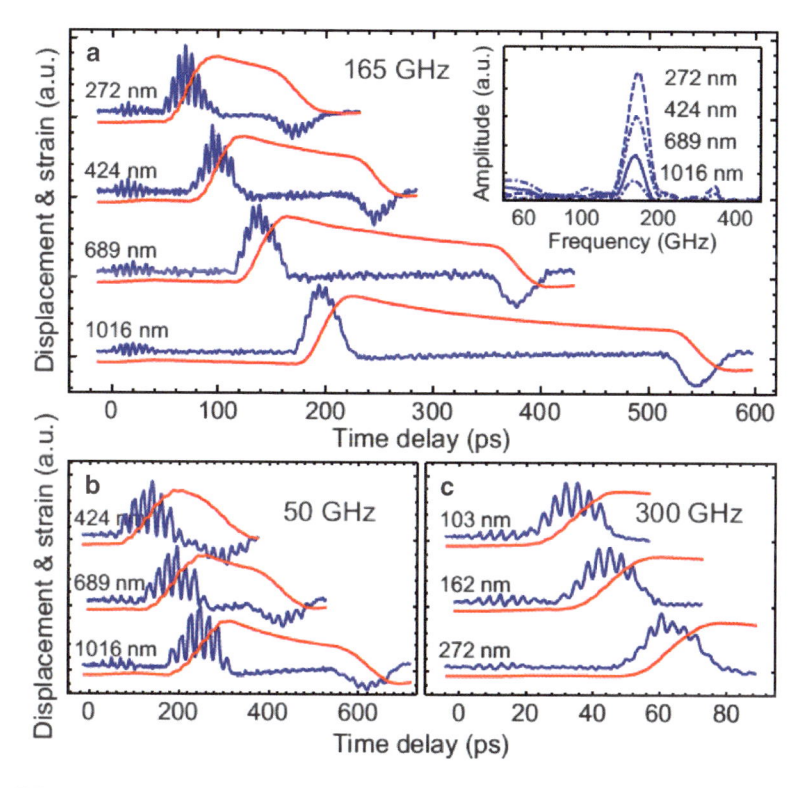

7 Time-Resolved Phonon Spectroscopy and Phonon Transport in Nanoscale Systems 221

Fig. 7.7 Raw data for different thickness of amorphous SiO$_2$ samples with frequency at 165 GHz (**a**), 50 GHz (**b**), and 300 GHz (**c**). Smooth curve is interferometrically measured displacement while strongly modulated curves show corresponding strain (time derivatives). The *inset* shows the Fourier spectrum of the transmitted acoustic signal. Reprinted from [64] with permission. Data displacement (*redline*) and strain (*blue line*)

7.5.1.2 Death Star-Signal Shape

Figure 7.7 shows an example of time-dependent surface displacement $u(t)$ of the Aluminum-air interface by the interferometer, together with the corresponding strain $\eta(t) = \frac{\partial u(t)}{\partial z} = \frac{1}{v_l}\frac{\partial u(t)}{\partial t}$, where v_l is the sound velocity of longitudinal acoustic phonon. Small acoustic signal near $t = 0$ is observed before the acoustic pulse travel through the sample. This small signal contribution is due to the displacement of transducer film. The probe pulses partially penetrate through the receiver film and acquire the signal from the transducer displacement because the penetration depth of the probe pulse is on order of the film thickness. This small signal plays a role in the precise time reference. Generated acoustic phonons at the transducer film propagate through the sample, and then reach to the detector transducer. The larger displacement and strain signals between 50 and 200 ps in Fig. 7.7 correspond to the arrival of the acoustic pulse to the receiver film. It is worth mentioning that the

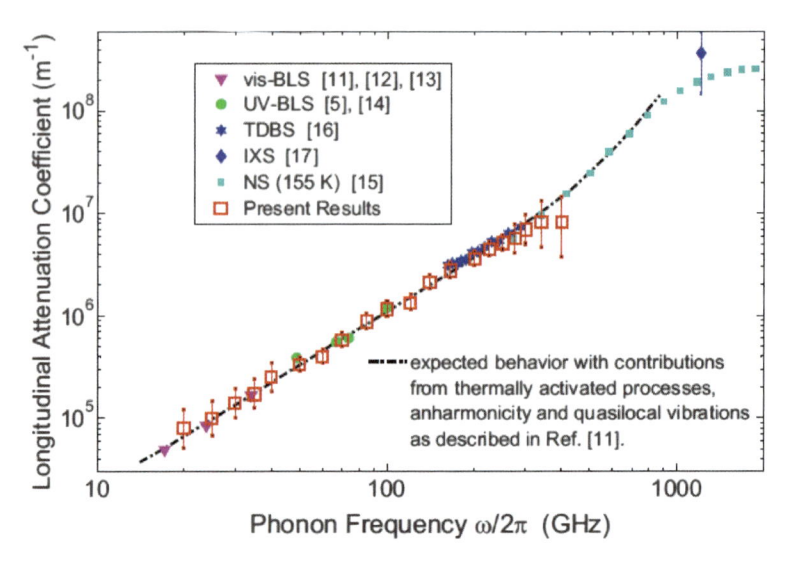

Fig. 7.8 Frequency-dependent acoustic attenuation in SiO_2 glasses at room temperature and the results from other spectroscopic techniques. Reprinted from [64] with permission

transducer configuration is the direct measurement of acoustic transport properties, in contrast to scattering experiments where localized vibration and propagating waves cannot be distinguished directly [55, 67, 68].

Phonon properties and their relation to the anomalous thermal properties in amorphous solids have been of considerable interest, and in particular, the transport properties of high frequency acoustic phonons (>100 GHz) have been a matter of controversial debates [58, 67, 69–77].

Amorphous materials show characteristic plateau region in their temperature dependence of thermal conductivity at low temperature (∼10 K) [69]. The feature is generally explained by the Ioffe–Regel crossover of acoustic phonon, where the mean-free-path of acoustic plane wave becomes shorter than the wavelength and acoustic phonons with above crossover frequency cease to propagate [67]. In other words, acoustic phonon above the crossover frequency does not have well-defined wave vector, and the crossover can be considered as the end point of acoustic branch in amorphous materials. It has been reported that the acoustic phonon branch merges to the Boson peak in densified silica glass in higher frequency range [78]. The crossover happens around the wavelength on the order of nanometer, and the mechanism of strong scattering of the acoustic phonons at high frequency regime is still under debate [72, 73]. Tunable acoustic spectroscopy is an acoustic transport experiment, unlike scattering-based experiments. The measurement of time of flight provides acoustic group velocity at the excitation frequency. The decrease in the amplitude gives the direct measure of phonon mean-free-path at the excitation frequency. It is a unique experiment since it can distinguish propagating and non-propagating vibrational modes. Figure 7.8 shows the frequency-dependent attenuation of longitudinal acoustic phonons in silica glass [64] from the tunable acoustics spectroscopy in the range of 20–400 GHz together with the results of Brillouin scattering with visible, UV, and X-ray excitation, and neutron inelastic scattering. The experiment covers the frequency range between optical Brillouin scattering and neutron inelastic scattering.

7.5.2 Summary and Future Perspective

We have reviewed time-resolved acoustic phonon measurements using ultrafast laser spectroscopy, with the particular emphasis on the methods that are relevant to the transport measurements. In nanoscale materials, the length scale-dependent transport property is the key to understanding the fundamental mechanisms of nanoscale transport phenomena. However, the lack of experimental means to directly measure acoustic transport properties, particularly in high frequency range, is a major obstacle. Tunable acoustic spectroscopy with a combination of picosecond acoustics and laser pulse shaping is one of the major efforts to overcome this difficulty. Current development of the spectrometer demonstrated direct measurement of the group velocity and mean-free-path of acoustic phonons directly at variable frequency up to about 400 GHz. The current upper frequency limit can be extended by the use of other acoustic generation mechanisms. In addition, the generation and detection of shear acoustic waves in tunable transport measurement geometry is expected to provide further details of nanoscale transport phenomena.

References

1. Cahill DG, Ford WK, Goodson KE, Mahan GD, Majumdar A, Maris HJ, Merlin R, Phillpot SR (2003) Nanoscale thermal transport. J Appl Phys 93:793
2. Schelling PK, Shi L, Goodson KE (2005) Managing heat for electronics. Mater Today 8:30
3. Hochbaum AI, Chen R, Delgado RD, Liang W, Garnett EC, Najarian M, Majumdar A, Yang P (2008) Enhanced thermoelectric performance of rough silicon nanowires. Nature 451:163
4. Boukai AI, Bunimovich Y, Tahir-Kheli J, Yu J-K, Goddard WA III, Heath JR (2008) Silicon nanowires as efficient thermoelectric materials. Nature 451:168
5. Dresselhaus MS, Chen G, Tang MY, Yang R, Lee H, Wang D, Ren Z, Fleurial JP, Gogna P (2007) New directions for Low-dimensional thermoelectric materials. Adv Mater 19:1043
6. Balandin A, Wang KL (1998) Significant decrease of the lattice thermal conductivity due to phonon confinement in a free-standing semiconductor quantum well. Phys Rev B 58:1544
7. Harman TC, Taylor PJ, Walsh MP, LaForge BE (2002) Quantum dot superlattice thermoelectric materials and devices. Science 297:2229
8. Stroscio M, Dutta M (2001) Phonons in nanostructure. Cambridge University Press, Cambridge
9. Chang C-M, Geller MR (2005) Mesoscopic phonon transmission through a nanowire-bulk contact. Phys Rev B 71:125304
10. Zou J, Balandin A (2002) Phonon heat conduction in a semiconductor nanowire. J Appl Phys 89:2932

11. Chen G (2005) Nanoscale energy transport and conversion. Oxford University Press, New York
12. Chen G (2000) Phonon heat conduction in nanostructures. Int J Therm Sci 39:471
13. Ziman JM (1960) Electrons and phonons. Clarendon, Oxford
14. Thomsen C, Grahan HT, Maris HJ, Tauc J (1986) Surface generation and detection of phonons by picosecond light pulses. Phys Rev B 34:4129
15. Sugawara Y, Wright OB, Matsuda O, Takigahira M, Tanaka Y, Tamura S, Gusev VE (2002) Watching ripples on crystals. Phys Rev Lett 88:1885504
16. Bartels A, Dekorsy T, Kurz K, Kohler K (1999) Coherent zone-folded longitudinal acoustic phonons in semiconductor superlattices: excitation and detection. Phys Rev Lett 82:1044
17. Gusev VE, Karabutov AA (eds) (1993) Laser optoacoustics. American Institute of Physics, New York
18. Kinoshita S, Shimada Y, Tsurumaki W, Yamaguchi M, Yagi T (1993) New high resolution phonon spectroscopy using impulsive stimulated Brillouin scattering. Rev Sci Instrum 64:3384
19. Yan YX, Nelson KA (1987) Impulsive stimulated light-scattering. 1. General-theory. J Chem Phys 87:6240
20. Yan YX, Nelson KA (1987) Impulsive stimulated light-scattering. 2. Comparison to frequency-domain light-scattering spectroscopy. J Chem Phys 87:6257
21. Zhu TC, Maris HJ, Tauc J (1991) Attenuation of longitudinal-acoustic phonons in amorphous SiO_2 at frequencies up to 440 GHz. Phys Rev B 44:4281
22. Sun CK, Liang JC, Yu XY (2000) Coherent acoustic phonon oscillations in semiconductor multiple quantum wells with piezoelectric fields. Phys Rev Lett 84:179
23. Stroscio MA, Kim KW, Yu S, Ballato A (1994) Quantized acoustic phonon modes in quantum wires and quantum dots. J Appl Phys 76:4670
24. Prasher R (2006) Thermal conductivity of tubular and core/shell nanowires. Appl Phys Lett 89:063121
25. Farhat H, Sasaki K, Kalbac M, Hofmann M, Saito R, Dresselhaus MS, Kong J (2009) Softening of the radial breathing mode in metallic carbon nanotubes. Phys Rev Lett 102:126804
26. Rego LGC, Kirczenow G (1998) Quantized thermal conductance of dielectric quantum wires. Phys Rev Lett 81:232
27. Nishiguchi N, Ando Y, Wybourne MN (1997) Acoustic phonon modes of rectangular quantum wires. J Phys Condens Matter 9:5751
28. Tian Z, Esfarjani K, Shiomi J, Henry AS, Chen G (2011) On the importance of optical phonons to thermal conductivity in nanostructures. J Appl Phys 99:053122
29. Sellan DP, Turney JE, McGaughey AJH, Amon CH (2011) Cross-plane phonon transport in thin films. J Appl Phys 108:113524
30. Broido DA, Malorny M, Birner G, Mingo N, Stewart DA (2007) Intrinsic lattice thermal conductivity of semiconductors from first principles. Appl Phys Lett 91:231922
31. Klemens PG (1958) Thermal conductivity of lattice vibrational modes. In: Seitz F, Turnbull D (eds) Book. Academic, New York
32. Casimir HBG (1938) Note on the conduction of heat in crystals. Physica 5:495
33. Majumdar A (1993) Microscale heat conduction in dielectric thin films. J Heat Transf 115:7
34. Swartz ET, Pohl RO (1989) Thermal-boundary resistance. Rev Mod Phys 61:605
35. Baldi G et al (2005) Brillouin ultraviolet light scattering on vitreous silica. J Non Cryst Solids 351:1919
36. Lindsay SM, Anderson MW, Sandercock JR (1981) Construction and alignment of a high performance multipass Vernier tandem Fabry–Perot interferometer. Rev Sci Instrum 52:1478
37. Benedek GB, Fritsch K (1966) Brillouin scattering in cubic crystals. Phys Rev 149:647
38. Wright OB, Gusev VE (1995) Ultrafast generation of acoustic waves in copper. IEEE Trans Ultrason Ferroelectr Freq Control 42:331
39. Eesley GL, Clemens BM, Paddock CA (1987) Generation and detection of picosecond acoustic pulses in thin metal films. Appl Phys Lett 50:717
40. Kashiwada S, Matsuda O, Baumberg JJ, Voti RL, Wright OB (2006) In situ monitoring of the growth of ice films by laser picosecond acoustics. J Appl Phys 100:073506

41. Thomsen C, Strait J, Vardeny Z, Maris HJ, Tauc J, Hauser JJ (1984) Coherent phonon generation and detection by picosecond light-pulses. Phys Rev Lett 53:989
42. Tas G, Maris HJ (1994) Electron diffusion in metals studied by picosecond ultrasonics. Phys Rev B 49:15046
43. Saito T, Matsuda O, Wright OB (2003) Picosecond acoustic phonon pulse generation in nickel and chromium. Phys Rev B 67:205421
44. Wright OB, Kawashima K (1992) Coherent phonon detection from ultrafast surface vibrations. Phys Rev Lett 69:1668
45. Kaganov MI, Lifshitz IM, Tanatarov LV (1957) Relaxation between electrons and the crystalline lattice. Sov Phys JETP 4:173
46. Wright OB (1994) Ultrafast nonequilibrium stress generation in gold and silver. Phys Rev B 49:9985
47. Cummings MD, Elezzabi AY (2001) Ultrafast impulsive excitation of coherent longitudinal acoustic phonon oscillations in highly photoexcited InSb. Appl Phys Lett 79:770
48. Makarona E, Daly B, Im J-S, Maris H, Nurmikko A, Han J (2002) Coherent generation of 100 GHz acoustic phonons by dynamic screening of piezoelectric fields in AlGaN/GaN multilayers. Appl Phys Lett 81:2791
49. Matsuda O, Wright OB, Hurley DH, Gusev V, Shimizu K (2008) Coherent shear phonon generation and detection with picosecond laser acoustics. Phys Rev B 77:224110
50. Hurley DH, Wright OB (1999) Detection of ultrafast phenomena by use of a modified Sagnac interferometer. Opt Lett 24:1305
51. Choi JD, Feurer T, Yamaguchi M, Paxton B, Nelson KA (2005) Generation of ultrahigh-frequency tunable acoustic waves. Appl Phys Lett 87:819071
52. Grahn HT, Maris HJ, Tauc J, Abeles B (1988) Time-resolved study of vibrations of amorphous hydrogenated silicon multilayers. Phys Rev B 38:6066
53. Chen W, Lu Y, Maris HJ, Xiao G (1994) Picosecond ultrasonic study of localized phonon surface modes in Al/Ag superlattices. Phys Rev B Condens Matter 50:14506–14515
54. Rossignol C, Perrin B, Bonello B, Djemia P, Moch P, Hurdequint H (2004) Elastic properties of ultrathin permalloy/alumina multilayer films using picosecond ultrasonics and Brillouin light scattering. Phys Rev B 70:9
55. Foret M, Courtens E, Vacher R, Suck JB (1996) Scattering investigation of acoustic localization in fused silica. Phys Rev Lett 77:3831
56. Benassi P, Krisch M, Masciovecchio C, Mazzacurati V, Monaco G, Ruocco G, Sette F, Verbeni R (1996) Evidence of high frequency propagating modes in vitreous silica – reply. Phys Rev Lett 78:4670
57. Alexander S, Entin-Wohlman O, Orbach R (1986) Phonon–fracton anharmonic interactions: the thermal conductivity of amorphous materials. Phys Rev B 34:2726
58. Nakayama T (2002) Boson peak and terahertz frequency dynamics of vitreous silica. Rep Prog Phys 65:1195
59. Yamaguchi M, Nakayama T, Yagi T (1998) Effects of high pressure on the Bose peak in a-GeS$_2$ studied by light scattering. Physica B 263–264:258
60. Yang B, Chen G (2003) Partially coherent phonon heat conduction in superlattices. Phys Rev B 67:195311–195314
61. Yoshino Y (2009) Piezoelectric thin films and their applications for electronics. J Appl Phys 105:061623
62. Morath CJ, Maris HJ (1996) Phonon attenuation in amorphous solids studied by picosecond ultrasonics. Phys Rev B 54:203
63. Yamamoto A, Mishina T, Masumoto Y, Nakayama M (1994) Coherent oscillation of zone-folded phonon modes in GaAs-AlAs superlattices. Phys Rev Lett 73:740
64. Klieber C, Peronne E, Katayama K, Choi J, Yamaguchi M, Pezeril T, Nelson KA (2011) Narrow-band acoustic attenuation measurements in vitreous silica at frequencies between 20 and 400 GHz. Appl Phys Lett 98:211908
65. Yu C-T, Lin K-H, Hsieh C-L, Pan C-C, Chyi J-I, Sun C-K (2005) Generation of frequency-tunable nanoacoustic waves by optical coherent control. Appl Phys Lett 87:093114

66. Glorieux C, Beers JD, Bentefour EH, van de Rostyne K, Nelson KA (2004) Phase mask based interferometer: operation principle, performance, and application to thermoelastic phenomena. Rev Sci Instrum 75:2906
67. Courtens E, Foret M, Hehlen B, Ruffl'e B, Vacher R (2003) The crossover from propagating to strongly scattered acoustic modes of glasses observed in densified silica. J Phys Condens Matter 15:S1279
68. Masciovecchio C, Ruocco G, Sette F, Krisch M, Verbeni R, Bergmann U, Soltwisch M (1996) Observation of large momentum phonon like modes in glasses. Phys Rev Lett 76:3356
69. Zeller RC, Pohl RO (1971) Thermal conductivity and specific heat of noncrystalline solids. Phys Rev B 4:2029
70. Hunklinger S, Arnold W (1976) In: Thurston RN, Mason WP (eds) Physical acoustics XII. Academic, New York, pp 155–215
71. Buchenau U (2001) Dynamics of glasses. J Phys Condens Matter 13:7827
72. Ruffle B, Parshin DA, Courtens E, Vacher R (2008) Boson peak and its relation to acoustic attenuation in glasses. Phys Rev Lett 100:015501
73. Schirmacher W, Ruocco G, Scopigno T (2007) Acoustic attenuation in glasses and its relation with the boson peak. Phys Rev Lett 98:025501
74. Masciovecchio C et al (2006) Evidence for a crossover in the frequency dependence of the acoustic attenuation in vitreous silica. Phys Rev Lett 97:035501
75. Rat E, Foret M, Courtens E, Vacher R, Arai M (1999) Observation of the crossover to strong scattering of acoustic phonons in densified silica. Phys Rev Lett 83:1355
76. Courtens E, Foret M, Hehlen B, Vacher R (2001) The vibrational modes of glasses. Solid State Commun 117:187
77. Ruocco G, Sette F (2001) High-frequency vibrational dynamics in glasses. J Phys Condens Matter 13:9141
78. Foret M, Vacher R, Courtens E, Monaco G (2002) Merging of the acoustic branch with the boson peak in densified silica glass. Phys Rev B 66:024204

Chapter 8
Semiconductor Superlattice Sasers at Terahertz Frequencies: Design, Fabrication and Measurement

A.J. Kent and R. Beardsley

Abstract This chapter describes the design, fabrication and measurement of sub-THz sound amplification by the stimulated emission of radiation (saser) devices based on semiconductor superlattices (SLs). The chapter begins with a review of the various methods of amplifying sound in the GHz–THz frequency range which have been explored during the past 50 years since the invention of the laser. This is followed by a detailed consideration of electrically pumped sasers using SLs as the acoustic gain medium and as acoustic mirrors. A theoretical model of the phonon amplification by stimulated emission in a weakly coupled SL is presented, and the experimental evidence for amplification is reviewed. Next, the principles of SL acoustic Bragg reflectors and the methods that can be used for their design are explained. Various prototype vertical cavity saser structures are described and experimental evidence for saser action reviewed. The chapter ends with a brief discussion of possible applications for sub-THz saser sound.

8.1 Introduction

The demonstration of sound amplification by the stimulated emission of radiation (saser) has been targeted by researchers ever since the invention of the laser just over 50 years ago. Like photons, the quanta of lattice vibrational energy, phonons, obey Bose–Einstein statistics and transitions of a quantum system, e.g. electronic or spin systems, between energy levels brought about by the stimulated absorption and emission of phonons can, if the levels are population inverted, give rise to an increase in the phonon occupation number at a particular frequency, i.e. phonon amplification.

A.J. Kent (✉) • R. Beardsley
School of Physics and Astronomy, University of Nottingham,
University Park, Nottingham NG7 2RD, UK
e-mail: Anthony.Kent@Nottingham.ac.uk; ryan.beardsley@nottingham.ac.uk

S.L. Shindé and G.P. Srivastava (eds.), *Length-Scale Dependent Phonon Interactions*,
Topics in Applied Physics 128, DOI 10.1007/978-1-4614-8651-0_8,
© Springer Science+Business Media New York 2014

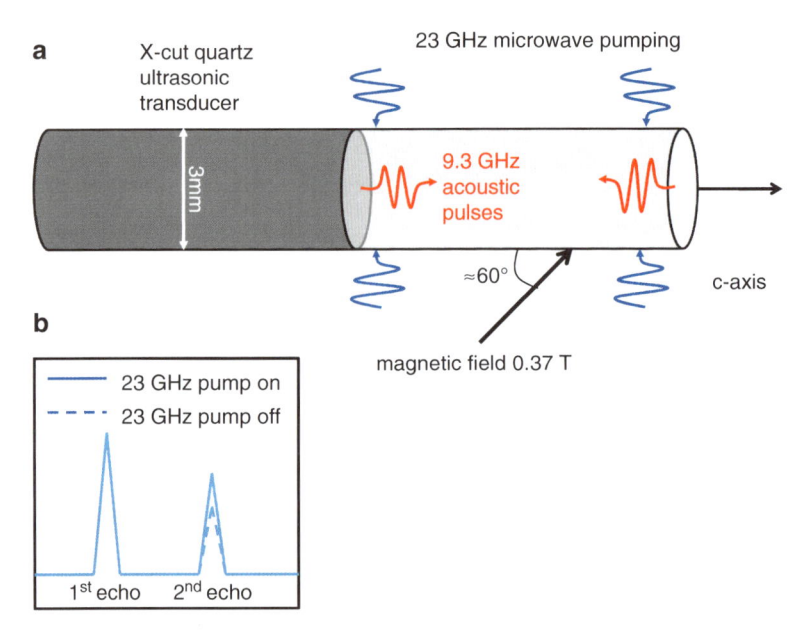

Fig. 8.1 (**a**) Sample arrangement used by Tucker [1] for measurement of the amplification of 9.3 GHz phonons due to spin–phonon interactions in microwave-pumped ruby; (**b**) schematic representation of the result: when the pumping and magnetic field are on, the acoustic echoes decay in intensity more slowly

One of the earliest observations of acoustic amplification in pink ruby was made by Tucker in [1] (1 year after the invention of the laser by Maiman [2]). Normally, due to the spin–phonon interaction with Cr^{3+} ions in Al_2O_3, longitudinal-polarized ultrasonic waves are attenuated on passage through the ruby [3]. To obtain amplification instead of attenuation, Tucker created a population inversion of the spin levels by microwave pumping using a frequency of 23.3 GHz. A schematic of the experimental arrangement is shown in Fig. 8.1. The experiment was conducted at a temperature of 1.5 K and using an ultrasonic frequency of 9.3 GHz. To obtain the appropriate value of spin Zeeman splitting a magnetic field of magnitude 0.37 T was applied. The ultrasonic pulse was propagated along the crystalline c-axis and the field was applied at an angle of 56° to the c-axis in order to achieve a compromise between the pumping efficiency and the spin–phonon coupling strength. Without the pumping applied, the ultrasonic echo train showed strong attenuation of echoes due to phonon absorption and only the first two or three echoes were detected. However, upon application of the pumping the attenuation decreased significantly and up to 18 echoes were detected. Tucker determined the gain for 9.3 GHz longitudinal ultrasonic waves in ruby as 0.12 cm^{-1}, and, taking account of the losses at 9.3 GHz, estimated that this could be sufficient to achieve "phonon maser" action.

Following on from Tucker's pioneering work, there have been further studies of phonon amplification in ruby. In 1980, using superconducting bolometer phonon

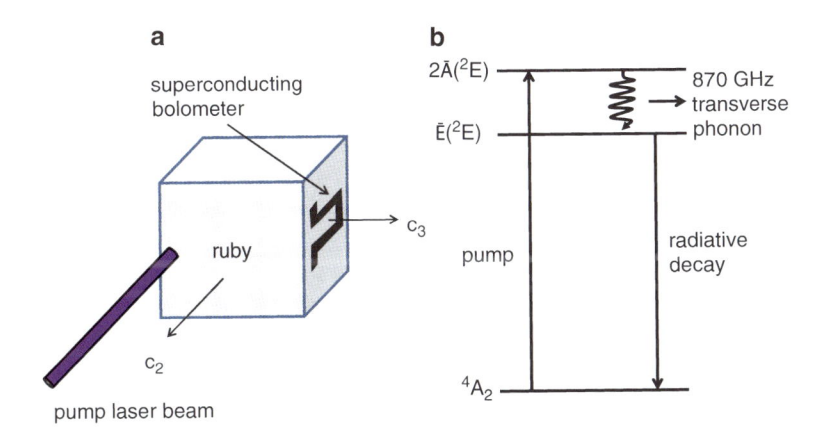

Fig. 8.2 (a) Experimental scheme for the measurement of amplification of 870 GHz transverse phonons in optically pumped ruby [4]; (b) Energy level diagram showing the pump and phonon emission transitions

detectors and optical pumping, Hu [4] claimed to observe stimulated emission of 29 cm^{-1} (870 GHz) transverse acoustic phonons. The 29 cm^{-1} phonons are emitted due to transitions between the $2\bar{A}(^2E)$ and the $\bar{E}(^2E)$ levels, see Fig. 8.2. Population inversion was produced by optically pumping with a dye laser tuned to the $^4A_2$–$2\bar{A}(^2E)$ optical transition. The geometry of the experiment was such that the phonons propagated along the c_3 axis of the ruby crystal to a granular aluminium bolometer (phonon detector) on the end face of the crystal. Evidence for stimulated emission was the narrow spectrum and directionality of the emitted phonons. However, when considering the latter, the effect of ballistic phonon focussing needs to be taken into account. On the basis of his measurements, Hu estimated the phonon gain to be of the order 10^5 cm^{-1} which is much larger than the estimated losses (~ 1 cm^{-1}), but no threshold was observed in the pump-power dependence of the emission. This was attributed to the system being in the high gain regime at even the lowest possible pump powers. Evidence for saturation of the population inversion was observed at very high optical pump powers, where the increase of the phonon signal levelled off. The sample did not provide an efficient cavity for 29 cm^{-1} phonons and continuous self-sustained oscillation was not possible. Overwijk et al. [5] also studied the emission of 29 cm^{-1} phonons by inverted Cr^{3+} ions in ruby. Depending on the Cr^{3+} concentration in the sample, they were able to identify two emission regimes: superfluorescence and amplified spontaneous emission. Tilstra et al. [6] studied the phonon emission due to transitions between the Zeeman split levels of the $\bar{E}(^2E)$ Kramers doublet. In a magnetic field of 3.48 T applied at 60° to the c-axis, the transverse-polarized phonons generated in the transition had a frequency of 50.4 GHz. The levels were inverted by pulsed optical pumping of the $^4A_2$ ground multiplet to $\bar{E}(^2E)$ transition and the population of the upper (E$^+$) and lower (E$^-$) states monitored by measuring the strengths of the luminescence lines due to transitions back to the ground multiplet, Fig. 8.3. The main evidence

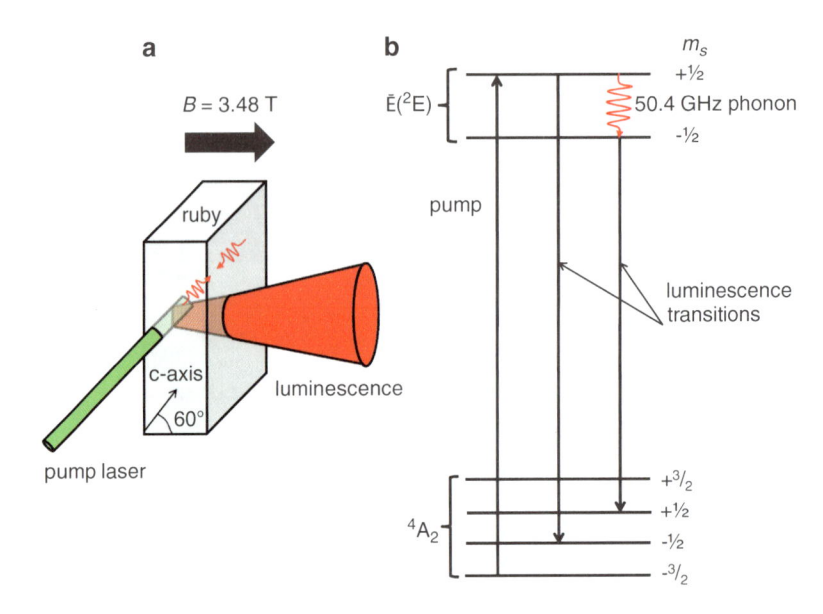

Fig. 8.3 (**a**) Experimental setup used by Tilstra et al. [6] to measure phonon avalanches in ruby; (**b**) Energy level diagram of Cr^{3+} in an applied magnetic field, with the pump, phonon emission and luminescence transitions indicated

for stimulated emission was the observation of "phonon avalanches," that is step decreases in the population of the E^+ level separated in time by the time taken for phonons to make the round trip from the inverted region to the end of the ruby crystal and back to the inverted region again. The avalanches continued until the inversion produced by the optical pump pulse had been fully depleted.

Bron and Grill [7] obtained evidence for acoustic amplification by the stimulated emission of phonons in another transition metal ion system: V^{4+} ions in Al_2O_3. The 24.7 cm^{-1} (740 GHz) phonons were emitted by the $_1E_{1/2}-_2E_{3/2}$ transition in V^{4+}, and these levels were inverted by pumping with IR light. Using superconducting bolometer phonon detectors, the authors noted a marked increase in the ballistic longitudinal-polarized phonon emission when the pumping exceeded the estimated threshold value necessary to create population inversion in the system. However, they were unable to obtain supporting evidence in the form of emission directionality measurements.

Acoustic amplification by stimulated emission has also been studied in a number of other systems: Prieur et al. [8] measured amplification of 545 MHz sound in BK7 glass. Acoustic phonons interact with localized two-level systems (TLS) present in glasses [9]. In the experiments, inversion of the TLS was created by an intense ultrasonic pump pulse, and this resulted in the observation of gain for a weaker probe pulse propagating behind the pump. Hutson et al. [10] observed acoustic amplification via the Cerenkov effect at frequencies of a few tens of MHz in the bulk piezoelectric semiconductor CdS when the drift velocity of the (optically excited) carriers exceeded the speed of sound in the material.

The experimental demonstrations of acoustic amplification discussed above fall into two groups: systems operating at relatively low frequencies, up to about 10 GHz, and those working in the technologically important sub-THz range. In most cases the object in which population inversion and acoustic amplification is achieved has macroscopic dimensions and it is difficult to form a high efficiency acoustic cavity for sub-THz phonons in such systems. Therefore a device exhibiting self-sustaining saser oscillation at such frequencies has proved elusive. At the lower frequencies, a cavity may be defined by the boundaries of the object as discussed in [1], but there are already well-established methods for generating coherent acoustic waves at these low frequencies.

Contemporary semiconductor laser devices are largely based on heterostructures having nano-scale feature sizes, which present many opportunities to engineer the electronic states to obtain strong population inversion under pumping. Such heterostructures may also be considered as the basis of a saser device. When considering the possibility of a saser, it is necessary to take account of the fact that, in the semiconductor crystals, the speed of sound is about five orders of magnitude smaller than the speed of light. As a result, for a given frequency, acoustic phonons have a much shorter wavelength than photons. In particular, for the THz frequency band, the phonon wavelength is only a few nanometers. This suggests that for a heterostructure-based saser, not only the electron states, but also the phonon states and their coupling with electrons can be tailored to facilitate its operation. On the other hand, the short wavelength of THz sound as well as its relatively strong scattering in comparison with light [11] brings about very severe requirements for design of the acoustic cavity. It has been demonstrated, however, that modern epitaxial semiconductor growth technologies enable the production of high-quality acoustic cavities for THz frequencies [12, 13].

The possibility of phonon amplification and sasing in semiconductor heterostructures has been considered in a number of theoretical works: transverse acoustic (TA) phonon amplification via the Cerenkov effect has been studied for the case of AlGaAs quantum well structures, it was predicted that significant gain, ~ 100 cm^{-1}, is achievable at sub-THz frequencies [14]. A source of THz coherent phonons based on resonant tunnelling in a double barrier resonant tunnelling diode (DBRTD) was proposed by Makler et al. [15]. The design of the device was such that the energy separation of the ground state and first excited state in the quantum well (QW) in the DBRTD structure was equal to the longitudinal optical (LO) phonon energy. A bias is applied to the device so that electrons resonantly tunnel into the excited state. When inside the well, the electrons rapidly relax to the ground state by emitting LO phonons. This LO phonon subsequently decays, resulting in the generation of THz TA phonons which escape the QW as an intense beam. The calculations showed that coherent amplification of the TA phonons via stimulated decay of the LO modes would be feasible in such a device. To date, there has been no convincing experimental demonstration of phonon amplification or sasing in either of these two systems.

Acoustic phonon amplification in semiconductor superlattice (SL) structures was considered by Glavin et al. [16–18]. An SL consists of alternating layers of two different semiconductor materials A and B, i.e. ABABAB..., with each layer being just a few nanometres thick. A single period, AB, of the SL may be repeated a few tens of times to build up the full structure. Using epitaxial crystal growth techniques such as molecular beam epitaxy (MBE) it is possible to fabricate SLs with atomically flat interfaces and sub nm control over the layer thicknesses. As we shall see in the following sections, the particular electronic and acoustic properties of superlattices are ideally suited to the realization of saser devices working in a wide range of sub-THz frequencies. Experimental evidence for coherent phonon amplification and sasing in GaAs/AlAs superlattices has recently been presented for the case of electrical pumping [19, 20], and optical pumping [21]. The electrical pumping results will be reviewed in this chapter.

8.2 Phonon Amplification due to Hopping Conduction in an SL

We will first discuss, from a theoretical point of view, the processes of phonon amplification in SLs. The GaAs/AlAs system will be considered as it is the subject of the experiments described in the following sections. The electronic properties of SLs have been extensively studied during the past 30 years, and many of the essential properties are discussed in the book by Grahn [22]. The electronic band structures of a SL, for the cases of zero and applied electrical bias, are illustrated in Fig. 8.4. Owing to the smaller band gap of the GaAs layers compared to AlAs, electrons are subject to a periodic confining potential in the SL growth direction, which we define here as the z-direction, but are free to move in the plane of the GaAs layers. Due to the process of quantum tunnelling, electrons can transfer between QWs and, if the barrier is sufficiently narrow, electronic minibands are formed. The width, Δ_m, of the minibands depends on the probability of tunnelling between adjacent quantum wells (QWs). Typically, in GaAs/AlAs SLs, Δ_m is of order a few meV. For $\Delta_m > \hbar/\tau$, where τ is the mean collision time, and for small values of the applied electric field, F, electrical conduction takes place through the minibands. At higher fields, such that $eFNd > \Delta_m$, where d and N are, respectively, the SL period and number of periods, the electron states become localized within a distance $\lambda \sim \Delta_m/eF$, each centred on a different QW. These states, known as the Wannier–Stark (WS) states, give rise to splitting of the minibands into a ladder of levels separated by $\Delta = eFd$ which are well defined if $eFd > \hbar/\tau$. Under these conditions miniband transport can no longer occur and electron transport between the WS states takes place primarily by quantum tunnelling. If F is sufficiently large that $\lambda \sim d$, tunnelling is between states centred on adjacent QWs. Miniband conduction is also broken if the scattering is sufficiently strong, i.e. $\Delta_m \leq \hbar/\tau$, in which case sequential tunnelling dominates the conduction even at low F.

Conservation of energy and momentum requires that tunnelling occurs either by an elastic process involving defect scattering with the subsequent emission of phonons, or by inelastic phonon-assisted tunnelling. In the latter case, tunnelling

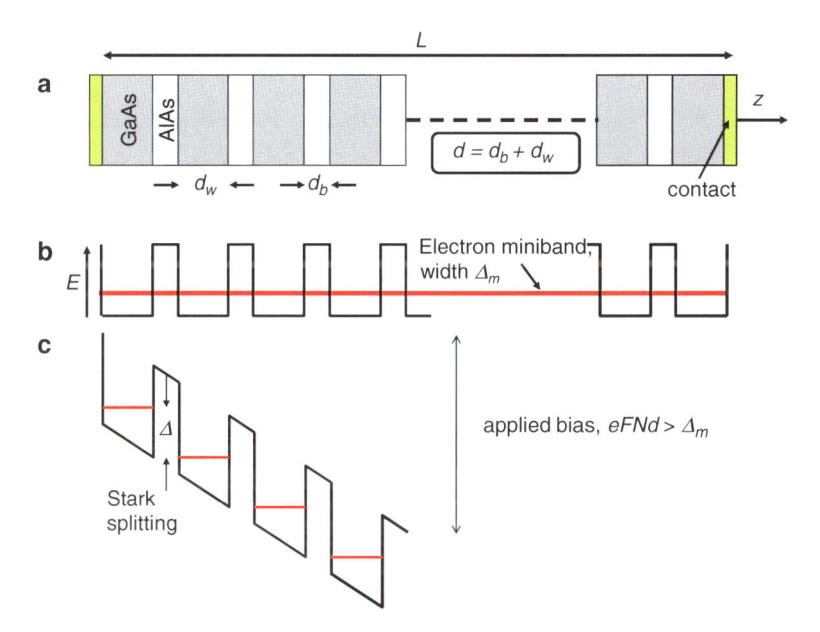

8 Semiconductor Superlattice Sasers at Terahertz Frequencies

Fig. 8.4 Schematic diagram of (**a**) GaAs/AlAs superlattice; (**b**) Electronic band structure with zero applied bias; (**c**) Band structure under applied bias electric field F such that $eFNd > \Delta_m$, the Stark splitting is Δ

of an electron through a barrier is accompanied by the absorption or emission of a phonon, either an optic (LO) or acoustic mode, depending on the magnitude of the Stark splitting, Δ. Now considering just the phonon-assisted processes, Fig. 8.5 shows in real and momentum space the possible electronic transitions between neighbouring wells involving phonon emission (1, 2) and absorption (3, 4). Taking account of all the possible phonon-assisted transitions, the kinetic equation for the population, $N_{\omega,\mathbf{q}}$, of the phonon mode, with frequency ω and wavevector $\mathbf{q}$, is given by:

$$\frac{dN_{\omega,\mathbf{q}}}{dt} = P_{\omega,\mathbf{q}}^{(em)}\left(N_{\omega,\mathbf{q}} + 1\right) - P_{\omega,\mathbf{q}}^{(ab)} N_{\omega,\mathbf{q}} - \beta_{\omega,\mathbf{q}} N_{\omega,\mathbf{q}}, \tag{8.1}$$

where $P_{\omega,\mathbf{q}}^{(em)}$ and $P_{\omega,\mathbf{q}}^{(ab)}$ are, respectively, the probabilities for phonon emission and absorption, and $\beta_{\omega,\mathbf{q}}$ accounts for phonon losses including phonon scattering and anharmonic decay. The first term on the right-hand side includes the two possible contributions to the transitions accompanied by emission of a phonon: spontaneous and stimulated, the latter being proportional to $N_{\omega,\mathbf{q}}$. Following the approach described in [16–18], we define the "phonon increment," $\gamma_{\omega,\mathbf{q}}$, as the difference of the probabilities of emission and absorption:

$$\gamma_{\omega,\mathbf{q}} = P_{\omega,\mathbf{q}}^{(em)} - P_{\omega,\mathbf{q}}^{(ab)}. \tag{8.2}$$

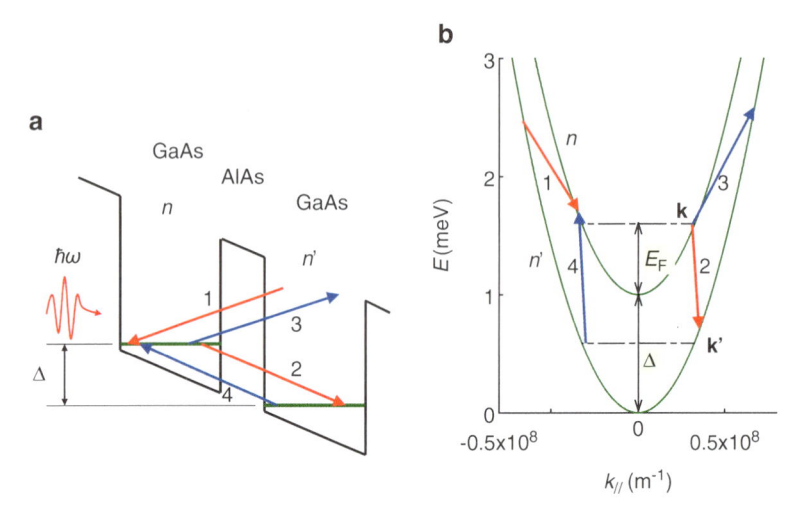

Fig. 8.5 Phonon-assisted inter-well transitions in (**a**) Real space and (**b**) Momentum space. Transitions due to stimulated emission (*1, 2*) and stimulated absorption (*3, 4*) are indicated

For $\gamma_{\omega,\mathbf{q}} > \beta_{\omega,\mathbf{q}}$, an exponential growth of the occupation of phonon mode $N_{\omega,\mathbf{q}}$ occurs, i.e. phonon instability. The emission and absorption probabilities may be written in terms of the rates of phonon-assisted transitions between neighbouring quantum wells n and n':

$$P^{(\text{em,ab})}_{\omega,\mathbf{q}} = \sum_{n,n',\mathbf{k},\mathbf{k}'} f_{n,\mathbf{k}} (1 - f_{n',\mathbf{k}'}) W^{(\text{em,ab})}_{n,n',\mathbf{k},\mathbf{k}',\omega,\mathbf{q}}, \quad (8.3)$$

where $\mathbf{k}$ and $\mathbf{k}'$ are the wave vectors of the electrons in wells n and n', respectively, and $f_{n,\mathbf{k}}$ are the Fermi functions for the electrons in the wells. The transition rates, $W^{(\text{em,ab})}_{n,n',\mathbf{k},\mathbf{k}',\omega,\mathbf{q}}$, are given by Fermi's golden rule

$$W^{(\text{em,ab})}_{n,n',\mathbf{k},\mathbf{k}',\omega,\mathbf{q}} = \frac{2\pi}{\hbar} \left| M^{n',\mathbf{k}'}_{n,\mathbf{k}} \right|^2 \delta(E_{n,\mathbf{k}} - E_{n',\mathbf{k}'} \mp \hbar\omega), \quad (8.4)$$

where $M^{n',\mathbf{k}'}_{n,\mathbf{k}}$ is the electron–phonon interaction matrix element. In the delta function term, which accounts for energy conservation, $E_{n,\mathbf{k}}$ and $E_{n',\mathbf{k}'}$ are the electron energies, (+) is for absorption and (−) is for emission. The matrix element for deformation potential electron–phonon coupling is given by Kent and Wigmore [23]:

$$\left| M^{n',\mathbf{k}'}_{n,\mathbf{k}} \right|^2 = \left| \langle n',\mathbf{k}' | V_{\mathbf{q}} | n,\mathbf{k} \rangle \right|^2 = \frac{\hbar D^2 q}{2\rho \Omega c_s} \delta_{\mathbf{k},\mathbf{k}',\mathbf{q}_\parallel} |F(q_z)|^2 S_{\mathbf{q}}^2. \quad (8.5)$$

8 Semiconductor Superlattice Sasers at Terahertz Frequencies

Here D is the deformation potential, ρ is the crystal density, c_s is the speed of sound and Ω is the interaction volume. The delta function accounts for momentum conservation in the plane of the QW, and screening of the electron–phonon interaction is taken into account by the term $S_\mathbf{q}$. The form factor, $F(q_z)$, was calculated by Tsu and Döhler [24]

$$F(q_z) = \int \varphi_z'^* \exp(iq_z z) \varphi_z dz = \left(\sqrt{\frac{T^2}{\Delta^2 + 4T^2}}\right) \frac{\sin(q_z d_b/2) \sin(q_z d_w/2)}{q_z d_w/2 \left[1 - (q_z d_w/2\pi)^2\right]}, \tag{8.6}$$

where φ_z are the QW bound-state wavefunctions, d_w and d_b are, respectively, the QW width and barrier width, and T is the tunnel coupling.

As well as inter-well transitions, phonons can be absorbed or emitted due to intrawell processes. For these we may write

$$W_{\mathbf{k},\mathbf{k}',\omega,\mathbf{q}}^{(\text{intra})} = \frac{2\pi}{\hbar} \left|M_\mathbf{k}^{\mathbf{k}'}\right|^2 \delta(E_\mathbf{k} - E_{\mathbf{k}'} \mp \hbar\omega), \tag{8.7}$$

where

$$\left|M_\mathbf{k}^{\mathbf{k}'}\right|^2 = |\langle \mathbf{k}'|V_\mathbf{q}|\mathbf{k}\rangle|^2 = \frac{\hbar D^2 q}{2\rho\Omega c_s} \delta_{\mathbf{k},\mathbf{k}',\mathbf{q}_{//}} |F(q_z)|^2 S_\mathbf{q}^2, \tag{8.8}$$

and, in the case of intrawell transitions,

$$|F(q_z)|^2 = \left(\frac{4\pi^2}{q_z^3 d_w^3 - 4\pi^2 q_z d_w}\right)^2 \left(\sin^2 q_z d_w - (1 - \cos q_z d_w)^2\right). \tag{8.9}$$

The above equations may be solved numerically for a given phonon mode $(\omega, \mathbf{q})$. Figure 8.6 shows the calculated increment for 620 GHz (2.6 meV) longitudinal acoustic (LA) phonons as a function of Δ and $q_{//}$ at a temperature of 15 K in an SL, with each period consisting of 5.9 nm of GaAs and 3.9 nm of AlAs, uniformly n-doped to a density of 10^{22} m^{-3}. For this SL, the Fermi energy $E_F \approx 0.6$ meV, and calculations of the miniband width using the Kronig–Penney model give $\Delta_m \approx 0.7$ meV. The maximum increment occurs at a Stark splitting near 2.6 meV (approximately equal to the phonon energy) and is $\sim 10^8$ s^{-1} per period of the SL. As $q_{//}$ increases, the maximum value of the increment is reduced and moves to larger Δ. This is due to efficient phonon absorption by intrawell transitions. For such processes, momentum conservation gives [23]

$$\mathbf{k}'^2 = \mathbf{k}^2 + q_{//}^2 - 2\mathbf{k} q_{//} \cos\varphi, \tag{8.10}$$

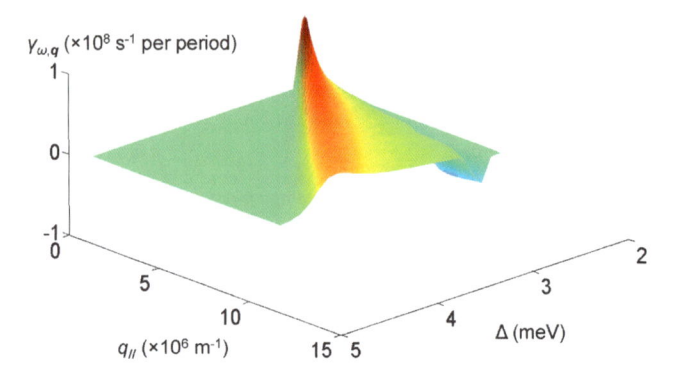

Fig. 8.6 Calculated increment per SL period for 620 GHz (2.6 meV) longitudinal acoustic (LA) phonons as a function of the Stark splitting, Δ, and in-plane component of the phonon wavevector, $q_{//}$, for an SL consisting of 50 periods of 5.9 nm GaAs and 3.9 nm of AlAs, uniformly doped with silicon to a density of 10^{22} m^{-3}, and at a temperature of 15 K

Fig. 8.7 Illustration of in-plane momentum conservation for intrawell transitions with emission or absorption of a phonon, q

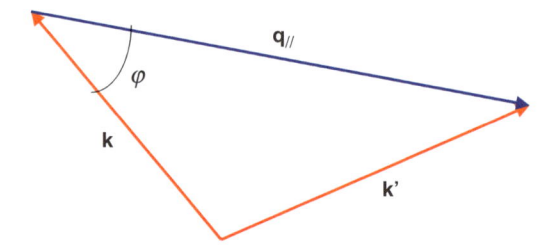

where φ is the angle between **k** and $q_{//}$, see Fig. 8.7. Including also energy conservation gives

$$\frac{c_s}{v_F}\frac{q}{q_{//}} - \frac{q_{//}}{2\mathbf{k}_F} = \cos\varphi \leq 1, \qquad (8.11)$$

where v_F and $\mathbf{k}_F$ are the Fermi velocity and wavevector, respectively. This condition means that the intrawell processes are effectively cut off for

$$q_{//} < q\frac{c_s}{v_F}. \qquad (8.12)$$

For the example chosen, $v_F = 56{,}000$ ms^{-1}, $c_s = 5{,}500$ ms^{-1}, and $q = \omega/c_s = 7.1 \times 10^8$ m^{-1}, which means intrawell transitions are cut off for $q_{//} < 7 \times 10^7$ m^{-1}, i.e. for phonons propagating at an angle of less than about 5° to the z-direction. It is possible to get a qualitative picture of how phonon amplification is possible for such near-vertical transitions by referring to Fig. 8.8. Consider a phonon with energy just slightly less than the Stark splitting, the transition due to stimulated emission is shown by the highlighted downward arrow and its rate is proportional to the probability of finding the initial state occupied multiplied by the probability of finding the final state empty. This is relatively high for the example shown, with the

Fig. 8.8 Illustration of quasi-population inversion for near-vertical inter-well transitions stimulated by phonons with energy $\hbar\omega < \Delta$

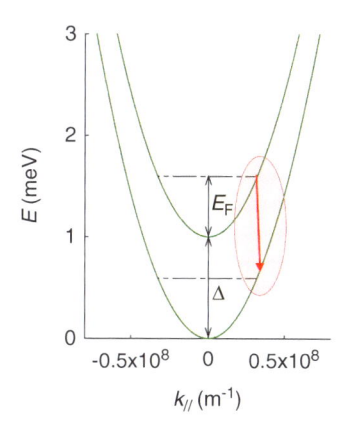

electron starting from the quasi-Fermi energy in the upper QW and ending just above the quasi-Fermi energy in the lower QW. However, absorption of the same energy phonon is much less probable, because the electron would either have to start from an empty state and/or end in a filled state. This means that there is an effective "inversion" for the phonon-assisted transitions, and it can be shown [16, 18] that this is a maximum for the phonon energy

$$\hbar\omega_q = \Delta - 2\sqrt{\frac{\hbar^2 q_{//}^2 E_F}{2m*}}, \tag{8.13}$$

where m^* the electron effective mass, i.e. for phonon energy slightly smaller than the Stark splitting.

Experimental measurements of the nonequilibrium acoustic phonon-induced transport in a 50-period SL with the key parameters, well and barrier width etc., as described above [25, 26], showed that for phonons incident close to the normal to the SL layers, stimulated emission was indeed the dominant phonon-assisted transport process at certain values of the Stark splitting. Thus, potentially, the SL could be used to amplify phonons. Based on the magnitude of the phonon-induced current, the phonon increment at $T = 4.2$ K could be estimated to be of the order 10^9 s^{-1}, which is in good agreement with the theoretical predictions.

To measure more accurately the acoustic gain of an SL, Beardsley et al. [27] devised an experiment in which a quasi-monochromatic beam of phonons was transmitted once through the SL and detected using a superconducting bolometer, Fig. 8.9. The phonons were generated by femtosecond pulsed laser excitation of a second, undoped, SL grown on top of the active SL. It had been previously shown [28] that this generates a propagating beam of quasi-monochromatic LA phonons of frequency $\approx c_s/d$, in this case about 650 GHz (2.7 meV). Figure 8.9c shows the bolometer signal as a function of the DC bias applied to the active SL, which consisted of 50 periods each made up from 5.9 nm GaAs and 3.9 nm of AlAs, uniformly doped with Si to 10^{22} m^{-3}. It is clear that application of the bias leads to

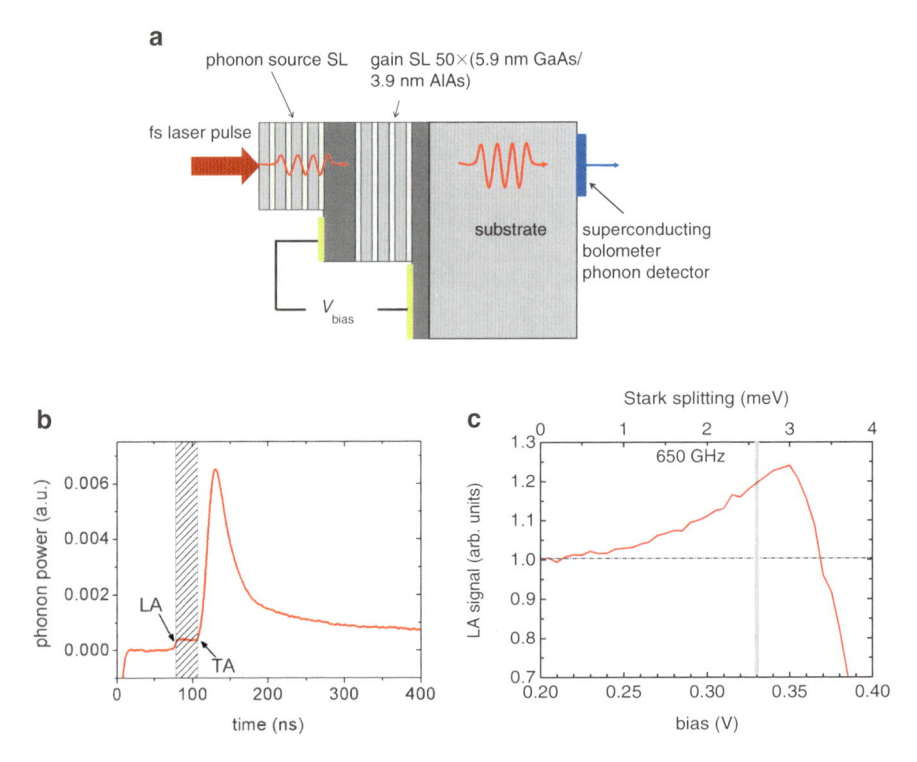

Fig. 8.9 (a) Sample details for single-pass phonon amplifier experiment; (b) Example bolometer signal; (c) Intensity of LA phonon signal as a function of the bias applied to the amplifier SL, normalized to the signal at zero applied bias

an increase of the detected LA phonon intensity, I, up to a peak at an applied bias corresponding to $\Delta \approx 3$ meV. For Stark splittings above about 3.4 meV, attenuation of the LA phonons is observed. From these results, the gain coefficient of the active SL, which is given by $g(\nu) = \frac{1}{l} \ln\left(\frac{I}{I_0}\right)$, where l is the length of the gain SL and I_0 the transmitted phonon intensity for zero applied bias, was determined to have a value of 4.5×10^5 m^{-1} at the maximum. Ignoring losses, this corresponds to an increment $\gamma = c_s g(\nu) = 2.5 \times 10^9$ s^{-1} at the temperature of the measurement, 2 K. The measured gain curve is very wide, at about 1 meV between the 50 % points. This is probably due to broadening of the QW levels due to scattering in the SL. It is necessary that the scattering is sufficiently strong that the SL is in the sequential tunnelling regime, even at biases below the current peak. Apparently, such scattering does not adversely affect the gain of the SL which is of the same order of magnitude as the theoretical predictions made assuming zero broadening.

Glavin et al. [17] calculated the effect of electron temperature on the increment. It was found that increasing the temperature to 77 K resulted in more than an order of magnitude decrease in the maximum increment and a significant broadening of the gain curve.

8.3 Threshold for Sasing

In the previous section we saw how a weakly coupled SL in the sequential tunnelling regime can amplify, by the process of stimulated emission, LA phonons of energy $\sim$ the Stark splitting, Δ, and propagating in a direction close ($< \sim 5°$) to the normal to the SL layers. Theoretical calculations and experimental measurements of the phonon increment at low, liquid helium, temperatures give a value of the order 10^9 s^{-1}, which corresponds to a gain coefficient of about 2×10^5 m^{-1} in a GaAs/AlAs SL.

In order to obtain saser oscillation, the gain SL needs to be within an acoustic cavity which provides phonon feedback. If the gain is sufficient to just compensate for the cavity losses, which must include the intended saser output, then self-sustaining saser oscillation is possible. Considering an acoustic cavity of length L defined by two acoustic mirrors having reluctances R_1 and R_2, see Fig. 8.10, the round trip gain is given by

$$G = R_1 R_2 \exp\{2(g(\nu)l - \beta L)\}, \quad (8.14)$$

where l is the length of the gain medium and β represents the losses per unit length of the cavity. The condition for saser oscillation to occur is $G = 1$; therefore, the threshold gain coefficient is obtained

$$\{g(\nu)\}_{th} = \frac{1}{l}\left\{\beta L + \frac{1}{2}\ln\left(\frac{1}{R_1 R_2}\right)\right\}. \quad (8.15)$$

As an example, we consider the superlattice used in the experimental measurements of the gain. This SL was about 0.5 μm in length, and if we include the electrical contact layers the cavity containing it would have to be at least 1 μm in length. The gain was measured as 4.5×10^5 m^{-1} at $\nu = 650$ GHz and in GaAs phonons of this frequency have a mean free path of 0.8 mm [29]. Using these values, we find the threshold for saser oscillation can be achieved if $R_1 R_2 \geq 0.64$. As we shall see in the next section, this is readily achievable in practice using acoustic

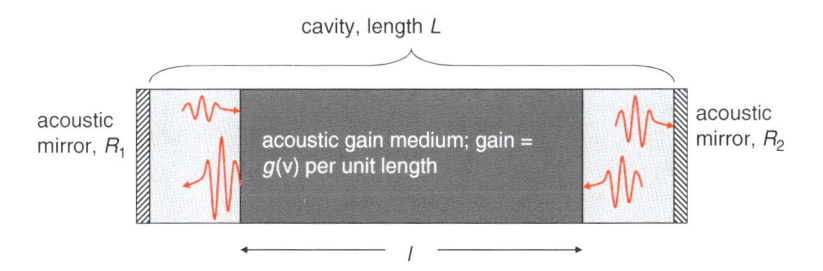

Fig. 8.10 Acoustic gain medium, length l, in a saser cavity of length L between two acoustic mirrors having reflectances R_1 and R_2

mirrors based on SLs. The threshold condition may also be expressed in terms of the quality, or Q-, factor of the cavity, which is defined as the ratio of the energy stored in the cavity to the energy lost per period of oscillation:

$$\{g(\nu)\}_{\text{th}} = \beta + \frac{\nu}{Qc_s}. \tag{8.16}$$

8.4 Superlattice Acoustic Bragg Mirrors

As well as providing the gain medium, SLs can be used as Bragg reflectors for sound and so form the basis of acoustic cavities for sub-THz phonons. Owing to the differences in the density, ρ, speed of sound, c_s, and, hence, the acoustic impedance Z of the two semiconductors making up the SL, an acoustic wave incident on an interface between the two materials is partly reflected and partly transmitted. For an acoustic wave travelling in a medium with acoustic impedance $Z_1 = c_{s1}\rho_1$, and incident normally on the boundary with a material of acoustic impedance $Z_2 = c_{s2}\rho_2$, the reflectance is given by:

$$R = \left(\frac{Z_1 - Z_2}{Z_1 + Z_2}\right)^2. \tag{8.17}$$

As an example, the acoustic impedances of GaAs and AlAs for LA waves are 2.52×10^7 and 2.12×10^7 kgm^{-2} s^{-1}, respectively, which gives $R = 0.086$. Even though the reflectance of an individual interface may be small, Bragg reflection may occur when the reflected waves from many such interfaces in an SL are in phase, i.e. for normal incidence $nc_s/\nu = 2d$, where n is an integer, c_s is the average speed of sound in the SL, ν is the frequency and d is the SL period.

Rytov [30] calculated analytically the acoustic dispersion of an infinite SL

$$\cos(qd) = \cos\left(\frac{\omega d_1}{c_{s1}}\right)\cos\left(\frac{\omega d_2}{c_{s2}}\right) - \frac{1+\kappa^2}{2\kappa}\sin\left(\frac{\omega d_1}{c_{s1}}\right)\sin\left(\frac{\omega d_2}{c_{s2}}\right), \tag{8.18}$$

where $\kappa = Z_1/Z_2$. This is shown in Fig. 8.11 for an SL with each period consisting of 6.2 nm of GaAs and 1.1 nm of AlAs. In this representation, the SL dispersion is folded into a mini-Brillouin zone which has its boundary at $q = \pi/d$. It is seen that stop gaps open at the mini-zone centre and edge, centred on frequencies corresponding to the conditions for Bragg reflections

$$\nu_n = n\left(\frac{d_1}{c_{s1}} + \frac{d_2}{c_{s2}}\right)^{-1} \tag{8.19}$$

8 Semiconductor Superlattice Sasers at Terahertz Frequencies

Fig. 8.11 Folded phonon dispersion for a 40-period GaAs/AlAs SL, with each period consisting of 6.2 nm of GaAs and 1.1 nm of AlAs

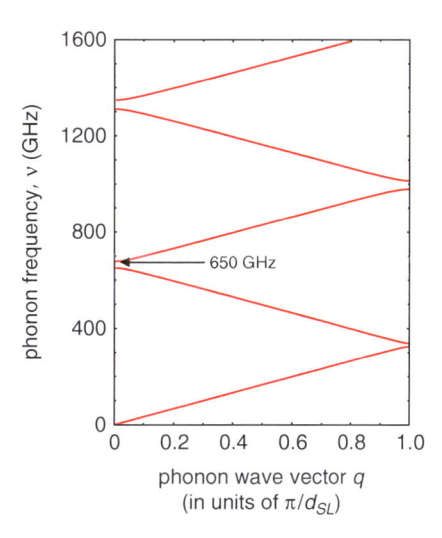

where d_1 and d_2 are the thicknesses of each of the layers making up a period of the SL ($d = d_1 + d_2$). The width of the stop gaps is given by

$$\Delta v_n = \frac{2v_n}{n\pi} \sqrt{\frac{(Z_1 - Z_2)^2}{Z_1 Z_2}} \left| \sin\left(n\pi \frac{d_1/c_{s1}}{d_1/c_{s1} + d_2/c_{s2}} \right) \right|. \quad (8.20)$$

In the stop gaps, the transmittance of the infinite SL is zero, and $R = 1$.

The acoustic properties of, more realistic, finite SLs may be calculated numerically using the transfer matrix method [31]. Boundary conditions applicable to sound propagation across the interface between two solids are that the lattice displacement U and the stress S are continuous across the boundary. Considering an acoustic wave of angular frequency ω, travelling in layer j, of thickness d_j, in the z-direction, towards the interface between layers j and $j+1$ as shown in Fig. 8.12

$$U_j(z) = a_j{}^f e^{iq_j z} + a_j{}^r e^{-iq_j z}, \quad (8.21)$$

and

$$S_j(z) = i\omega Z_j \left(a_j{}^f e^{iq_j z} - a_j{}^r e^{-iq_j z} \right). \quad (8.22)$$

Here $a_j{}^f$ and $a_j{}^r$ are, respectively, the amplitudes of the forward and reverse waves and Z_j is acoustic impedance of layer j. These equations can be written in a, more convenient, matrix form

$$\mathbf{W}_j(z) = \underline{\mathbf{h}}_j(z) \mathbf{A}_j, \quad (8.23)$$

Fig. 8.12 Acoustic wave normally incident on the boundary between the jth and $(j+1)$th layers of an SL. The wave is partly reflected and partly transmitted at the interface

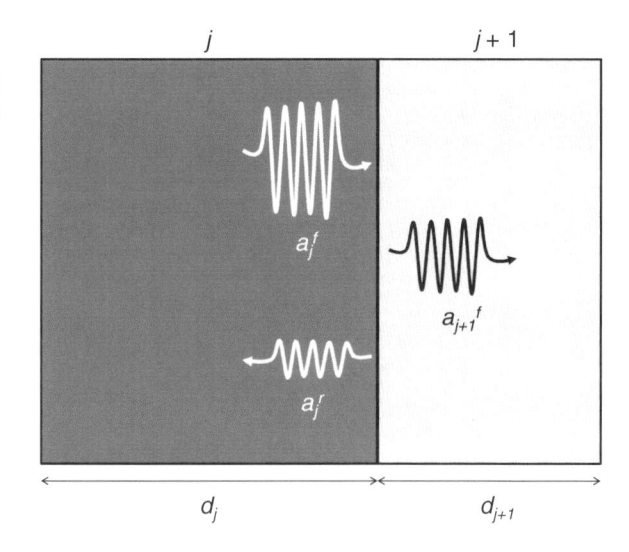

where

$$\mathbf{W}_j(z) = \begin{pmatrix} U_j(z) \\ S_j(z) \end{pmatrix} ; \quad \mathbf{A}_j = \begin{pmatrix} a_j^{\text{f}} \\ a_j^{\text{r}} \end{pmatrix} \qquad (8.24)$$

and

$$\underline{\mathbf{h}}_j(z) = \begin{pmatrix} e^{iq_j z} & e^{-iq_j z} \\ i\omega Z_j e^{iq_j z} & -i\omega Z_j e^{-iq_j z} \end{pmatrix}. \qquad (8.25)$$

The boundary conditions applied at the interface between layer $j-1$ and layer j require

$$\mathbf{W}_{j-1}(z) = \mathbf{W}_j(z), \qquad (8.26)$$

and, at the interface between layers j and $j+1$

$$\mathbf{W}_j(z+d_j) = \mathbf{W}_{j+1}(z+d_j). \qquad (8.27)$$

Therefore, the change in $\mathbf{W}$ after passing through layer j is given by

$$\mathbf{W}_{j+1}(z+d_j) = \underline{\mathbf{t}}_j(z+d_j)\mathbf{W}_{j-1}(z), \qquad (8.28)$$

where

$$\underline{\mathbf{t}}_j(z+d_j) = \underline{\mathbf{h}}_j(z+d_j)\left[\underline{\mathbf{h}}_j(z)\right]^{-1}. \qquad (8.29)$$

Setting $z = 0$ at the $j - 1/j$ interface, we obtain

$$\underline{t}_j(d_j) = \begin{pmatrix} \cos(q_j d_j) & \frac{1}{\omega Z_j}\sin(q_j d_j) \\ -\omega Z_j \sin(q_j d_j) & \cos(q_j d_j) \end{pmatrix}. \tag{8.30}$$

A similar 2×2 matrix, $\underline{t}_{j+1}(d_{j+1})$, can be defined for propagation through the $(j+1)$th layer. These two matrices can be used to give the change in **W** after propagation through a single period of the SL

$$\mathbf{W}_{j+2}(d) = \underline{T}\mathbf{W}_j, \tag{8.31}$$

where $\underline{T} = \underline{t}_{j+1}\underline{t}_j$ and $d = d_j + d_{j+1}$. Finally, after passing through p equal periods of the SL,

$$\mathbf{W}_p(pd) = \left(\underline{T}\right)^p \mathbf{W}(0). \tag{8.32}$$

Figure 8.13a shows the reflectance of a 40-period SL with each period consisting of 6.2 nm of GaAs and 1.1 nm of AlAs, calculated using the transfer matrix method. For an SL with 40 periods, the reflectance at the first mini-zone edge and centre stop gaps is very close to unity, and for most purposes the infinite SL approximation may be used for the position and width of the stop gaps. The reflectance at these two gaps as a function of the number of periods for an SL with the same layer thicknesses is shown in Fig. 8.13b. In this case, for the first mini-zone centre gap, the reflectance only deviates significantly from unity when the number of periods is reduced below 20.

Experimental observation of the zone-folded LA phonon modes in GaAs/AlAs SLs has been made in Raman scattering measurements [32, 33]. In the back scattering geometry, pairs of modes are observed at frequencies corresponding to $q = 2k_{\text{laser}}$, where k_{laser} is the magnitude of the laser photon wave vector. Resonant enhancement of the Raman cross section was observed when the photon energy matched the E1-HH1 transition in the SL [32].

The phononic properties of SLs have been exploited in phonon optics elements. For example, band-stop phonon filters [34–36], phonon mirrors and phonon cavities [12, 13, 37, 38]. In the experiments described by Narayanamurti et al. [34], superconducting tunnel junctions were used to generate quasi-monochromatic phonons and also to detect them after propagation through the SL. Dips in the phonon transmission at frequencies corresponding to the SL stop bands were observed. However, due to the spectral characteristics of the tunnel junction emitters and detectors, the dips were weak and superimposed on a large background. In the work described in [35], quasi-monochromatic phonons generated by femtosecond laser excitation of an SL were propagated via a second, filter, SL to a superconducting bolometer detector. Strong attenuation of the phonon pulses was observed when the

Fig. 8.13 (a) Reflectance as a function of phonon frequency for a 40-period GaAs/AlAs SL, with each period consisting of 6.2 nm of GaAs and 1.1 nm of AlAs; (b) Reflectance of the SL in (a) as a function of the number of periods for the lowest two modes

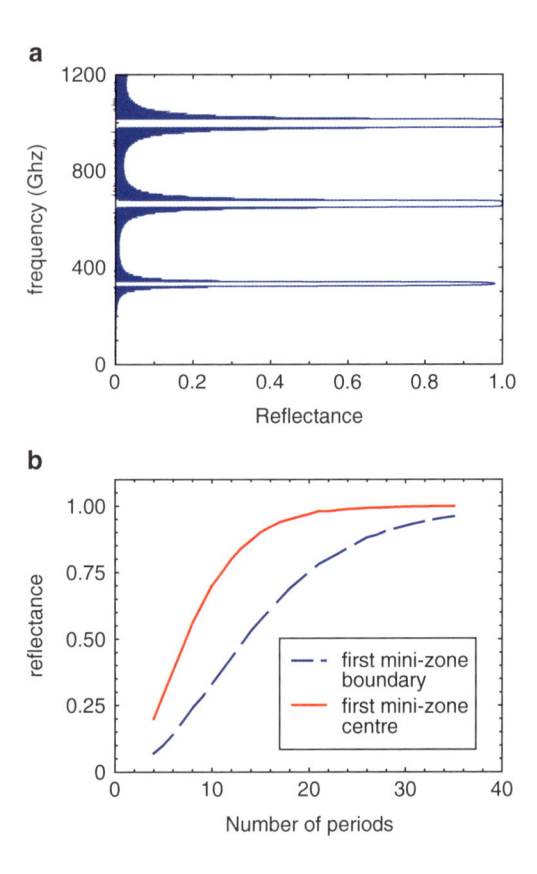

stop band of the filter SL matched the frequency of the phonon generated. Phonon propagation through superlattice structures has also been studied using phonon imaging [39]. The experimental measurements were found to be consistent with Rytov's analytical model and/or the transfer matrix calculations.

To summarize, using the transfer matrix method it is possible to calculate the parameters of SL acoustic Bragg mirrors, thicknesses of layers and number of periods, to give the desired value of reflectance at the intended frequency of operation a saser device.

8.5 Electrically Pumped SL Saser Device

The essential elements of a sub-THz saser: a gain medium based on a weakly coupled and lightly doped SL, and SL acoustic Bragg mirrors to form an acoustic cavity, are in place. It now remains to assemble these together to make a prototype saser device. Since both key elements are based on GaAs/AlAs SLs it is straightforward

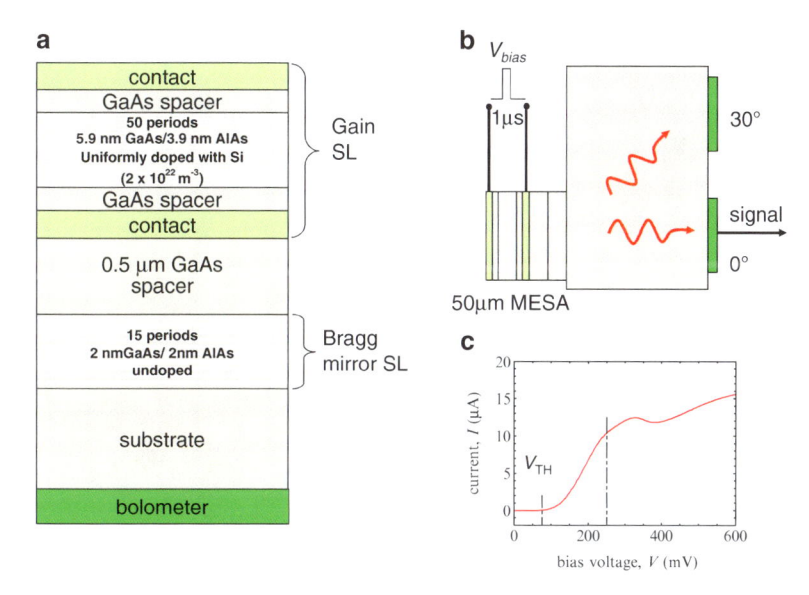

Fig. 8.14 (a) Structure of the conventional cavity saser device: the gain SL is in a cavity defined by the output coupler Bragg mirror and the top surface of the MESA; (b) The experimental arrangement: the structure was fabricated into a 50 μm-diameter mesa and electrically pumped with 1 μs-duration pulses. Two superconducting bolometers (phonon detectors) were fabricated on the opposite side of the substrate, one directly opposite to the device and one subtending an angle of 30°; (c) The current–voltage characteristics of the device: the operating point is in the region between the two *dashed lines*, the lower of which corresponds to the device threshold

to fabricate the complete structure by epitaxial growth using, e.g. MBE [40]. We consider two types of device working at different frequencies: one is a conventional cavity device consisting of the gain SL sandwiched between two acoustic mirrors; the other is a distributed feedback (DFB) device using the same SL to provide both the gain and the phonon confinement.

8.5.1 Conventional Cavity Device

Figure 8.14a shows the structure of a device consisting of the amplifying SL, the parameters of which were as described in Sect. 8.2, above a single SL Bragg mirror structure grown on a 380 μm-thick semi-insulating GaAs substrate. The Bragg mirror contained 15 periods, each consisting of 2 nm of GaAs and 2 nm of AlAs. Calculations using the transfer matrix method give $R = 0.97$ at a phonon energy of 2.6 meV (630 GHz). In this device the high reflector is the free surface of the sample. Ideally, such a vacuum interface would have a reflectance of 100 % independent of phonon frequency. However, in practice, there is phonon scattering at the surface and the specular reflectance is somewhat less than 100 %. Experimental measurements

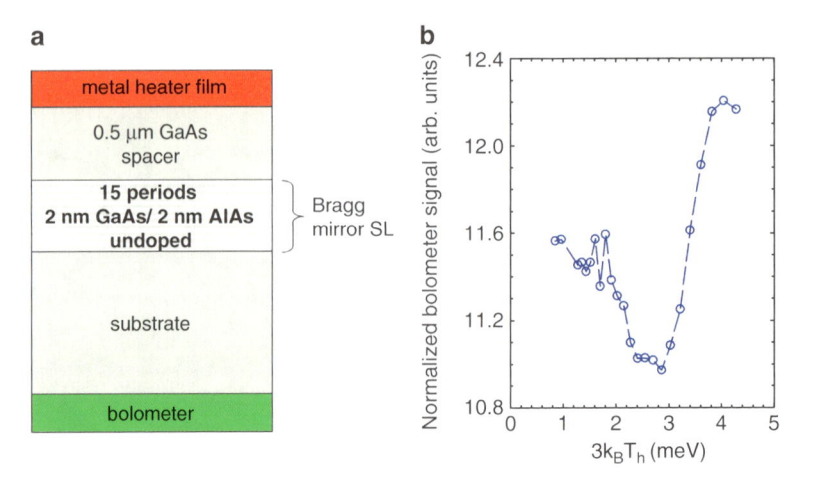

Fig. 8.15 (a) Arrangement for testing the SL Bragg mirror using the ballistic heat pulse technique; (b) Heat pulse transmission normalized to the total heater power as a function of the energy of the peak of the Planck phonon spectrum. A broad dip in the transmission centred on about 2.7 meV is observed, which corresponds to the design frequency of the mirror

of the fraction of sub-THz phonons that are specularly reflected at a surface typically give values in the region of 80 % [41, 42], which gives $R_1R_2 = 0.78$ (>0.64) so the conditions for oscillation should be achieved in the device.

First, the mirror SL was characterized by means of a heat pulse [43] measurement. The amplifying SL was removed by etching and a metal film resistive heater fabricated in its place. Opposite to the heater and on the other side of the substrate a superconducting aluminium bolometer was fabricated for phonon detection. The sample was placed in a helium cryostat and cooled to a temperature T_0 of about 2 K, which is on the superconducting transition edge of the bolometer. Under these conditions, and with a small constant current bias applied to the bolometer, incident phonons cause a change in bolometer resistance and a corresponding voltage transient which can be amplified and detected using a digital averaging oscilloscope. The heater was excited by a ~10 ns-duration electrical pulse and generates nonequilibrium phonons with an approximately Planckian spectrum which propagate ballistically through the substrate. The peak in the phonon spectral energy density occurs at $\nu_m = 3k_B T_h / h$, where T_h is the temperature of the heater. The heater temperature is proportional to the electrical power P_h dissipated in the heater and given by $T_h = \sqrt[4]{\frac{P_h}{\varsigma} - T_0^4}$, where the constant ς is determined from acoustic mismatch theory [44]. In order to reach the bolometer, the nonequilibrium phonons emitted by the heater had to pass through the mirror SL. Thus we could study the transmission of the mirror SL as a function of phonon frequency by measuring the bolometer signal as a function of the power applied to the heater. Figure 8.15 shows the result of this measurement. The bolometer signal has been normalized to the power supplied to the heater. A dip in the transmission is seen,

8 Semiconductor Superlattice Sasers at Terahertz Frequencies

with a minimum at $h\nu_m = 2.6$ meV ($\nu_m \approx 630$ GHz), which is due to the reflection of phonons by the mirror. The dip is not very deep and is wide due to the broad spectrum of the nonequilibrium phonons, but it does confirm that the mirror is working at its design frequency.

The sample arrangement for measuring the phonon emission by the saser structure is shown in Fig. 8.14b: the structure was processed into an active device of diameter 50 μm using standard photolithography and wet etching methods, and electrical contacts were formed at the ends of the amplifying SL. On the surface of the substrate opposite to the device, superconducting aluminium bolometers were fabricated for detection of the emitted phonons. One bolometer was placed directly opposite to the device (0°) and one subtending an angle of about 30° to the SL axis. The DC current–voltage, I–V, characteristics of the amplifying SL are shown in Fig. 8.14c. It is seen that above the turn-on voltage, $V_T \approx 75$ mV, which is the voltage required to align the Fermi energy in the emitter contact with the energy level in the first quantum well, the current increases monotonically. At higher bias ($V > 250$ mV), the current increases more slowly and regions of negative differential resistance are seen. This behaviour is attributed to the formation of space charge domains in the structure. At applied biases between the turn-on and the onset of domain formation, the potential drop across the active SL is assumed to be uniform, and the Stark splitting is given by $\Delta \approx \frac{e(V-V_T)}{50}$. The amplifying SL was excited by 1 μs duration voltage pulses, short enough to enable the bolometer measurement to be carried out, but long enough for the device to reach steady-state conditions, which takes a few nanoseconds. The emitted phonons propagated ballistically across the substrate at the speed of sound and were incident on the bolometer.

Figure 8.16a shows the signals on the two bolometers as a function of the magnitude of the Stark splitting. Because the total power dissipated in the device increases strongly with the amplitude of the applied pulse, the signals have been normalized to the total power dissipated in the device as determined from the I–V measurements. The important result is the peak in the normalized signal for the 0° bolometer occurring at a Stark splitting of about 2.6 meV. This broad peak corresponds to the spectral gain profile of the amplifying SL exceeding the threshold gain condition in the range of Stark splitting between 2 and 3 meV, as shown in Fig. 8.16b. No such peak is observed on the 30° bolometer, which is consistent with the prediction of theory that the increment is only observed for phonons propagating close to the SL growth axis.

8.5.2 DFB Device

As well as providing acoustic gain, the amplifying SL also acts to confine phonons which have frequencies corresponding to its own acoustic stop gaps. Thus the amplifying SL can be its own DFB cavity. Such an arrangement is analogous to a DFB laser. Figure 8.17 shows the folded acoustic dispersion of the amplifying SL, which consisted of 50 periods of 6.6 nm GaAs and 4.4 nm of AlAs, uniformly

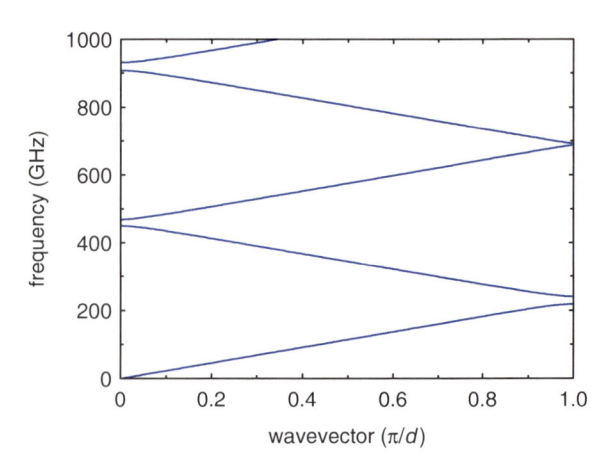

Fig. 8.16 (**a**) Signals at the bolometer directly opposite to the saser device (*squares*), and at the 30° bolometer (*circles*). Both have been normalized to the total power dissipated in the device. A peak in the 0° bolometer signal occurs when the gain spectrum of the active SL is tuned to resonance with the cavity mode at 2.6 meV; (**b**) Gain coefficient of SL as a function of the bias, determined from the result shown in Fig. 8.9, the *horizontal line* shows the threshold gain condition calculated for the cavity

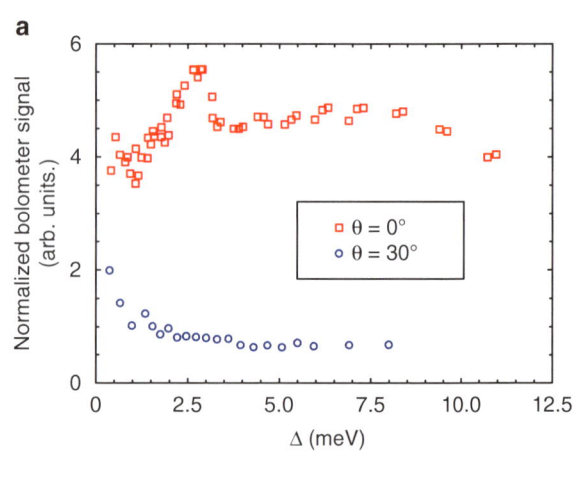

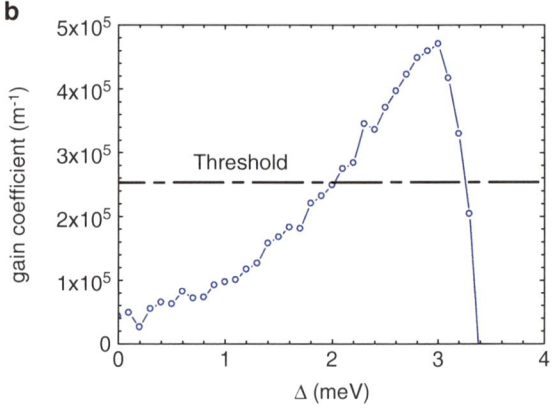

Fig. 8.17 Folded longitudinal phonon dispersion of the active SL. The first mini-Brillouin zone centre mode is at about 440 GHz (1.9 meV)

Fig. 8.18 Phonon signal at bolometer directly opposite to the device MESA normalized to the total power input (*points*), showing a clear peak when the gain spectrum is tuned to the 1.9 meV mode of the distributed feedback cavity. Also shown (*line*) is the differential conductance of the device

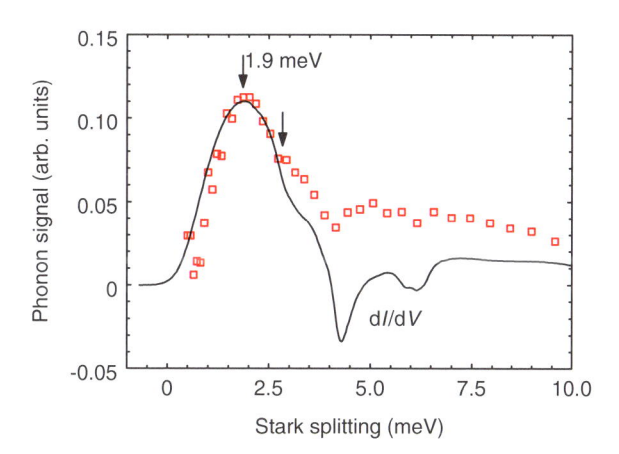

doped with Si to a density of $\sim 10^{21}$ m^{-3}. The first mini-zone centre stop gap occurs at a frequency of about 450 GHz (phonon energy ≈ 1.9 meV). At this frequency the predicted gain of the SL is about 4×10^5 m^{-1}, and Eq. (8.16) gives that saser oscillation is theoretically possible if the DFB cavity has a Q-factor of at least 230. This corresponds to a lifetime of 450 GHz phonons in the SL of about 0.5 ns. High resolution measurements of the phonon modes of the SL (see the next section for details) show that this condition can be met in practice.

The structure of the DFB saser is the same as the conventional cavity device described in the previous section, except without the Bragg mirror SL. For the emission measurements the same arrangement as shown in Fig. 8.14b was used. The signal on the 0° bolometer is shown in Fig. 8.18. There is a clear peak seen at a Stark splitting of about 1.9 meV, corresponding to the gain spectral profile overlapping the high-Q cavity mode. Just as in the case of the conventional cavity device, this "resonance" provides evidence that saser oscillation is achieved in the device [19]. Also shown in Fig. 8.18 is the differential conductance of the device, which also peaks at a Stark splitting of 1.9 meV. This shows that there is an increase of current in the device as a result of the phonon-stimulated inter-well transitions. Using the conductance data, the acoustic emission was estimated to be in the class mW/cm^{-2}. However, bolometers are sensitive to the total phonon flux and give no information about the spectral characteristics of the emitted phonons which, as a result of the stimulated emission process, should exhibit a higher degree of coherence and a reduced spectral linewidth compared to spontaneously emitted phonons.

8.5.3 Injection Seeding and Line Narrowing

To obtain information about the spectral characteristics of the acoustic oscillations in the DFB saser cavity [20] the ultrafast optical pump-probe technique of coherent phonon measurement was employed. This technique had been extensively used to study coherent acoustic phonons in superlattices and other nanostructures, see for example [45–53], and involves generating zone-folded modes by absorbing in the SL a femtosecond duration optical pulse. The acoustic vibrations in the SL give rise to a modulation of the optical properties of the SL which may be strong near to an optical, e.g. exciton, resonance. Therefore, by measuring the time dependence of the reflectance for a femtosecond probe pulse, which is delayed in time with respect to the pump, the acoustic oscillations of the SL may be detected. The basic idea here is to use a femtosecond optical pulse to injection seed the DFB cavity. The intense acoustic pulse generated in the SL by this method would have the effect of phase locking the saser device and then it is possible to observe the oscillations in the cavity using the optical probe pulse.

The experimental arrangement used is fairly typical and is shown in Fig. 8.19: the pump and probe pulses are derived from the output of a femtosecond oscillator ($\lambda \approx 770$ nm; pulse width $= 100$ fs; repetition rate $= 82$ MHz and pulse energy ≈ 12 nJ/pulse). The laser pulses are split with about 90 % of the energy to the pump beam and 10 % to the probe. The pump is focussed to a spot of about 100 μm diameter on the top of the saser structure, which has a special ring-shaped contact metallization to allow optical access to the SL, Fig. 8.19b. The probe is passed through an optical delay line focussed to a spot of about 50 μm diameter within the pump spot. The reflected probe is detected by a photodiode and lock-in amplifier referenced to either the modulation of the pump or the bias applied to the device. By adjusting the length of the optical delay line the dependence on time, t, of the reflected probe intensity can be measured with femtosecond resolution.

Figure 8.20 shows the reflectance pump-probe signal from the DFB device with zero applied bias. Zero time in this trace corresponds to the pump and probe being coincident in time. The large step at $t = 0$ followed by the slow decay is due to the relaxation of photoexcited carriers. Superimposed on this background signal are oscillations having frequencies of about 40 and 440 GHz. The former are Brillouin backscattering oscillations and the latter are due to the SL mode near the first mini-Brillouin zone stop gap. We focus attention on the higher frequency mode: Fig. 8.21 shows the decay of the mode following the pump excitation. There are two frequencies beating and their amplitudes are decaying exponentially, possibly due to leakage of the modes from the SL or phonon scattering in the structure. The solid line is a fit of the equation

$$F(t_0) = A_0 \exp(-\gamma_0 t_0) \left[1 + B^2 + 2B \cos(2\pi \Delta \nu \, t_0)\right]^{1/2} \quad (8.33)$$

where $\gamma_0 = (2.1 \pm 0.1)\,\text{ns}^{-1}$, $B = (-0.43 \pm 0.03)$, $|\Delta \nu| = (4.5 \pm 0.03)\,\text{GHz}$ and A_0 is a normalization factor The two frequencies arise from quantization of the cavity vibrational modes due to the finite length of the SL, and the beat frequency is given by $\Delta \nu \approx c_s/2L$, where L is the total length of the SL. The characteristic decay time of the SL acoustic modes is about $\gamma_0^{-1} = 0.5$ ns.

Application of bias to the SL results in an increase of the amplitude of the stronger of the two modes. Figure 8.22 shows the amplitude of the mode at a time

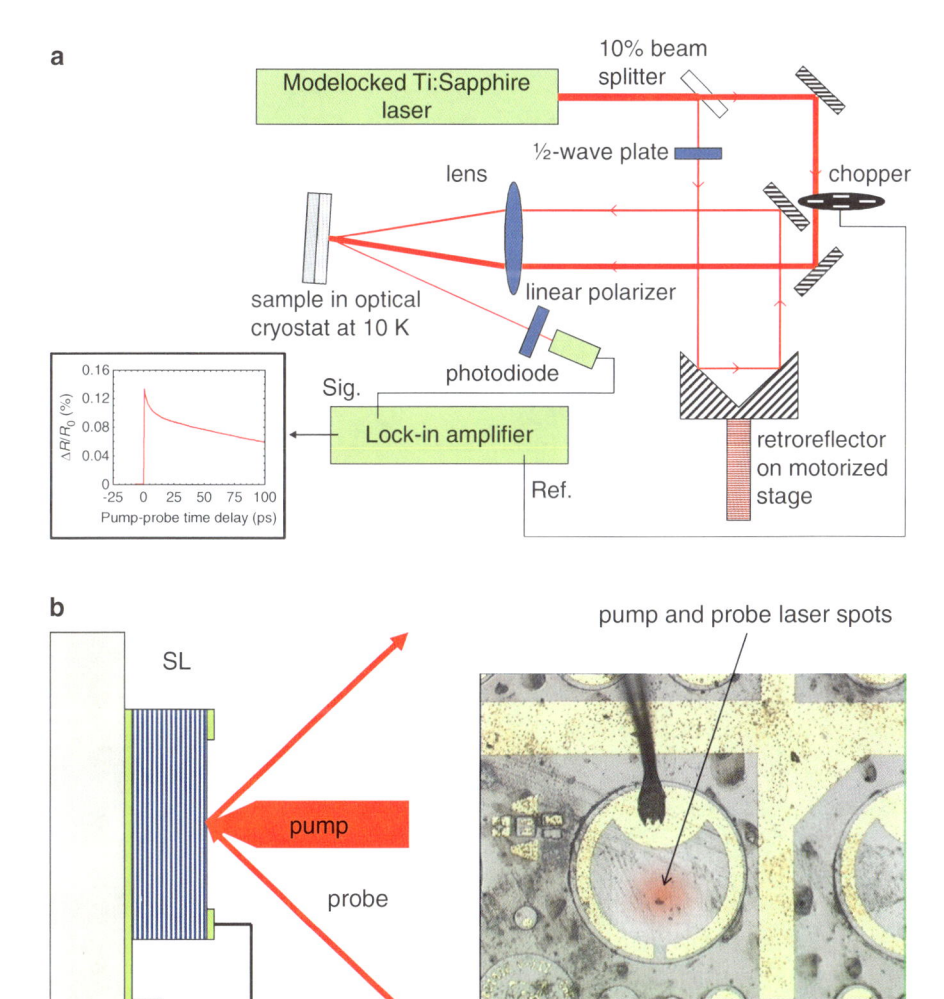

8 Semiconductor Superlattice Sasers at Terahertz Frequencies

Fig. 8.19 (**a**) Experimental arrangement for pump-probe measurement of the acoustic vibrations in the DFB saser cavity (see text for a description); (**b**) Sample arrangement for the pump-probe measurement showing the electrical contact arrangement

delay of one nanosecond as a function of the applied bias voltage. A peak at about 140 mV corresponds to the Stark splitting being coincident with the energy of the 440 GHz phonon mode, the condition for maximum amplification. At this bias, the increase of the signal relative to zero bias is about 40 %, which is consistent with the measurement of the gain using the single-pass amplifier (Sect. 8.2).

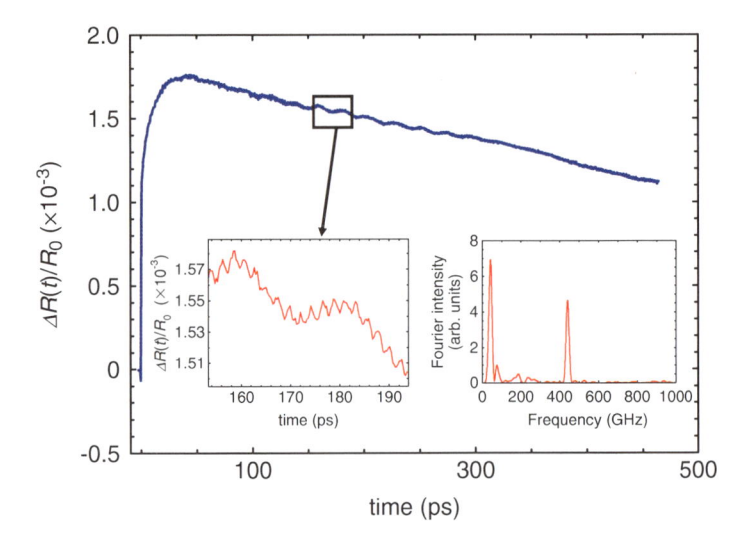

Fig. 8.20 Pump-probe signal, the 440 GHz oscillations and Fourier spectrum are shown in the *insets*

Fig. 8.21 Amplitude of the 440 GHz acoustic oscillations as a function of pump-probe delay time. The oscillations are due to beating of two exponentially decaying modes which are different in frequency by about 5 GHz

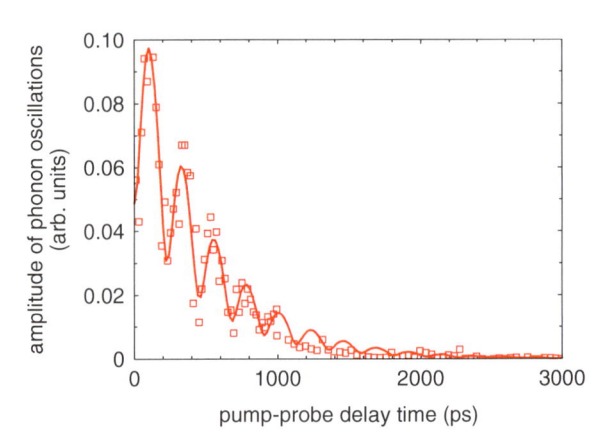

Using the technique of Stark modulation we are able to resolve the bias-dependent part of the signal from any background. The temporal dependence of this signal is shown in Fig. 8.23. The signal builds up to a peak at about 0.4 ns and then decays more slowly than the zero bias signal. The Fourier spectrum of the modes at zero and 140 mV bias is shown in Fig. 8.24. At zero bias the full width at half maximum (FWHM) line width of the Lorentzian peak is about 1.2 GHz. The bias-induced signal has a Gaussian shape with FWHM = 0.7 GHz. Therefore application of bias to the SL results in acoustic spectral line narrowing, which is a signature of amplification by stimulated emission. The Gaussian line shape implied that the spectrum is broadened due to dephasing, which could be due to timing jitter in the measurement system or inhomogeneity of the device structure in the probed area.

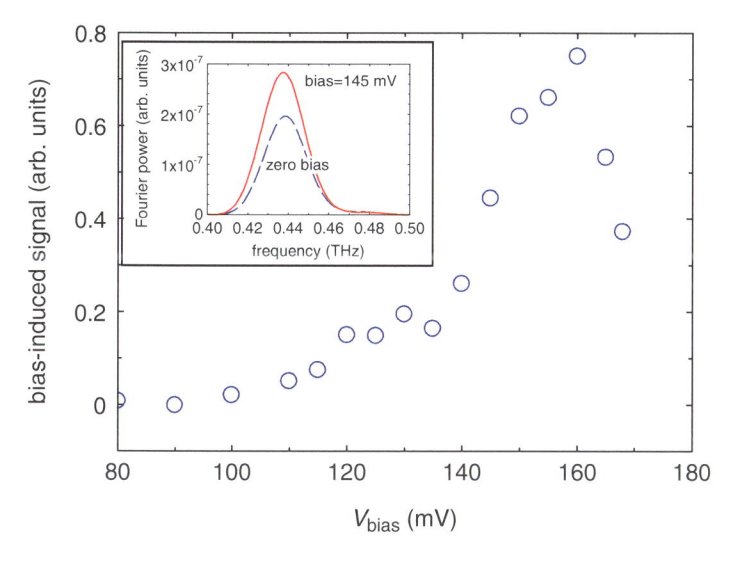

8 Semiconductor Superlattice Sasers at Terahertz Frequencies

Fig. 8.22 Amplitude of the stronger of the two beating modes at a pump-probe time delay of 1 ns as a function of the bias applied to the device. The *inset* shows that at a bias of 145 mV (corresponding to $\Delta \approx 1.9$ meV) the amplitude of the oscillations is increased by about 40 % relative to at zero bias

Fig. 8.23 Amplitude of the bias-induced increase in the acoustic oscillations for a bias of 145 mV as a function of pump-probe delay time

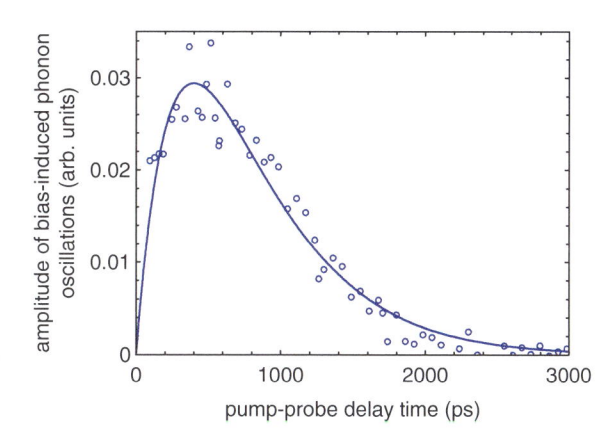

8.6 Summary and Outlook

In this chapter we have discussed the design criteria for the SLs to be used as the gain medium and acoustic mirrors in saser devices. We have also reviewed recent work which has demonstrated that coherent phonon amplification occurs in a weakly coupled superlattice under applied electrical bias which puts it into the Wannier–Stark ladder regime. Experimentally measured values of the acoustic gain were shown to be consistent with theoretical calculations and sufficient to meet the threshold for sasing in a device consisting of an acoustic cavity having a readily achievable Q-factor. The evidence for saser oscillation in such devices was a peak

Fig. 8.24 Spectrum of the acoustic oscillations at zero bias (*dashed line*) and for the bias-induced signal (*solid line*)

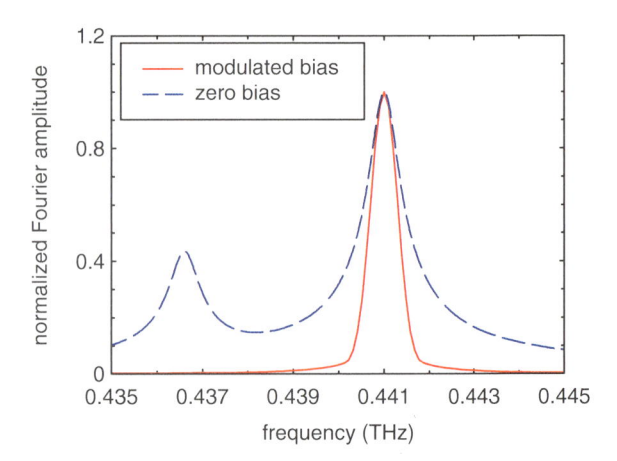

in the acoustic emission normal to the superlattice when the Stark spitting was tuned into resonance with a the high-Q cavity mode, thus achieving the sasing threshold condition.

The advantage of the superlattice device compared with some other methods of achieving sub-THz acoustic amplification, reviewed in Sect. 8.1, is that the gain spectrum is tuneable with the applied bias. However, the superlattice devices have so far only been demonstrated to work at low temperatures (<20 K). Theoretically, operation at up to liquid nitrogen temperatures might be possible, which will enhance the prospects for practical application of the devices.

Such a source of intense coherent THz and sub-THz sound as a saser has numerous potential applications. The obvious application in science and technology is nanometre-resolution acoustic probing and imaging of nanoscale structures and devices. This can be regarded as an extension of established ultrasonic techniques to the nanoscale. Acoustic techniques allow non-destructive probing of structures buried in an optically opaque matrix. A number of other applications for sasers can be envisaged, and some of these are discussed in [14]. However, more recently, there have been developments which point to applications in the area of THz acousto-electronics. One of these is the conversion of sub-THz acoustic impulses to sub-THz electromagnetic waves using piezoelectric materials [54]. Thus the saser could form the basis of a stable, narrow band continuous wave THz electromagnetic source. It has also been shown that sub-THz acoustic impulses can be converted into electrical signals using electronic devices such as *p–n* and Schottky diodes [55, 56], and active devices like heterojunction bipolar transistors and mesfets. This raises the prospect of sub-THz acoustical "clocking" of electronic circuits and the possibility to use sasers as highly stable local oscillators for heterodyne detection of THz signals.

Current work is focussed on improving the efficiency of the superlattice saser through enhancements of the gain and the cavity Q-factor, and to develop the technologies to exploit the THz sound waves generated. It remains to be seen if, during the next 50 years, saser applications become as widespread as have the uses of lasers in the past 50 years since their invention.

References

1. Tucker EB (1961) Amplification of 9.3-kMc/sec ultrasonic pulses by maser action in ruby. Phys Rev Lett 6:547–548
2. Maiman TH (1960) Stimulated optical radiation in ruby. Nature 187:493–494
3. Tucker EB (1961) Attenuation of longitudinal ultrasonic vibrations by spin-phonon coupling in ruby. Phys Rev Lett 6:183–185
4. Hu P (1980) Stimulated emission of 29-cm^{-1} phonons in ruby. Phys Rev Lett 44:417–420
5. Overwijk MHF, Dijkhuis JI, de Wijn HW (1990) Superfluorescence and amplified spontaneous emission of 29-cm^{-1} phonons in ruby. Phys Rev Lett 65:2015–2018
6. Tilstra LG, Arts AFM, de Wijn HW (2003) Coherence of phonon avalanches in ruby. Phys Rev B 68:144302
7. Bron WE, Grill W (1978) Stimulated phonon emission. Phys Rev Lett 40:1459–1463
8. Pieur J-Y, Devaud M, Joffrin J, Barre C, Stenger M, Chapellier M (1996) Sound amplification by stimulated emission of phonons using two-level systems in glasses. Phys B Condens Matter 219–220:235–238
9. Phillips WA (1987) Two-level states in glasses. Rep Prog Phys 50:1657
10. Huston AR, McFee JH, White DL (1961) Ultrasonic amplification in CdS. Phys Rev Lett 6:547–548
11. Srivastava GP (1990) The physics of phonons. A. Hilger, Bristol
12. Trigo M, Bruchhausen A, Fainstein A, Jusserand B, Thierry-Mieg V (2002) Confinement of acoustical vibrations in a semiconductor planar phonon cavity. Phys Rev Lett 89:227402
13. Huynh A, Lanzillotti-Kimura ND, Jusserand B, Perrin B, Fainstein A, Pascual-Winter MF, Peronne E, Lemaitre A (2006) Subterahertz phonon dynamics in acoustic nanocavities. Phys Rev Lett 97:115502
14. Komirenko SM, Kim KW, Demidenko AA, Kochelap VA, Stroscio MA (2000) Generation and amplification of sub-THz coherent acoustic phonons under the drift of two-dimensional electrons. Phys Rev B 62:7459
15. Makler SS, Vasilevskiy MI, Anda EV, Tuyarot DE, Weberszpil J, Pastawski HM (1998) A source of terahertz coherent phonons. J Phys Condens Matter 10:5905
16. Glavin BA, Kochelap VA, Linnik TL, Kim KW, Stroscio MA (2002) Generation of high-frequency coherent acoustic phonons in superlattices under hopping transport. I. Linear theory of phonon instability. Phys Rev B 65:085303
17. Glavin BA, Kochelap VA, Linnik TL, Kim KW, Stroscio M (2002) A generation of high-frequency coherent acoustic phonons in superlattices under hopping transport. II. Steady-state phonon population and electric current in generation regime. Phys Rev B 65:085304
18. Glavin BA, Kochelap VA, Linnik TL (1999) Generation of high-frequency coherent acoustic phonons in a weakly coupled superlattice. Appl Phys Lett 74:3525–3527
19. Kent AJ, Kini RN, Stanton NM, Henini M, Glavin BA, Kochelap VA, Linnik TL (2006) Acoustic phonon emission from a weakly coupled superlattice under vertical electron transport: observation of phonon resonance. Phys Rev Lett 96:215504
20. Beardsley RP, Akimov AV, Henini M, Kent AJ (2010) Coherent terahertz sound amplification and spectral line narrowing in a Stark ladder superlattice. Phys Rev Lett 104:85501
21. Walker PM, Kent AJ, Henini M, Glavin BA, Kochelap VA, Linnik TL (2009) Terahertz acoustic oscillations by stimulated phonon emission in an optically pumped superlattice. Phys Rev B 79:245313
22. Grahn HT (1994) Semiconductor superlattices: growth and electronic properties. World Scientific, Singapore
23. Kent AJ, Wigmore JK (2003) Energy relaxation by hot two-dimensional carriers in zero magnetic field. In: Challis L (ed) Electron–phonon interactions in low-dimensional structures. Oxford University Press, Oxford, pp 5–59
24. Tsu R, Döhler G (1975) Hopping conduction in a "superlattice". Phys Rev B 12:680–686

25. Cavill SA, Challis LJ, Kent AJ, Ouali FF, Akimov AV, Henini M (2002) Acoustic phonon-assisted tunnelling in GaAs/AlAs superlattices. Phys Rev B 66:235320
26. Kini RN, Kent AJ, Stanton NM, Henini M (2005) Angle dependence of phonon-assisted tunnelling in a weekly coupled superlattice: evidence for terahertz phonon amplification. J Appl Phys 98:033514
27. Beardsley RP, Campion RP, Glavin BA, Kent AJ (2011) A GaAs/AlAs superlattice as an electrically pumped THz acoustic phonon amplifier. New J Phys 13:073007
28. Hawker P, Kent AJ, Challis LJ, Bartels A, Dekorsy T, Kurz H, Kohler K (2000) Observation of coherent zone-folded acoustic phonons generated by Raman scattering in a superlattice. Appl Phys Lett 77:3209
29. Kent AJ, Stanton NM, Challis LJ, Henini M (2002) Generation and propagation of monochromatic acoustic phonons in gallium arsenide. Appl Phys Lett 81:3497
30. Rytov SM (1956) Acoustical properties of a thinly laminated medium. Sov Phys Acoust 2:68–80
31. Tamura S, Hurley DC, Wolfe JP (1988) Acoustic-phonon propagation in superlattices. Phys Rev B 38:1427–1449
32. Colvard C, Merlin R, Klein MV, Gossard AC (1980) Observation of folded acoustic phonons in a semiconductor superlattice. Phys Rev Lett 45:298
33. Popovic ZV, Spitzer J, Ruf T, Cardona M, Nötzel R, Ploog K (1993) Folded acoustic phonons in GaAs/AlAs corrugated superlattices grown along the [3 1 1] direction. Phys Rev B 48:1659
34. Narayanamurti V, Störmer HL, Chin MA, Gossard AC, Wiegmann W (1979) Selective transmission of high-frequency phonons by a superlattice: the "dielectric" phonon filter. Phys Rev Lett 43:2012–2016
35. Stanton NM, Kini RN, Kent AJ, Henini M (2003) Terahertz phonon optics in GaAs/AlAs superlattice structures. Phys Rev B 68:113302
36. Lanzillotti-Kimura ND, Fainstein A, Lemaître A, Jusserand B (2006) Nanowave devices for terahertz acoustic phonons. Appl Phys Lett 88:083113
37. Lanzillotti-Kimura ND, Perrin B, Fainstein A, Jusserand B, Lemaître A (2010) Nanophononic thin-film filters and mirrors studied by picosecond ultrasonics. Appl Phys Lett 96:053101
38. Rozas G, Pascual Winter MF, Jusserand B, Fainstein A, Perrin B, Semenova E, Lemaître A (2009) Lifetime of THz acoustic nanocavity modes. Phys Rev Lett 102:015502
39. Hurley DC, Tamura S, Wolfe JP, Morkoç H (1987) Imaging of acoustic phonon stop bands in superlattices. Phys Rev Lett 58:2446
40. Vvedensky DD (2001) Epitaxial growth of semiconductors. In: Barnham K, Vvedensky D (eds) Low dimensional semiconductor structures: fundamentals and device applications. Cambridge University Press, Cambridge, pp 1–55
41. Wybourne MN, Wigmore JK (1988) Phonon spectroscopy. Rep Prog Phys 51:923–987
42. Northrop GA, Wolfe JP (1984) Phonon reflection imaging: a determination of specular versus diffuse boundary scattering. Phys Rev Lett 52:2156–2159
43. Wolfe JP (1998) Imaging phonons: acoustic wave propagation in solids. Cambridge University Press, Cambridge
44. Rösch F, Weis O (1977) Phonon transmission from incoherent radiators into quartz, sapphire, diamond, silicon and germanium within anisotropic continuum acoustics. Z Phys B 27:33
45. Yamamoto A, Mishina T, Masumoto Y, Nakayama M (1994) Coherent oscillation of zone-folded phonon modes in GaAs–AlAs superlattices. Phys Rev Lett 73:740
46. Bartels A, Dekorsy T, Kurz H, Köhler K (1999) Coherent zone-folded longitudinal acoustic phonons in semiconductor superlattices: excitation and detection. Phys Rev Lett 82:1044
47. Sun C-K, Liang J-C, Yu X-Y (2000) Coherent acoustic phonon oscillations in semiconductor multiple quantum wells with piezoelectric fields. Phys Rev Lett 84:179
48. Matsuda O, Wright OB, Hurley DH, Gusev VE, Shimizu K (2004) Coherent shear phonon generation and detection with ultrashort optical pulses. Phys Rev Lett 93:095501
49. Trigo M, Eckhause TA, Reason M, Goldman RS, Merlin R (2006) Observation of surface-avoiding waves: a new class of extended states in periodic media. Phys Rev Lett 97:124301

50. Devos A, Poinsotte F, Groenen J, Dehaese O, Bertru N, Ponchet A (2007) Strong generation of coherent acoustic phonons in semiconductor quantum dots. Phys Rev Lett 98:207402
51. Moss DM, Akimov AV, Kent AJ, Glavin BA, Kappers MJ, Hollander JL, Moram MA, Humphreys CJ (2009) Coherent terahertz acoustic vibrations in polar and semipolar gallium nitride-based superlattices. Appl Phys Lett 94:011909
52. Lanzillotti-Kimura ND, Fainstein A, Perrin B, Juserrand B, Mauguin O, Largeau L, A L (2010) Bloch oscillations of THz acoustic phonons in coupled nanocavity structures. Phys Rev Lett 104:197402
53. Bruchhausen A, Gebs R, Hudert F, Issenmann D, Klatt G, Bartels A, Schecker O, Waitz R, Erbe A, Scheer E, Huntzinger J-R, Mlayah A, Dekorsy T (2011) Subharmonic resonant optical excitation of confined acoustic modes in a free-standing semiconductor membrane at GHz frequencies with a high-repetition-rate femtosecond laser. Phys Rev Lett 106:077401
54. Armstrong MR, Reed EJ, Kim K-Y, Glownia JH, Howard WM, Piner EL, Roberts JC (2009) Observation of terahertz radiation coherently generated by acoustic waves. Nat Phys 5:285
55. Moss DM, Akimov AV, Campion RP, Kent AJ (2011) Ultrafast strain-induced electronic transport in a GaAs p–n junction diode. Chin J Phys 49:499
56. Moss DM, Akimov AV, Glavin BA, Henini M, Kent AJ (2011) Ultrafast strain-induced current in a GaAs Schottky diode. Phys Rev Lett 106:066602

Chapter 9
Acoustic Carrier Transport in GaAs Nanowires

Snežana Lazić, Rudolf Hey, and Paulo V. Santos

Abstract Present semiconductor technologies allow the growth of different types of nanostructures, such as quantum wells, wires, and dots on the surface of a single semiconductor crystal. The piezoelectric field of surface acoustic waves (SAWs) propagating on the crystal surface provides an efficient mechanism for the controlled exchange of electrons and holes between these nanostructures. In this review, we explore this ability of dynamic SAW fields to demonstrate acoustically driven single-photon sources using coupled quantum wells and dots based on (Al,Ga)As (311)A material system. We address the growth of the coupled nanostructures by molecular beam epitaxy, the dynamics of the acoustic carrier transfer between them, as well as the acoustic control of recombination in quantum dots. The latter provides the basis for the operation of the acoustically driven single-photon sources, which are characterized by a low jitter and repetition frequency close to 1 GHz.

9.1 Introduction

Surface acoustic waves [8, 48, 52] are long-wavelength elastic vibrations confined to the surface of a material. The presence of the surface allows for elastic modes consisting of admixture of bulk modes with lower phase velocity propagating close to the surface. The surface region acts, therefore, as an acoustic waveguide, which prevents the spreading of the SAW fields into the bulk. The propagation velocity

---

S. Lazić (✉)
Paul-Drude-Institut für Festkörperelektronik, Hausvogteiplatz 5–7, 10117 Berlin, Germany

Pres. Address: Departamento de Física de Materiales, Universidad Autónoma de Madrid, 28049 Madrid, Spain
e-mail: lazic.snezana@uam.es

R. Hey • P.V. Santos
Paul-Drude-Institut für Festkörperelektronik, Hausvogteiplatz 5–7, 10117 Berlin, Germany
e-mail: hey@pdi-berlin.de; santos@pdi-berlin.de

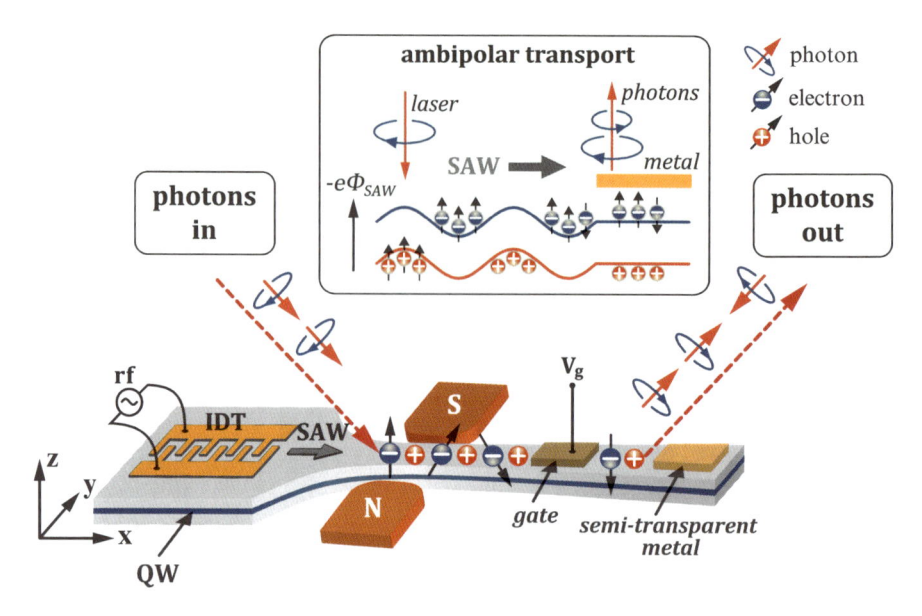

Fig. 9.1 Processing of quantum information in the form of polarized photons and electronic wave packets using the moving piezoelectric potential Φ_{SAW} created by a surface acoustic wave (SAW). The *inset* shows the modulation of the conduction and valence band induced by Φ_{SAW}, which captures and transports the photoexcited carriers. IDT denotes an interdigital transducer for SAW excitation. V_g is a control-gate voltage on the carrier transport path [55]

v_{SAW} of SAW modes lies normally below the ones for the bulk longitudinal and transverse modes. In the case of GaAs, $v_{SAW} \approx 3,000$ m/s= 3 μm/ns. The SAW acoustic field is typically confined within a depth corresponding to one acoustic wavelength λ_{SAW}. For modes with frequency close to 1 GHz, the wavelength $\lambda_{SAW} \approx$3 μm is comparable to the typical depth ranges and dimensions of planar electronic devices.

SAWs have become particularly important in piezoelectric materials, where they can be electrically generated via the inverse piezoelectric effect using interdigitated transducers (IDTs) deposited on the sample surface [65, 66]. A typical configuration for IDTs is illustrated in Fig. 9.1. They consist of a metal finger grating with a periodicity equal to λ_{SAW}. The application of a radio-frequency (rf) voltage with frequency $f_{SAW} = v_{SAW}/\lambda_{SAW}$ launches a SAW, which propagates to the region outside the transducer. The IDT itself is an acoustic resonator, whose frequency response can be controlled by changing the dimensions of the finger electrodes. The dimensions of a resonator are normally comparable to the wavelength of the excitation. Since the acoustic velocity is over 4 orders of magnitudes lower than the electromagnetic one, the size of an acoustic resonator for a given resonance frequency is several orders of magnitude smaller than the one for an electromagnetic resonator. This feature has been exploited to realize compact electronic filters with well-defined frequency characteristics, which constitute a main application area for SAW devices.

The electric-to-acoustic energy conversion in an IDT is reversible: by placing a second IDT in front of the first one, it is thus possible to retrieve electrical energy from the propagating SAW. Due to the low acoustic velocity (compared to the electromagnetic one), this configuration forms a delay line for electric signals. Furthermore, the propagation velocity can be modified, for instance, by exposing the surface to adsorbates. This property has been exploited for the realization of sensitive acoustic sensors.

The strain field of a SAW induces, through the deformation potential mechanism, a periodic modulation of the materials band gap as well as of the refractive index. The diffraction of light in the refractive index grating induced by acoustic waves forms the basis for different acousto-optical SAW devices [37]. In contrast to conventional Bragg cells based on bulk acoustic waves, the planar distribution of the SAW fields makes them compatible with waveguide optics [63].

9.1.1 Surface Acoustic Waves in Semiconductors

Most of the previously mentioned applications of SAW in electronics and optics have been carried out using piezoelectric insulators such as quartz and lithium niobate. In these materials, the piezoelectric modulation has a minor effect on the electronic properties since it only interacts with bounded charge. The situation becomes different in semiconductors, where free charges may be induced by doping or illumination. Here, the SAW piezoelectric potential Φ_{SAW} introduces a type-II modulation of the materials band gap, which can reach hundreds of meV. The amplitude of this modulation is normally much larger than the one induced by the SAW strain field and represents the main interaction mechanism with free charges. These charges can be captured at the positions of minimum (in the case of electrons) and maximum (in the case of holes) electronic energy, as illustrated in the inset of Fig. 9.1. The trapped charge can then be transported with the SAW propagation velocity, provided that the carrier mobility (μ) is sufficiently high to follow the moving SAW field. The last condition can be expressed as $\mu |\partial \Phi_{SAW}/\partial x|_{max} > v_{SAW}$, where $|\partial \Phi_{SAW}/\partial x|_{max}$ is the amplitude of the longitudinal component of the SAW piezoelectric field (i.e., along the propagation direction x-direction).

Charge transport by SAWs was originally reported for electrons in bulk III-V semiconductors [31] and, subsequently, in III-V heterostructures [62, 68]. An important landmark in this field has been the demonstration of the transport of single electrons by SAWs by Shilton et al. [56]. A second important landmark has been the observation of simultaneous acoustic transport of photoexcited electrons and holes [51], which is illustrated in the inset of Fig. 9.1. The type-II piezoelectric potential of a SAW is sufficiently strong to ionize photoexcited excitons—the resulting free electrons and holes can then be captured at the minima and maxima of the electronic potential energy ($-e\Phi_{SAW}$), respectively. The spatial separation prevents the recombination of the carriers, thereby increasing their radiative lifetime. The electrons and holes can then be transported with the acoustic velocity over several hundreds of μm. Photons can be retrieved if the carriers are forced to recombine after transport. As shown in Fig. 9.1, recombination can be induced by using a thin semi-transparent metal layer to quench the SAW piezoelectric potential close to the surface [51]. The low acoustic velocity together with the reversible inter-conversion between photons and carriers provides a way to effectively store photons over times up to a few μs [51].

A third important development in the field of acoustic transport in semiconductors has been the demonstration of enhanced electron spin coherence during acoustic transport. Spin coherence lengths ℓ_s up to several tens of μm have been reported, thus implying spin coherence times ℓ_s/v_{SAW} in excess of 20 ns [16, 59, 60] (hole spins normally live only for a short time compared to the SAW period of a few ns). The long spin coherence times have been attributed to the fact that the spatial electron–hole separation by the type-II potential not only prevents recombination but also, simultaneously, reduces electron spin scattering through the electron–hole exchange interaction (BAP-mechanism) [59]. The spatial confinement of the carriers also reduces spin scattering processes associated with the spin–orbit interaction [60]. Furthermore, during transport, the spin vector can be manipulated by an external field (magnetic or electric) [16, 60] (as illustrated in Fig. 9.1) or by the effective spin–orbit magnetic field generated by the SAW strain and piezoelectric fields [60]. Finally, the spin-polarized electrons can be brought to recombine with the holes after transport, leading to the emission of polarized photons. The latter has been used to determine the spin state by optical means.

The scheme of Fig. 9.1 provides a simple quantum processor for polarized photons based on the acoustic transport and manipulation of electrons spins. For applications in quantum communication, the outgoing photons can be coupled to an optical fiber for long range communication. The overall efficiency of the process can be very high. By embedding the GaAs transport channel in a microcavity, it has recently been shown that over 90% of the incoming photons can be converted into electron–hole pairs. The latter can be transported over 200 μm with a transport efficiency exceeding 85% [9, 34].

Most of the functionalities required for spin transport and manipulation using the scheme in Fig. 9.1 have been demonstrated over the past years. There are, however, important missing links. First, although there are proposals for photon detectors with photon number discrimination based on SAWs [61], the detection of single photons using a process compatible with acoustic transport has so far not been demonstrated. The second are acoustically based single-photon sources (SPSs), which are required for the emission of single photons after the manipulation of the electrons spins. Progress in the realization of these SPSs will be the main subject of this work. Finally, large efforts will be required to combine these functionalities on a single semiconductor chip.

9.1.2 Outline

Present semiconductor technologies allow the growth of different electronic systems, such as quantum wells (QWs), wires (QWRs), and dots (QDs) on the surface of a single semiconductor crystal. The functionalities that can be implemented using these structures depend on the ability to transfer charges between them. The moving character of the SAW fields provides a powerful tool for controlling the charge transfer without requiring the structuring of the interaction region.

In this review, we exploit the ability of the SAW fields for the controlled manipulation of carriers for the realization of single-photon sources (SPSs). The general requirements for efficient SPSs together with concepts for their implementation using acoustic transport will be reviewed in Sect. 9.2. In Sect. 9.2.3 we introduce the approach used by us, which is based on the acoustic transfer of charge between a QW and a QWR grown epitaxially on GaAs (311)A substrates. The structures are grown via molecular beam epitaxy (MBE) of (Al,Ga)As layers on a (311)A substrate patterned with shallow (approx 35 nm high) mesas. As will be discussed in Sect. 9.3, the special growth mode during the deposition of a QW on the patterned surface leads to material accumulation along one of the mesa edges, which creates a sidewall QWR along it [42]. These QWRs are almost planar and electrically connected to the surrounding QW. SAW modes can be easily excited on these surfaces, as demonstrated by the SAW propagation studies in Sect. 9.3.3.

The sidewall QWRs act as efficient channels for the ambipolar transport of photo-excited electrons and holes by a SAW. Optical studies of carrier transport along the QWRs are presented in Sect. 9.4. Time-resolved investigations (Sect. 9.4.2) show that the electrons and holes are captured close to the minima and maxima, respectively, of $-e\Phi_{SAW}$, which move with the acoustic velocity.

Potential fluctuations within the sidewall QWRs create QD-like states with narrow photoluminescence (PL) linewidths. By reducing the QWR length, it thus becomes possible to create arrays of a few QDs embedded at well-defined positions within a QW. These arrays can be dynamically populated via the acoustic transport of carriers from the surrounding QW. The experimental results reviewed in Sect. 9.5.2 show that the SAW controls not only the transport but also the subsequent recombination of carriers within the array. In particular, by appropriately selecting the acoustic power it becomes possible to limit the emission to a few QD centers. The photon anti-bunching experiments described in Sect. 9.5.4 demonstrate the emission of non-classical light by these centers, which form the basis for acoustically driven SPSs. Finally, the main conclusions and future perspectives in the field of acoustic manipulation of carriers in nanostructures are reviewed in Sect. 9.6.

9.2 Single-Photon Sources Based on Acoustic Transport

Single-photon emitters operating at optical frequencies are essential components for quantum information processing, where they find application in quantum key distribution, quantum teleportation, and linear optics quantum computation [13, 39, 69]. In this section, we first review processes for the generation of single photons using attenuated laser beams or two-level systems (Sect. 9.2.1). We then discuss

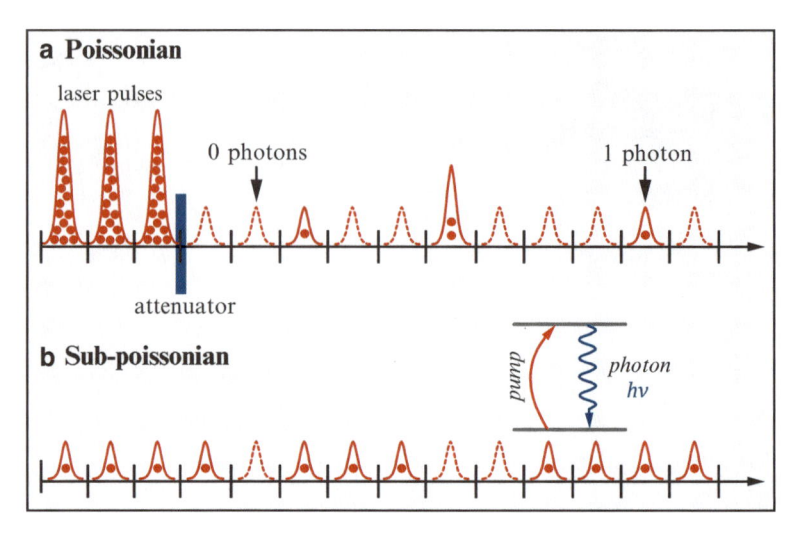

Fig. 9.2 (a) Poissonian single-photon source (SPS) based on attenuated laser pulses. (b) Sub-Poissonian SPS based on the two-level quantum system sketched in the inset

concepts for the implementation of SPSs using the acoustic transport of charge by SAWs (Sect. 9.2.2). Section 9.2.3 introduces the approach used by us to realize an acoustically driven SPS, which will be discussed in detail in the subsequent sections.

9.2.1 Photon Statistics and Single-Photon Sources

SPSs for applications in quantum information should ideally deliver a train of light pulses, each containing a single photon. The information (the secret key) is encoded in the photon polarization. The inherent security of the transmission channel is ensured by the fact that any attempt to detect the photon state will invariably change it. By using well-defined information exchange protocols [28], it then becomes possible to check the integrity of the transmission channel.

A major challenge for the realization of efficient SPSs is associated with the statistics of the light emission process. An SPS can be approximated by a train of attenuated laser pulses, as illustrated in Fig. 9.2a. The attenuation should be large enough to reduce the probability of having more than one photon per pulse. The laser emission process, however, follows a Poissonian statistics. As a result, the high attenuation level required to avoid multiple occupancy also implies that many of the pulses will have zero occupancy, thereby reducing the information transmission rate.

SPSs with much higher efficiency than attenuated laser pulses can be realized using sub-Poissonian emitters based on two-level systems. Due to Pauli's exclusion principle, a two-level system has a single excited state, which cannot be re-excited before it relaxes emitting a single photon (cf. inset of Fig. 9.2b). If the characteristic time for photon emission is shorter than the time interval between excitation pulses (but much longer than the pumping time), the two-level system will behave as a turn-style source emitting a single photon per excitation cycle [11, 35]. In this way, one can obtain light pulses with the the sub-Poissonian distribution presented in Fig. 9.2b, where each light pulse ideally contains a single photon.

During the last years, SPSs based on different types of two-level systems have been demonstrated including atomic and molecular levels, lattice defects, and semiconductor quantum dots [39]. SPSs in solid state systems are of particular interest because of their potential for scalability as well as for the integration with other electronic components. The key to generating single photons on demand in the solid state is the control of the recombination process. As discussed in more detail below, the modulation caused by a SAW provides an effective method to control the charge flow and recombination, which will be exploited for the generation of single photons.

9.2.2 Concepts for SAW-Based Single-Photon Sources

Two different concepts for the generation of single photons on demand using the carrier transport by SAWs have been proposed. The first of these concepts [22] is based on the unipolar acoustic transport of a single electron per SAW cycle originally demonstrated by Shilton et al. [56]. One possible realization is illustrated in Fig. 9.3a. Here, a SAW extracts electrons from an n-type doped region and drives them through a narrow constriction (point contact), which restricts the number of transported carriers to a single one per SAW cycle. The carrier is subsequently brought by the SAW to a p-doped region, where the SAW potential becomes screened by the high carrier density. The transported electron can then recombine, thus leading to just one electron–hole recombination event per SAW cycle. A similar procedure using hole transport can also be realized. The process can be extended to the generation of a well-defined number of photons per SAW period by controlling the number of carrier within each SAW cycle. Also, the rate at which photons are emitted can be controlled by adjusting the SAW frequency. Although different experiments have demonstrated the acoustic transport of carriers between in-plane n-i-p junctions followed by photon emission [25, 26, 30], SPSs based on this approach have so far not been realized.

The second approach for SAW-based SPS relies on the ambipolar transport of carriers by a SAW to a QD [67], as sketched in Fig. 9.3b Here, the charge carriers are optically generated in the QW and then acoustically transported to an embedded QD [14]. The discrete QD states are then populated by a periodic train of electrons and holes delivered with the acoustic velocity v_{SAW}. These carriers then recombine emitting single photons [32, 67]. In principle, since the QDs are periodically pumped, the photon emission rate can be easily varied by tuning the frequency of the driving SAW.

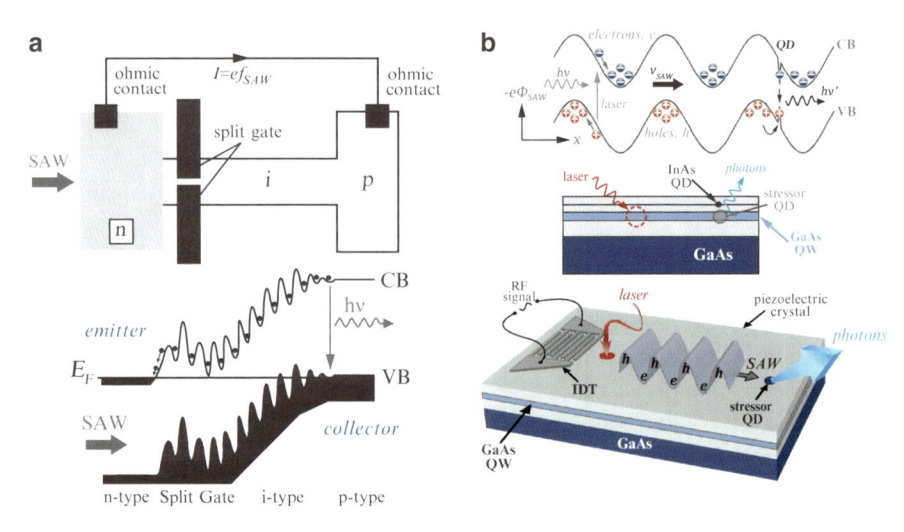

Fig. 9.3 Proposals for single-photon sources based on carrier transport by a SAW. (**a**) Unipolar transport [22][1]: a SAW extracts electrons from an n-type doped region and transports them to a p-type contact, where they recombine emitting photons. A split-gate in the transport path ensures that only one electron is transported in each SAW minima [56], thus leading to the emission of a single-photon per SAW cycle. [Reprinted figure with permission from C. L. Foden, V. I. Talyanskii, G. J. Milburn, M. L. Leadbeater and M. Pepper, Physical Review A **62**, 011803(R) (2000), http://pra.aps.org/abstract/PRA/v62/i1/e011803. Copyright (2000) by the American Physical Society.] (**b**) Ambipolar transport [67]: a controllable SAW-induced transport of electrons and holes optically excited in a quantum well (QW) to an embedded strain-induced quantum dot (stressor QD). The discrete energy levels of the QD ensure that at most one recombination event occurs per SAW cycle

The approach described in the previous paragraph requires a high-mobility acoustic transport channel (usually a QW) electrically connected to the QD. Conventional self-assembled QDs obtained by the Stranski–Krastanow growth mode are normally connected to a thin QW-like wetting layer. The mobility of carriers in the wetting layer, however, is expected to be too low for efficient acoustic transport. In order to overcome this difficulty, the authors in [67] proposed the use of stressor dots. As shown in Fig. 9.3b, the latter consists of a dot-like local potential minima in the QW layer induced by the strain generated by an InAs QD grown over the QW.

The SAW-based sources for single photons on demand have several potential advantages: (i) The SPS in Fig. 9.3c does not require a pulsed laser since the

---

[1]Due to the use of APS copyrighted material, this contribution is subjected to the following restriction: Readers may view, browse, and/or download material for temporary copying purposes only, provided these uses are for noncommercial personal purposes. Except as provided by law, this material may not be further reproduced, distributed, transmitted, modified, adapted, performed, displayed, published, or sold in whole or part, without prior written permission from the American Physical Society.

repetition rate is determined by the SAW frequency, which can easily reach the GHz range. Such high repetition rate is desirable for practical applications. (ii) The acoustically driven SPSs can be combined with SAW-based spin transport [16, 60] for the control of the polarization of the emitted photons. The ability to manipulate the polarization of a photon state leads to applications that cannot be achieved in classical context. One example is the unconditionally secure transmission of information using quantum key distribution [15]. Finally, as will be discussed in Sect. 9.5.2, the SAW-driving mechanism can be used to reduce the time jitter for photon emission [10], which is a crucial requirement for the implementation of quantum key distribution protocols.

Even though no working devices based on the approaches described in Fig. 9.3 have been realized, the idea of using solid-state systems to generate single photons opens promising perspectives towards practical on-chip photon-based quantum light applications. QDs are especially appealing due to their narrow emission linewidth, long-term stability, and the advantage of tailorable emission energies. Also, the fabrication of QDs is compatible with mature semiconductor technologies allowing them to be integrated with other on-chip components. They have emerged as promising platforms for single-photon emitters [40, 53] and quantum logic elements [18, 47] in quantum information processing. Finally, the band gap modulation by the strain field of a SAW provides an interesting mechanism to tune the emission energy of SPSs based on InAs QDs, as recently demonstrated by Gell and co-workers in [27].

9.2.3 Single-Photon Source Based on GaAs (311)A

We have recently demonstrated a SAW-based SPS [17] similar to the one presented in Fig. 9.3b. The structure of this SPS is sketched in Fig. 9.4. The recombination leading to single-photon emission takes place in short (approximately 2 μm long) QWRs embedded within well-defined positions in a high-mobility QW transport channel. The QWRs are formed during the MBE growth of (Al,Ga)As layers on patterned GaAs (311)A substrates. The particular growth mode on the patterned substrate leads to the formation of QWRs along one of the mesa edges, as sketched in the inset of Fig. 9.4 (details of the growth mechanism are given in Sect. 9.3.1). The QD-like recombination centers for the generation of single photons are induced by the fluctuations in the QWR thickness [12, 29]. As illustrated in Fig. 9.4, the carriers are optically generated in the QW and transported by a SAW as successive trains of positive and negative charges. The transported carriers are then captured by localized QD states within the short QWRs, where they recombine emitting single photons.

Experimental analyses presented in Sect. 9.5 show that the driving SAW fields govern not only the carrier transport but also the recombination from the emitting QD states within the short QWRs. Thus, by adjusting the SAW intensity it becomes

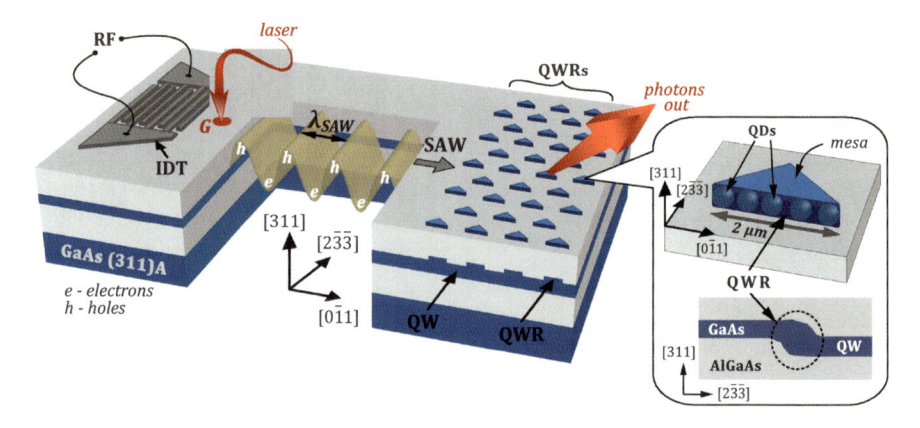

Fig. 9.4 (a) Acoustically driven single-photon source using GaAs/(Al,Ga)As (311)A heterostructures with an IDT for SAW excitation. Electron–hole pairs are optically excited in the quantum well (QW) at the generation spot G and transported by the SAW towards the embedded quantum wires (QWRs), where they recombine radiatively in QD emission centers formed by potential fluctuations along the QWR axis. Inset shows a 2 μm wide triangular mesa together with a cross-sectional view of the QW and the QWR overgrown on its edge

possible to restrict the number of emitting QDs within the QWR. This ability of the SAW to control the recombination of transported carriers yields a high-repetition-rate source of single photons with tunable emission energy.

9.3 Coupled Nanostructures for Acoustic Transport

The epitaxial growth of semiconductor layers on patterned substrates has proven to be a powerful technique for the formation of low-dimensional structures like quantum wires and quantum dots—reviews of the different growth approaches can be found in [64] and [21]. The acoustically driven SPSs investigated in this work employ quasi-planar QWRs embedded in a high-mobility GaAs QW. The QWRs were fabricated via MBE through the overgrowth of (Al,Ga)As layers on GaAs (311)A substrates patterned with shallow mesas, following the procedure originally introduced in [42]. In this section, we first describe the MBE deposition procedure and then address the electrical generation of SAWs on the surface of the samples using interdigital transducers.

9.3.1 Sidewall Quantum Wires on GaAs (311)A

The formation of sidewall QWRs during MBE growth on patterned GaAs (311)A substrates is schematically illustrated in Fig. 9.5a. Prior to the growth, ridge- or

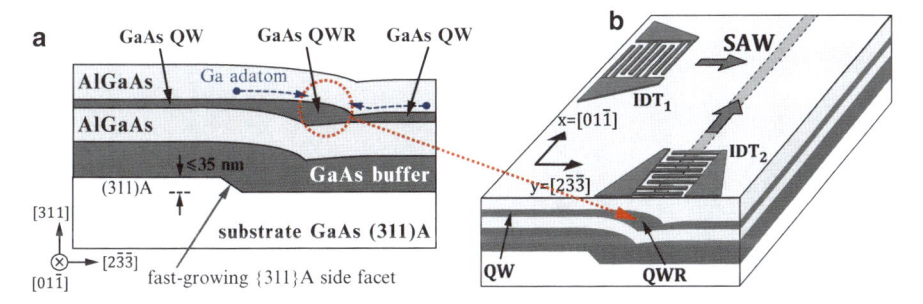

Fig. 9.5 (**a**) Formation of GaAs/(Al,Ga)As QWRs during the growth of a QW on a patterned GaAs (311)A surface (adapted from [45]). The QWR is formed due to the enhanced growth rate near the fast-growing sidewall aligned along the [01$\bar{1}$] surface direction. The *dashed arrows* indicate the migration of Ga adatoms during growth (see [45]). (**b**) Structure of sidewall QWR sample with IDTs for the generation of SAWs along the [01$\bar{1}$] (IDT$_1$) and [2$\bar{3}\bar{3}$] (IDT$_2$) main axes of the (311)A surface [4]

trench-type structures are prepared on the sample surface using wet chemical etching. The etching depths range typically from 20 to 35 nm. The dimensions of the structures and their location on the surface are defined by optical lithography. During the growth of GaAs on the patterned structure, the preferential migration and attachment of gallium adatoms at the mesa sidewall aligned along the [01$\bar{1}$] direction gives rise to the formation of [01$\bar{1}$]-oriented GaAs QWR (cf. Fig. 9.5a) [42, 45]. The samples investigated in this work consist of one to three GaAs QWs, with a nominal thickness of 6 nm, separated by 10 nm-thick Al$_x$Ga$_{1-x}$As barriers (x≈0.3). The sidewall QWRs are about twice as thick as the adjacent QW layer and have a lateral extension of approximately 50 nm. Further details of the QWR fabrication process are given in [23, 42, 44, 45].

The previously described process allows the fabrication of QWRs oriented along the [01$\bar{1}$] direction at well-defined positions on the sample surface. The QWRs are electrically connected to the neighboring QW, so that charges can easily be transferred between them. Further investigations of the growth process have shown that the addition of atomic hydrogen during the growth creates a periodic surface corrugations with nanometer scale along the QWR axis, which leads to the formation of an arrays of QDs [43]. Finally, the process can also be used to produce InGaAs QWRs by introducing indium into the QW structure [46].

9.3.2 Structural Characterization

The formation of QWRs through the selective growth on patterned substrates is demonstrated by the cathodoluminescence (CL) measurements displayed in Fig. 9.6. Figure 9.6a shows a scanning electron micrograph (SEM) of a region of the sample

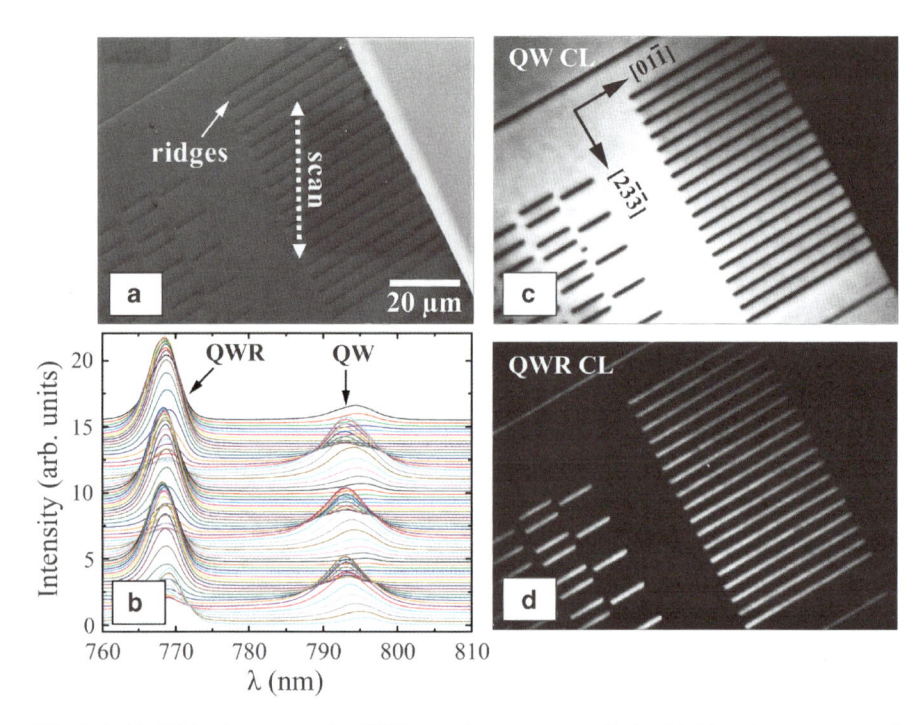

Fig. 9.6 (a) SEM micrograph of a QWR sample grown on a GaAs (311)A substrate patterned with an array of 5.6 µm-wide ridges separated by 1-µm-wide trenches. (b) CL spectra recorded along the scan line indicated in (a) show the spectral features of the QW and QWRs. CL images of the region in (a) obtained by integrating the emission of the (c) QW and (d) QWR [SOURCE: Alsina, F., Stotz, J. A. H., Hey, R., Jahn, U. and Santos, P. V. (2008), Acoustic charge and spin transport in GaAs quantum wires. physica status solidi (c), 5: 2907–2910. Copyright Wiley-VCH Verlag GmbH & Co. KGaA. Reproduced with permission]

surface patterned with an array of 5.6 µm-wide ridges separated by 1-µm-wide trenches. Figure 9.6b displays a sequence of CL spectra recorded while scanning the electron beam along the line indicated in Fig. 9.6a over three ridges. The emission lines centered at 768 and 793 nm are attributed to the QW and QWR, respectively. Note that the 793 nm emission line only appears near the ridges parallel to the $[01\bar{1}]$ surface direction, where the QWR forms. The QWR confinement energy is given by the energetic separation between the two lines of approx. 25 meV, which is comparable to the thermal energy at room temperature.

Figure 9.6c,d display CL images recorded by integrating the CL intensity around the emission lines of the QW and QWRs, respectively. The two images are complementary since the emission regions of the QW and QWR do not overlap. In addition, since the QW and QWR are interconnected, carriers generated in the QW at positions close to the QWR are efficiently collected by QWR, thereby further reducing the CL emission intensity at the QW energy.

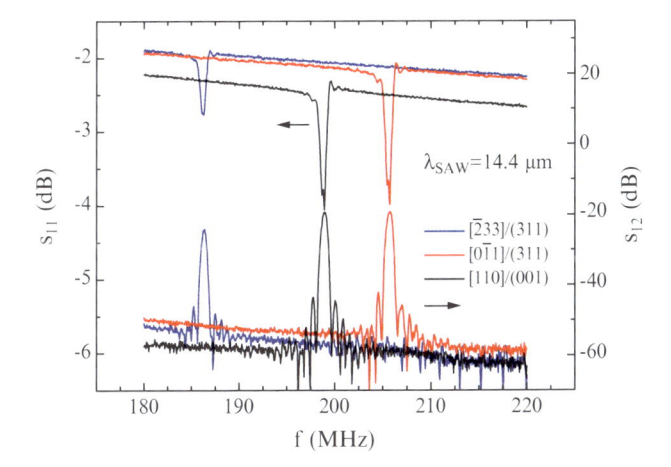

Fig. 9.7 rf-Power reflection (s_{11}) and transmission (s_{12}) coefficients measured on a SAW delay line for an acoustic wavelength $\lambda_{SAW} = 14.4$ μm on GaAs (311) (oriented along the [$2\bar{3}\bar{3}$] and [$01\bar{1}$] surface directions) and GaAs (001) (oriented along the [110] surface direction)

9.3.3 SAW Propagation on (311) Surfaces

The use of SAWs for the controlled transfer of carriers between QWs and QWRs in Fig. 9.5a presupposes that one can electrically excite acoustic modes with strong longitudinal piezoelectric fields both along ($x = [01\bar{1}]$ direction) and perpendicular ($y = [2\bar{3}\bar{3}]$ direction) to the QWR axis using IDTs. Fortunately, SAWs with relatively high intensities can be electrically excited along these two directions. The two main surface axes of the (311) surfaces are not equivalent with respect to SAW propagation, giving rise to acoustic modes with different strain fields and propagation velocities [41, 70]. We will denote the modes along x and y as [$01\bar{1}$]/(311) and [$2\bar{3}\bar{3}$]/(311), respectively. As for the conventional SAW mode propagating along the [110] direction of (001) GaAs substrates, the [$2\bar{3}\bar{3}$]/(311) mode is a true surface Rayleigh wave. In contrast, the [$01\bar{1}$]/(311) mode is a leaky surface mode with a very small attenuation coefficient (of approx. 2.5×10^{-3} dB/λ_{SAW}), as demonstrated by Zhang and co-workers using an elastic model [70].

In order to investigate SAW propagation on the (311) surface, SAW delay lines consisting of two IDTs facing each other were fabricated along the two main surface axes indicated in Fig. 9.5b [4]. We present here results for structures containing IDTs designed for an acoustic wavelength $\lambda_{SAW} = 14.4$ μm. The IDTs are of the split-finger type, where each period consists of 4 fingers, each with a width $\lambda_{SAW}/8$ and a thickness of 40 nm. The total length and width of the IDTs are 350 λ_{SAW} and 21 λ_{SAW}, respectively. Figure 9.7 displays the rf-power reflection (s_{11}) and transmission (s_{12}) coefficients for the delay lines, which were recorded

Table 9.1 Propagation properties of SAWs along the $x = [01\bar{1}]$ and $y = [2\bar{3}\bar{3}]$ directions of the GaAs (311)A surface obtained from rf-transmission and reflection measurements. v_{SAW} is the SAW propagation velocity. r_a is the measured ratio between the SAW acoustic power and the corresponding one for the [110]/(001) mode on GaAs (001), which was determined from the dips of the s_{11} resonances in Fig. 9.7. r_f denotes the calculated ratio between the amplitude of the longitudinal piezoelectric field at the surface to the one for the [110]/(001) mode (from [4])

Propagation direction/ surface	v_{SAW} (meas.) (m/s)	v_{SAW} (calc.) (m/s)	r_a	r_f
$[01\bar{1}]/(311)$	2,932	2,961	1.0	1.03
$[2\bar{3}\bar{3}]/(311)$	2,677	2,683	0.42	0.785
$[110]/(001)$	2,862	2,863	1.0	1.0

using a network analyzer. For comparison, the figure also includes results obtained on a similar delay line deposited on GaAs (001) substrates (the [011]/(100) mode).

The dips (maxima) in the s_{11} (s_{12}) curve indicate the frequency bands for SAW excitation. The acoustic velocity $v_{SAW} = \lambda_{SAW} f_{SAW}$ can be directly determined from the resonance frequencies f_{SAW} and is summarized for the different configurations discussed above in the second column of Table 9.1 [4]. The measured velocities agree well with the ones calculated using an elastic model [20], which are listed in the third column of the figures.

In addition to the frequency, the strength of the electromechanical coupling, which allows for the electrical SAW excitation, is another important parameter for the acoustic transport of carriers. The effective electromechanical coupling can be quantified from the depth of the s_{11} dips. The coupling r_a normalized to the one for the [011]/(100) mode is listed in the fourth column of Table 9.1. r_a is approximately the same for the $[01\bar{1}]/(311)$ and [011]/(100) modes, but considerably smaller for the $[2\bar{3}\bar{3}]/(311)$ one.

The efficiency of the acoustic transport depends on the amplitude of the longitudinal piezoelectric field F_x generated by the SAW. The last column in Table 9.1 displays the ratio r_f between the amplitude of F_x near the surface and the corresponding one for the [100]/(001) mode. r_f has comparable values for the $[01\bar{1}]/(311)$ and [011]/(100) modes but is much smaller for SAWs along the y direction of the (311) surface. The longitudinal piezoelectric field F_y for the $[2\bar{3}\bar{3}]/(311)$ mode is then expected to be a factor $r_f\sqrt{r_a} \sim 0.51$ smaller than in the other cases.

9.4 Acoustic Carrier Transport in Quantum Wires

The optical and structural properties of sidewall QWRs have been the subject of different investigations. Extensive PL studies have been carried out to address optical nonlinearities [46] and anisotropic effects [54] in these QWRs. By using scanning near-field microscopy with a spatial resolution below 200 nm, Richter and co-workers [49] have successfully established the one-dimensional quantum states of single QWRs. They have also investigated the transfer of electron–hole pairs between the QWR and the surrounding QW. These microscopic studies were subsequently extended to probe the carrier transfer between these nanostructures in the time domain [50]. They have revealed that the lateral material flow responsible for the QWR formation leads to the thinning of the QW in the regions close to the QWR. These regions act, therefore, as shallow lateral barriers for the diffusive transfer of electrons and holes from the QW to the QWR [38]. The near-field studies have also revealed that the broad PL band of the wires (approx. 10 meV at 10 K) arises from the superposition of closely spaced emission lines from states created by potential fluctuations along the wire axis [33]. In addition to these states, sharp luminescence peaks were also detected and attributed to excitons localized on a sub-150-nm length scale. As will be discussed in detail in Sect. 9.5, the acoustically driven SPS is based on the controlled population of these states using acoustic transport.

In the present section, we review the acoustic transport properties of sidewall QWRs as probed by spatially and time-resolved PL spectroscopy. This technique, which will be introduced in Sect. 9.4.1, shows that a SAW can efficiently transport electrons and holes over several tens of micrometer. During the transport, the carriers are enclosed within moving QDs created by the combination of the lateral confinement in the QWR with the longitudinal confinement induced by the SAW piezoelectric field. Time-resolved measurements are then used to prove that the carriers are transported as successive packets of electrons and holes stored at a well-defined phase of the SAW piezoelectric potential (Sect. 9.4.2).

9.4.1 Ambipolar Carrier Transport

The experimental configuration used to optically probe acoustic transport in QWRs is illustrated in Fig. 9.8. The sample in this case contains arrays of straight (upper part of Fig. 9.8a) and zig-zag (lower part of Fig. 9.8a) ridges, which have been overgrown in order to form QWRs along their edges. A SAW with a wavelength $\lambda_{SAW} = 4$ μm is generated by an IDT placed on the upper part of the sample. The experiments were carried out at low temperatures (10–20 K) in an optical cryostat with electrical radio-frequency (rf) connections for the excitation of the IDTs. The intensity of the SAWs will be specified in terms of the nominal electric power P_{rf} applied to the IDT.

The carriers are generated in the undoped nanostructures using a laser beam focused onto a 2 μm spot on the SAW path, as indicated by spots G_1 and G_2 in Fig. 9.8a. Selective excitation of the QWR can be achieved by using a laser energy between the band gaps of the QW and the QWR. Alternatively, carriers generated in the QW can be efficiently transferred to the QWR if the excitation spot is close to it. In order to probe the acoustic transport, the PL emitted along the QWR axis is collected and imaged using a CCD-camera. In the experiments, optical filters

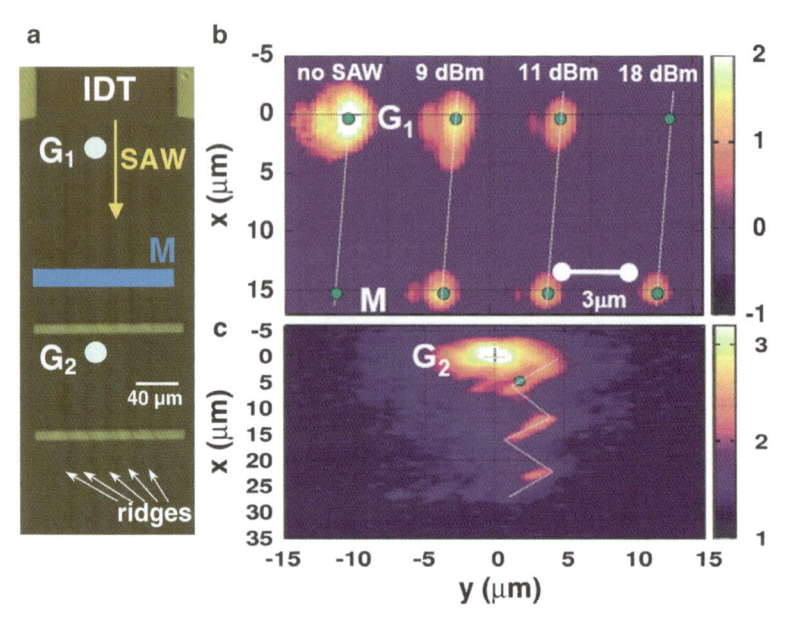

Fig. 9.8 (a) Optical micrograph of a patterned (311)A samples illustrating the acoustic carrier transport in QWRs detected using spatially resolved PL. A laser spot focused at G_1 or G_2 generates carriers. A SAW produced by the IDT transports the carriers from generation spot towards the metal stripe M, which is used to quench the piezoelectric field and induce recombination. (b) Sequence of PL images recorded for different acoustic powers P_{rf} (*from left to right*) showing the reduction of the PL at G_1 and its increase near M due to the recombination of transported carriers. [SOURCE: Alsina, F., Stotz, J. A. H., Hey, R., Jahn, U. and Santos, P. V. (2008), Acoustic charge and spin transport in GaAs quantum wires. physica status solidi (c), 5: 2907–2910. Copyright Wiley-VCH Verlag GmbH & Co. KGaA. Reproduced with permission.] (c) Corresponding PL images for the acoustic transport along the zig-zag wires, recorded by exciting carriers at spot G_2 in (a)

can be used to select the QWR emission and to reduce stray light from the laser. Figure 9.8b displays a sequence of PL images recorded for increasing nominal rf-power levels P_{rf}. In the absence of a SAW (left image), emission is only observed near the excitation spot G_1. When a SAW is excited, its piezoelectric potential ionizes the photogenerated excitons and transports the resulting free electrons and holes along the wire axis. As a consequence, the PL close to G_1 reduces significantly with increasing P_{rf}. In order to detect the remote PL induced by the transported carriers, we have deposited a semi-transparent metal stripe on the sample surface (indicated by M in Fig. 9.8a). Since the distance of the QWRs from the surface (of approx. 100 nm) is much smaller than λ_{SAW}, the stripe short-circuits the SAW piezoelectric potential, thus forcing carrier recombination [51]. The bright spot near M in Fig. 9.8b thus gives direct evidence for the acoustic transfer of electrons and holes along the QWR.

The acoustic transport takes place at the SAW velocity, which is normally much smaller than the typical electronic velocities. As a result, when the transport

9 Acoustic Carrier Transport in GaAs Nanowires   275

is carried out in a QW, there is considerable diffusion of the carriers in the
perpendicular direction (i.e., along the SAW wave fronts). This lateral transport is
inhibited in the case of QWRs: in fact, the dimensions of the emission region near
M in Fig. 9.8b is limited by the spatial resolution of the setup. The QWRs provide,
therefore, a very narrow channel for acoustic transport.

The QWR confinement energy (corresponding to the band gap difference to the
surrounding QW) reduces when the orientation of the mesa edges deviates from the
$[01\bar{1}]$ surface directions [44]. For small deviations, however, the confinement is not
significantly reduced, thus making it possible to create transport channels for other
orientations. One example is illustrated in Fig. 9.8c, which shows acoustic transport
along the zig-zag QWRs imaged in the lower part of Fig. 9.8a [44]. It is interesting
to note that the PL emission from zig-zag segments with one orientation is much
higher than for the other orientation. The difference is attributed to a much higher
density of trapping centers for one of the orientations. During one half-cycle of
the SAW, these traps capture carriers with one polarity, which then recombine with
carriers with the opposite polarity in a subsequent SAW half-cycle.

Further information about the trapping centers can be obtained by spectrally
analyzing the emission along the transport channel. Figure 9.9a displays the PL
micrograph of the acoustic transport in a QWR excited under conditions that lead
to enhanced recombination along the transport path [5]. The recombination sites
appear as pronounced hot spots in the PL image. Enhanced recombination during
transport has been observed for either a very low or a very high acoustic power. In
the former case, the SAW piezoelectric field is not sufficient to overcome potential
fluctuations along the QWR, which trap the carriers [33, 64]. Carrier trapping may
also be mediated by the SAW fields [7]: in this case (which corresponds to the
situation in Fig. 9.9a), the PL along the channel becomes enhanced at high acoustic
powers.

The plots in Fig. 9.9b–d show PL spectra recorded at different positions on the
SAW path. The PL spectrum close to the generation spot G is characterized by a
series of closely spaced sharp lines, which superimpose to form a broad PL emission
line with a full width at half maximum (FWHM) of approx. 10 meV. As one moves
away from G, the sharp lines (FWHM of approx. 0.4 meV) become much better
resolved. From the line widths we estimate a density of trapping centers of 1–
2 states/meV over the 10 meV spectral band width in Fig. 9.9c–d. By taking into
account the spatial resolution of the setup of approx. 2 μm, we calculate an average
separation between the trap centers of 100–200 nm. This estimate, which is only a
factor of 2 to 4 larger than the QWR width, agrees well with the results of [33]. We
will show in Sect. 9.5 that the sharp lines in Fig. 9.9c arise from quantum dot-like
emission centers, which can be used for the generation of single photons.

9.4.2 Real-Time Transport Dynamics

During transport by a SAW, the electrons and holes are expected to be kept close to
the minima and maxima of the piezoelectric potential (cf. inset of Fig. 9.1). In this

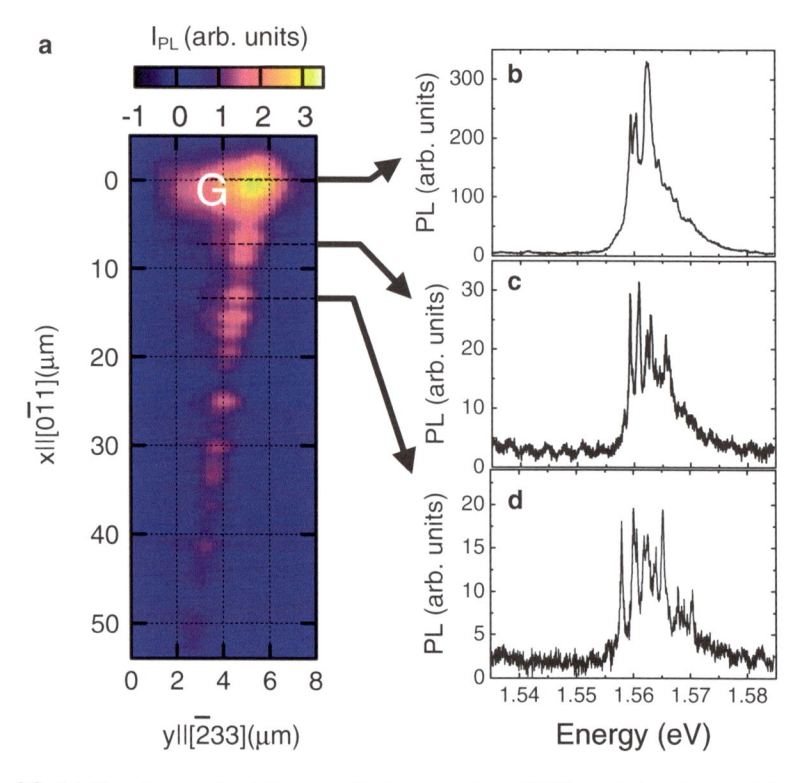

Fig. 9.9 (a) PL micrograph of the acoustic transport in a QWR excited under conditions of enhanced recombination along the transport path, which was induced by using a high SAW intensity ($P_{rf} = 20$ dBm). The PL results from the recombination of electrons and holes generated at G and transported by the SAW to trap centers along the QWR axis. (**b**)–(**d**) displays PL spectra recorded at different wire cross sections [SOURCE: Alsina, F., Stotz, J. A. H., Hey, R., Jahn, U. and Santos, P. V. (2008), Acoustic charge and spin transport in GaAs quantum wires. physica status solidi (**c**), 5: 2907–2910. Copyright Wiley-VCH Verlag GmbH & Co. KGaA. Reproduced with permission]

section, we address the dynamic carrier distribution during transport, which has been accessed using time-resolved PL measurements carried out with a spatial resolution less than half the SAW wavelength.

The time-dependent optical measurements of the acoustic transport have been carried out in the same configuration of Fig. 9.10a using a continuous wave (cw) laser beam for carrier excitation. The PL emission along the QWR transport channel was imaged with high spatial (of approx. 2 μm) and time resolution (< 0.5 ns). The latter can be achieved by using a streak camera [57] or a gated intensified ICCD camera [3] synchronized with the rf-signal used for SAW excitation.

The gray scale plot of Fig. 9.10b displays the spatial (horizontal axis) and temporal (vertical axis) dependence of the integrated PL intensity from a QWR in the presence of a SAW. The spatial and temporal coordinates are expressed in terms

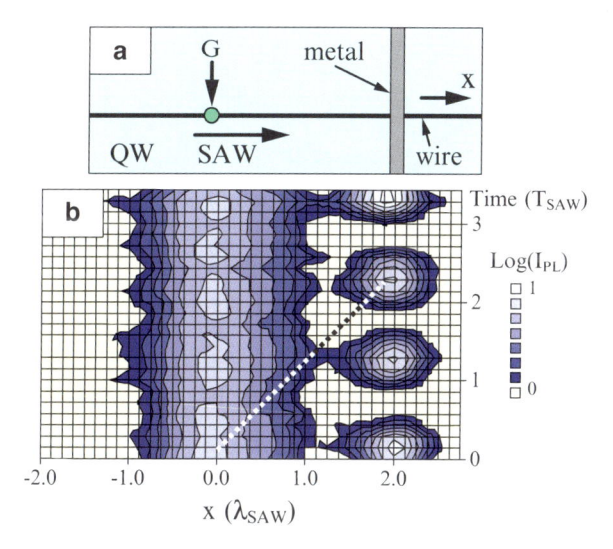

Fig. 9.10 (**a**) Setup for spatially and time-resolved optical detection of acoustic transport in QWRs [3]. The carriers are generated at spot G using a 680 nm cw laser diode. The PL along the transport path is detected by an intensified CCD-camera synchronized with the rf-signal used for SAW excitation. (**b**) Spatial (horizontal scale, in units of the SAW wavelength λ_{SAW}) and temporal (vertical scale, in units of the SAW period T_{SAW}) dependence of the QWR PL under a SAW. The *dotted line* indicates the trajectory of an electron package as it is transported by the moving SAW potential from the generation point G to the edge of the metal stripe, where it recombines with transported holes [Reprinted figure with permission from F. Alsina, P. V. Santos, H.-P. Schönherr, R. Nötzel and K. H. Ploog, Physical Review B 67, 161305(R) (2003), http://prb.aps.org/abstract/PRB/v67/i16/e161305. Copyright (2003) by the American Physical Society]

of the SAW wavelength λ_{SAW} and period $T_{SAW} = 2\pi/\omega_{SAW}$, respectively. Vertical cross sections of the plots yield, therefore, the PL time dependence at the spatial position indicated by the horizontal axis. Note that despite the continuous laser excitation, the PL exhibits a strong time dependence both close to the generation point (G) and to the semi-transparent metal stripe used to block the transport (position M). More important, the emission at M takes place as well-defined pulse bursts with the SAW repetition frequency. The last results show that the carriers are synchronously transported by the SAW fields.

A time-dependent emission near the photoexcitation spot has also been reported for microscopic PL measurement in QWs under acoustic fields [1, 2, 60] and attributed to the different mobilities for electrons and holes. Due to the efficient exciton ionization by the longitudinal component F_x of the SAW piezoelectric field, a sizeable PL intensity is only expected when the carriers are generated at the instants of time when F_x vanishes. At these time instants, Φ_{SAW} is either a maximum (corresponding to the storage site for electrons) or a minimum (storage site for holes). When the light spot is close to the hole storage sites (minima of Φ_{SAW}), the excited electrons can quickly move to the neighboring regions of high Φ_{SAW}, thereby preventing recombination. In contrast, when carriers are created close to the electron storage sites (maxima of Φ_{SAW}), holes have to be quickly extracted to avoid recombination. This process, however, is less efficient due to the lower hole mobility, thus leading to higher PL intensities at these instances of time, which repeat with the SAW frequency. This phase relationship between the PL emission and the SAW strain field has been corroborated by interferometric measurements of the SAW displacement field reported in [3].

The trajectory of the carriers during acoustic transport follows lines of slope v_{SAW}^{-1} in Fig 9.10b. In particular, the dashed line with slope v_{SAW}^{-1} and intersecting the maximum PL at G describe the trajectory of the electron packet generated at $t = 0$ as it moves from G to the recombination spot at the metal edge M while trapped in a minimum of $-e\Phi_{SAW}$. Note that a maximum PL at G corresponds to a maximum PL at M: based on the previous discussion, we then conclude that the PL bursts at M appear at the arrival time of the electrons at the recombination spot. This behavior is also associated with the higher electron mobility as compared to holes. As a consequence, the electron packets remain highly concentrated close to the minima of $-e\Phi_{SAW}$ during the transport: their recombination with holes trapped at M leads to the emission of the short light burst displayed in Fig 9.10b [2].

9.5 Carrier Dynamics and Photon Anti-Bunching in (311)A Quantum Dots

We have seen in the previous section that potential fluctuations in sidewall QWRs on GaAs (311)A create QD-like potentials. These potentials can trap and induce the recombination of acoustically transported carriers, giving rise to the sharp PL lines in Fig. 9.9b–d. In this section, we will show that the population of these QD states by a SAW can be used for the generation of single-photon pulses. As indicated in Fig. 9.4, we use for that purpose short QWRs (approximately 2 μm long) formed at the edges of triangular mesas patterned on the substrate surface. The fabrication of these structures using MBE is described in Sect. 9.5.1. Studies of the acoustic transport of charge from the surrounding QW to the short QWR presented in Sect. 9.5.2 show that the SAW not only transports carriers to the short QWRs but also controls the recombination within the QD states. By appropriately selecting the acoustic intensity, it becomes possible to restrict the emission to very few QD state as well as to control the photon emission energy. Investigations of the photon emission statistics described in Sect. 9.5.4 demonstrate a reduced probability for multiple emission events (photon anti-bunching). The latter provides the basis for the realization of acoustically driven single-photon sources.

9.5.1 Short Quantum Wires on GaAs (311)A

The fabrication of the short QWRs closely follows the one for QWRs described in Sect. 9.3.1. Instead of long ridges or trenches, the substrate surface was etched prior

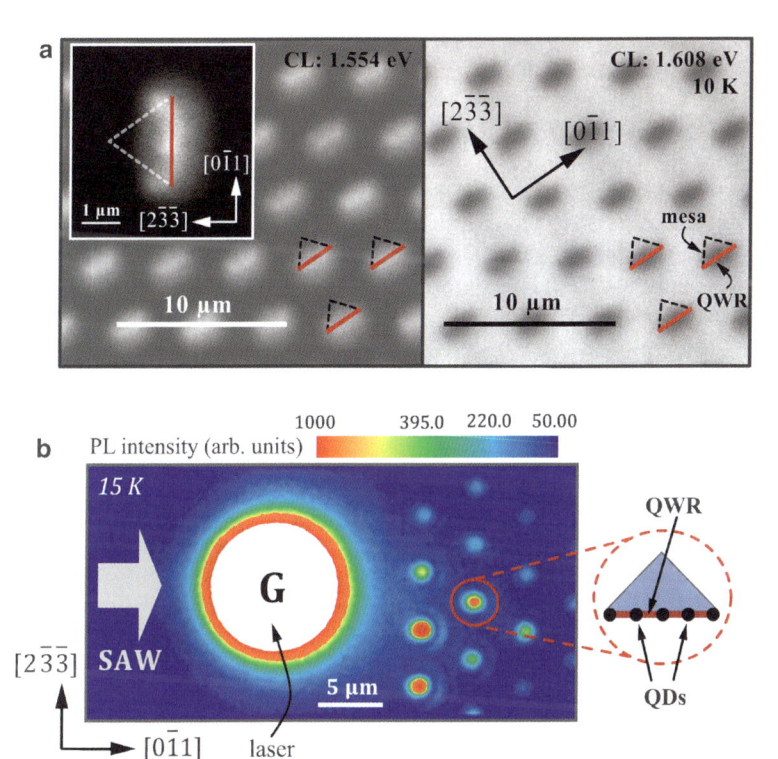

Fig. 9.11 (a) CL images recorded at 10 K showing the emission at 1.554 eV from the QWRs at mesa edges along the [0$\bar{1}$1] direction (*left panel*) as well as the QW emission at 1.608 eV in the unpatterned areas (*right panel*). Dashed and thick solid lines indicate the etched mesa edges. The inset displays a magnified CL image of a single QWR. (b) PL micrograph of QWRs at 15 K excited by the acoustic transport of carriers generated at spot G in the QW. The bright spot at the G position corresponds to the emission from the QW. The QWRs appear as the bright recombination sites along the SAW propagation path

to the MBE growth to form a matrix of triangular mesas, each with a side length of 1–2 μm. The formation of the short QWRs is demonstrated in Fig. 9.11a, which displays CL images recorded by detecting the emission at the QWR (left side) and at the QW energy (right side). Note that the short QWR forms only along one of the mesa edges (the one along the [0$\bar{1}$1]-direction) indicated by thick solid lines in Fig. 9.11a. Due to carrier diffusion, the width of the QWR emission region in the image is much larger than the physical width of the QWRs. This is because in the CL image recording process the spatial coordinates for each pixel correspond not to the exact position of the emission center but rather to the location of the electron beam. As a result, the diffusion of carriers excited in the QW to the QWRs gives rise to a luminescence halo around them. As discussed in Sect. 9.3.2, carrier diffusion also reduces the PL emission at the QW energy in the regions close to the QWRs.

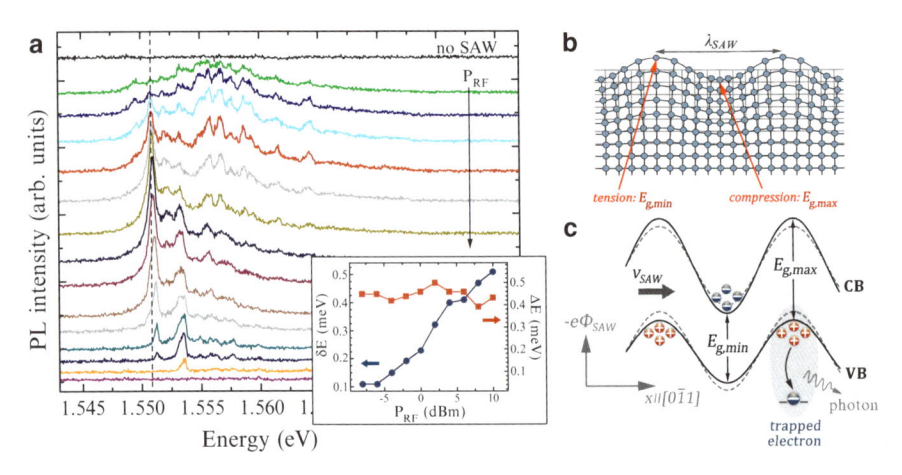

Fig. 9.12 (a) P_{rf} dependence of the PL intensity from a single remotely excited short QWR. The light was collected from the QWR indicated by the *circle* in Fig. 9.11b. The inset displays the P_{rf} dependence of the energy blueshift (δE, *circles*) and linewidth (ΔE, *squares*) measured for the PL line at 1.551 eV. (b) Strain field and (c) type-II modulation of the conduction band (CB) and valence band (VB) edges of a GaAs crystal by a SAW. The bandgap energy oscillates between maximum and minimum values given by $E_{max,g}$ and $E_{min,g}$, respectively. The diagram in (c) also illustrates the spatial separation of photo-generated carriers by the SAW piezoelectric field, as well as the recombination process following capture by discrete QD states

The recombination centers for the generation of single photons are probably induced by monolayer fluctuations in the QWR thickness as a result of step bunching [12, 29]. The latter also causes the fluctuations in the CL intensity along the QWRs illustrated in the inset of Fig. 9.11a. Note that the density of emission centers is highest at the corners of the mesa and at the center of the QWR. Due to carrier diffusion, however, it is not possible to spatially resolve the individual QD emission centers in the CL images.

In the CL experiments, the diffusive carrier transfer between the QW and the short QWRs only takes place within a small region around the QWRs. Transfer over much longer distances can be achieved by using a SAW to transport the carriers. Figure 9.11b displays a PL image obtained by exciting electron–hole pairs at a spot G in the QW approximately 10 μm away from the QWR matrix. The carriers are then acoustically transported to the short QWRs, where they recombine leading to the bright PL spots.

9.5.2 Recombination Control

Further information about the carrier recombination process in the short QWRs can be obtained from the spectral analysis of the PL emitted from a single QWR under remote optical excitation. Figure 9.12a displays PL spectra recorded under different

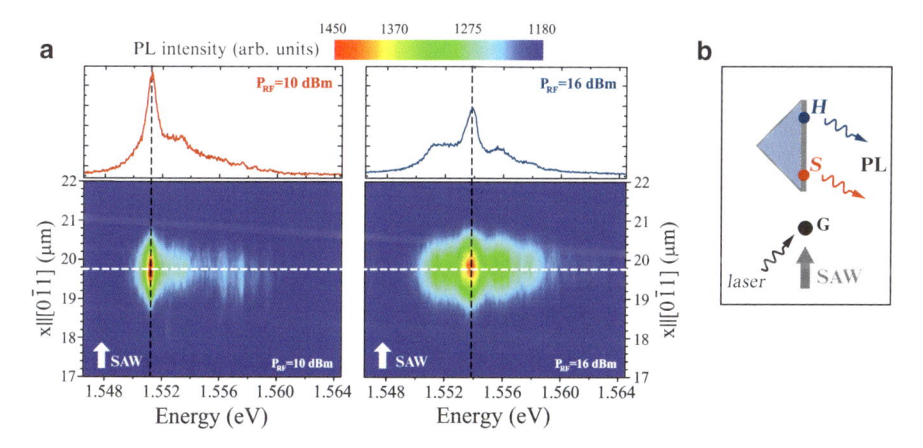

Fig. 9.13 (**a**) *Lower panels*: Time-integrated PL images of a single QWR recorded for P_{rf}=10 dBm (*left*) and 16 dBm (*right*). The *horizontal and vertical scales* represent the energy and the distance (along the SAW propagation direction x=[0$\bar{1}$1]) from the generation spot, respectively. *Upper panels*: The corresponding PL intensity profiles along the *dashed horizontal line*. The sketch in (**b**) shows a QWR which contains QDs with shallow (S) and high (H) trapping barriers

SAW intensities (described in terms of the rf-power P_{rf} applied to the IDT). The generation spot G (cf. Fig. 9.11b) was placed ~17 µm away from the selected short QWR indicated by the red circle in Fig. 9.11b. This distance is much larger than the typical carrier diffusion lengths. As a result, the remotely excited PL essentially vanishes in the absence of a SAW (upper curve in Fig. 9.12a). For low P_{rf} values, the PL spectra are similar to the one recorded for optical excitation directly on the QWR (similar to Fig. 9.9b), which consists of broad spectral features spread over a range of 10 meV. The energetic spacing between the lines (~2.7 meV) suggests that the emission arises from regions in the QWR differing in thickness by approximately two monolayers (~0.34 nm) on the stepped (311)A surface [33]. With increasing the P_{rf}, the broad PL peaks split into a series of sharp lines, which are attributed to the capture and recombination of carriers in the isolated QD states within the QWR.

The acoustic selection of a single QD within a short QWR relies on the fact that the SAW can not only inject but also extract carriers from the QD array. Namely, as the SAW intensity rises, the carriers impelled by a strong piezoelectric field can escape over the barriers imposed by the potential fluctuations in the QWR. This process eventually prevents capture and recombination even in the deeper QD trapping centers. As a result, the PL from the short QWRs disappears for very high acoustic intensities (lowest curve in Fig. 9.12a). By adjusting the acoustic power, one of the QDs within the short QWR can be selectively populated, giving rise to a PL spectrum consisting of a single emission line (spectrum for the second to the highest power in Fig. 9.12a). Such a behavior was observed in approximately 10% of probed short QWRs indicating the acoustic control of the recombination dynamics.

Further evidence for recombination control by the acoustic intensity is given by Fig. 9.13a, which displays the spectral distribution of the PL emitted from a single

QWR remotely excited by a pulsed diode laser (with wavelength of 635 nm, <90 ps pulse width, and 20.31 MHz repetition rate). The laser pulses were synchronized with the rf generator employed to drive the IDT in order to create carriers at a well-defined phase of the SAW field. The two pronounced peaks at 1.5511 and 1.5539 eV are associated with the recombination of acoustically transported carriers in quantized potentials within the single QWR. For the lower SAW power P_{rf} (upper left panel in Fig. 9.13a), the PL intensity is dominated by the single QD line at 1.5511 eV. With increasing P_{rf}, the SAW piezoelectric potential becomes strong enough to overcome the barriers induced by this QD. The recombination then takes place at a second QD emitting at 1.5539 eV (upper right panel). A possible mechanism for the change in the emission energy with acoustic power is illustrated in Fig. 9.13b. Here, the acoustically transported carriers first pass a QD with shallower trapping barrier (denoted as S) before reaching another one with a higher trapping barrier (H). For low acoustic powers, recombination is expected to take place at S, whereas emission from D only occurs for high SAW amplitudes.

Another important observation, which unveils information about the recombination mechanism, is the blueshift of the QD emission lines with increasing acoustic power. This blueshift (δE) is highlighted by the vertical dashed line in Fig. 9.12a and plotted as function of P_{rf} in the inset (circles). Note that for high acoustic powers δE exceeds the PL line width (ΔE), which is displayed by squares in the plot. As illustrated in Fig. 9.12b,c, the SAW creates regions of maximum compression and tension separated by $\lambda_{SAW}/2$, which induce a deformation potential modulation of the band gap energy with maximum and minimum values given by $E_{max,g}$ and $E_{min,g}$, respectively [27, 58]. In a time-integrated measurement (as in Fig. 9.12a), this periodic band gap modulation should result in a strain-induced broadening of the PL lines without a shift of the mean recombination energy [58]. In contrast to this expectation, data in the inset show an energetic shift with broadening of the PL line.

An increasing blueshift δE with P_{rf} together with a constant linewidth ΔE requires that the recombination process responsible for the appearance of the sharp PL lines takes place at a well-defined phase of the SAW field. This phase determines the relative amplitudes of the SAW piezoelectric potential energy and the strain-induced band gap modulation at the recombination time. Calculations of SAW fields using an elastic model show that the regions of maximum band gap (where the material is under maximum compressive strain) correspond to the sites of maximum electronic energy ($-e\Phi_{SAW}$) [17], where the holes are stored during acoustic transport. Based on this conclusion the recombination process in the QDs proceeds as sketched in Fig. 9.12c. Here, electrons transported in a SAW half-cycle with minimum of the electronic energy are initially captured by the QDs. The sharp PL lines are emitted in a subsequent SAW half-cycle, when moving holes, which are stored close to the positions of the maximum band gap, reach the QDs and recombine with the trapped electrons.

The recombination model introduced in the previous paragraph can explain the P_{rf} dependence of the remotely excited PL spectra in Fig. 9.12a. The broad PL lines observed for low acoustic powers are due to the recombination of transported

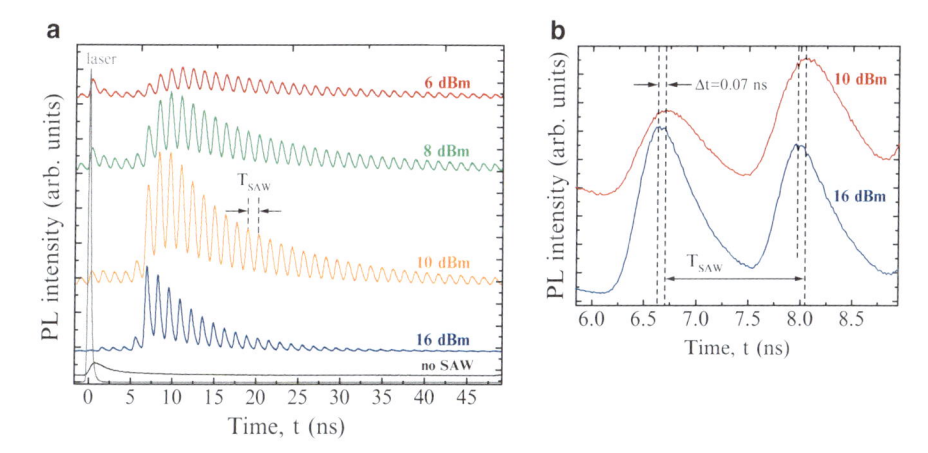

Fig. 9.14 (a) Time-resolved PL spectra of the QWR from Fig. 9.13a recorded under different acoustic powers (indicated in the figure). The trace labeled "laser" displays the reflectivity of the laser excitation pulse. The lowest trace was recorded in the absence of a SAW (trace labeled "no SAW"). The traces are vertically offset for clarity. (b) Enlarged portion of the PL decay curves recorded for P_{rf}=10 and 16 dBm

holes with a high density of electrons trapped within the short QWRs. The density of trapped carriers reduces for high acoustic powers, thereby leading to fewer recombination events.

9.5.3 Time-Resolved Carrier Transfer

The recombination dynamics is further supported by the time-resolved PL measurement presented in Fig. 9.14a, which were recorded under a SAW with frequency $f_{SAW} = 750$ MHz. Here, the PL from a single QWR in Fig. 9.13a was detected by an avalanche photodiode (APD). The time delay measuring electronics was triggered by a signal derived from the pulsed excitation and stopped on a detection of the photon. The precise excitation time and the time resolution of the detection setup (approximately 400 ps) were established by measuring the reflectance of the pulsed laser (trace labeled "laser"). Under acoustic excitation, the SAW field modulates the PL yielding pulses separated by the SAW period $T_{SAW} = 1/f_{SAW} = 1.33$ ns. The delay of approximately 6.5 ns between the laser excitation and the PL onset in Fig. 9.14a corresponds to the carrier transit time over the transport distance of ~19 μm between the excitation spot and the selected QWR. Both results attest that the carriers are periodically transported with the SAW velocity v_{SAW} and that the recombination takes place synchronously with the driving acoustic field. The repetition rate of the PL pulses is compatible with the recombination model presented above, which predicts that photons are emitted only once in a SAW cycle when transported holes recombine with trapped electrons.

The small PL peak observed in Fig. 9.14a at $t = 0$ (corresponding to a very short transit time) is an artifact due to imperfect filtering of the QW emission at the excitation spot. This peak has also been observed in the absence of a SAW (trace labeled "no SAW"). It is interesting to note that a single laser excitation pulse gives rise to multiple PL pulses in the time-resolved traces. The appearance of several pulses is attributed to the formation of a pool of long-living carriers close to the excitation spot. For weak SAWs (upper curves in Fig. 9.14a), the piezoelectric field becomes screened by the high carrier density generated by the laser pulse. The screened field can still increase the carrier lifetime by spatially separating electrons and holes. It can, however, only extract and transport away carriers from the borders carrier pool during the first SAW cycles after the laser pulse. Subsequent carrier extraction leads to the PL oscillations persisting over times much longer than the laser repetition period. The number of SAW cycles required to deplete the carrier density reduces at G with the SAW amplitude.

The rise and the decay time of the PL pulses are limited by the finite timing jitter of the detector (of approx. 400 ps), thus indicating that recombination takes place within a time scale substantially shorter than the detector resolution. The short recombination times in the discrete QD states results from the fact that the spatial location of the transport holes overlap with electron trapping sites only within a time window much shorter than $T_{SAW}/2$. The emission process has intrinsically a small jitter, which is an important feature for high-frequency single-photon sources.

The time-resolved results in Fig. 9.14a also allow us to extract information about the position of the QD emission centers within the short QWRs, thereby adding sub-micrometer spatial resolution to the conventional optical spectroscopy. The dimensions of the localized emission centers presented in the two-dimensional plots of Fig. 9.13a are below the optical spatial resolution, so that their exact location cannot be precisely determined from spatially resolved PL studies. Furthermore, we have seen that the precise location of the emission centers cannot be determined from the CL images presented in Sect. 9.5.1 due to carrier diffusion. However, by taking into account that the carriers are transported with the well-defined SAW velocity v_{SAW}, the average separation between the emission regions in Fig. 9.13a can be estimated from the time shift between the PL pulses for different P_{rf} in Fig. 9.14b. The observed maximum time shift of 70 ps thus yields an upper limit for the separation between the emission regions for the two lines in Fig. 9.13a of approximately 210 nm. This result compares well with near-field PL studies on similar QWR structures [33], where the average length of the emission region of about 150 nm was estimated. It is interesting to note that the emission region is much shorter than the length of the short QWRs.

9.5.4 Photon Anti-Bunching

The results presented in the previous section indicate the possibility of populating single QD states using SAWs. In order to demonstrate the discrete character of

Fig. 9.15 Photon correlation histograms of the remotely excited QWR of Fig. 9.13a recorded for $P_{rf}=10$ dBm (*middle trace*) and 16 dBm (*upper trace*). The traces are shifted vertically for clarity. The short and long oscillations correspond to the periodicity of the SAW ($T_{SAW} = 1.33$ ns) and laser pulses ($T_{laser} = 27.94$ ns). The lower histogram was measured from the same QWR for local optical excitation and in the absence of a SAW

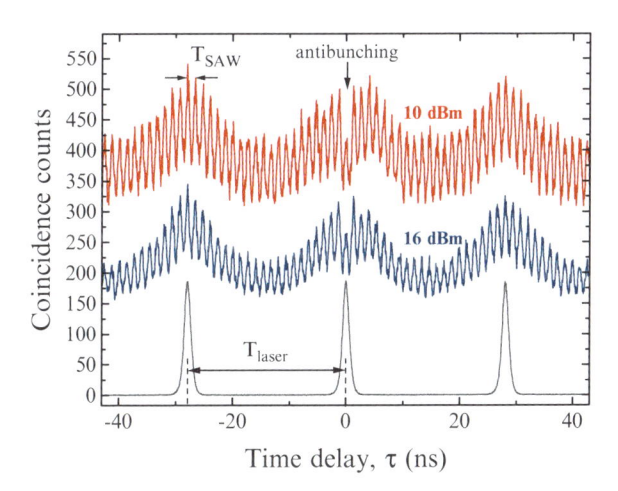

these states, we have investigated the photon statistics of the emitted PL using a Hanbury–Brown and Twist (HBT) setup. The latter consists of a 50:50 non-polarizing beam splitter, which directs the photon flux to two APDs connected to a time-correlated single-photon counting electronics. In the experiments, the number of photon pairs from an acoustically pumped short QWR with arrival-time separations of τ is summarized in the form of a photon correlation histogram. Due to the low intensity of the remotely excited PL, the autocorrelation measurements could not be performed by spectrally resolving a single PL line. Instead, the whole emission band of a single QWR was collected and analyzed. The selected QWR contains in general more than one emission center. In the experiments, the acoustic power was chosen as in Fig. 9.13a in order to reduce the total number of emission lines.

The two upper curves in Fig 9.15 display photon autocorrelation histograms recorded for a short QWR under the same excitation conditions as for the PL spectra in the left and right sides of Fig. 9.13a, respectively. The histograms include photon emission events separated in time τ by as much as 40 laser cycles: only the three cycles close to $\tau = 0$ are displayed in Fig. 9.15. For comparison, the lower trace shows a similar histogram recorded in the absence of a SAW for optical excitation directly on the short QWR. In the latter, several recombination centers contribute to the emission and only correlation events at the laser repetition rate ($1/T_{laser}$) are observed. Under remote acoustic excitation, in contrast, the recombination of transported carriers in the QWR is effectively frequency-locked to the driving acoustic filed: the correlation measurements deliver a number of coincidence peaks spaced by a SAW period $T_{SAW} = 1.33$ ns, corresponding to the maxima in the recombination probability. The overlap of the adjacent peaks is due to the finite time resolution of the correlation setup.

Figure 9.15 shows a clear reduction of the coincidence rate at zero-time delay (photon anti-bunching, indicated by the arrow) for both acoustic excitation powers (P_{rf}). These anti-bunching minima reflect the reduced probability for simultaneous emission of multiple photons, thus proving the non-classical character of the light emitted from the acoustically excited QWR. The residual counts at $\tau = 0$ are attributed to contributions from additional emission centers within the detection area, which also accounts for the wide background in the PL spectra of Fig. 9.13a. Note that the two different P_{rf} values lead to different photon emission energies (cf. Fig. 9.13a). Thus, in the single-photon source presented here the photon energy can be changed by controlling the acoustic power.

As for the time-resolved PL results in Fig. 9.14a, the amplitude of the oscillations in the HBT histograms is limited by the time-resolution of the APDs used in the HBT setup rather than by the carrier recombination time. The photon emission region must, therefore, be much smaller than the half the acoustic wavelength. Under these conditions, the acoustic transport reduces the time available for recombination to values below half of the acoustic period as well as the jitter in photon emission time. In agreement with these results, the photon anti-bunching only takes place at $\tau = 0$. Reduced jitter and short recombination times are important consideration for the realization of high frequency single-photon sources. The repetition ratio of 750 MHz for the source in Fig. 9.15 is comparable to those achieved using high-frequency sources based on InAs QDs [10, 27]. Higher rates can be obtained by increasing the SAW frequency.

Information about the number of emission centers can be obtained from the suppression level of the $\tau = 0$ peak in the HBT histograms. In general, the residual coincidence rate at $\tau = 0$ is proportional to $1 - 1/n$, where n denotes the number of independent emission centers present in the emission region. The determination of the suppression level (and, therefore, of n) from the histograms of Fig. 9.15 becomes complicated due to the finite time resolution of the HBT setup. In order to estimate the suppression, we used the construction sketched in the inset of Fig. 9.16. Here, the intensity of the $\tau = 0$ coincidence peak (indicated by $h_c(i = 0)$) is compared to the averaged intensity ($\bar{h}_c$ in Fig. 9.16b) of the central peaks at $\tau = iT_{laser}$ ($h_c(i)$ for all laser cycles $i \neq 0$). The latter is represented by the amplitude of the dashed envelope at $\tau = 0$ in Fig. 9.16. The amplitudes of the central peaks $h_c(i)$ were determined by using an extrapolation procedure (indicated by the dashed lines) to subtract the background.

In analogy with a quantum emitter probed by pulse laser excitation [24], the ratio $h_c(i = 0)/\bar{h}_c$ is a measure of the probability of detecting photon pairs per SAW cycle and monitors, therefore, deviations from a classical light source. For photon correlation histograms in Fig. 9.15, the estimated ratio $h_c(i = 0)/\bar{h}_c$ is 0.72 ± 0.05 for $P_{rf} = 10$ dBm and 0.74 ± 0.05 for $P_{rf}=16$ dBm, as indicated, respectively, by the circled star and square symbols in Fig. 9.16. The deviation of these points from the average values (represented by the dot-dashed line) is more than 5 standard deviations (dashed lines). These anti-bunching levels suggest that less than four emission centers are present in the emission area ($n < 4$). Similar degrees of anti-bunching (not shown) have been obtained for other short QWRs. The probability of the emission of multiple photons from these remotely excited QWRs is, therefore, reduced by approximately 25–30% in comparison with a coherent source of the same intensity.

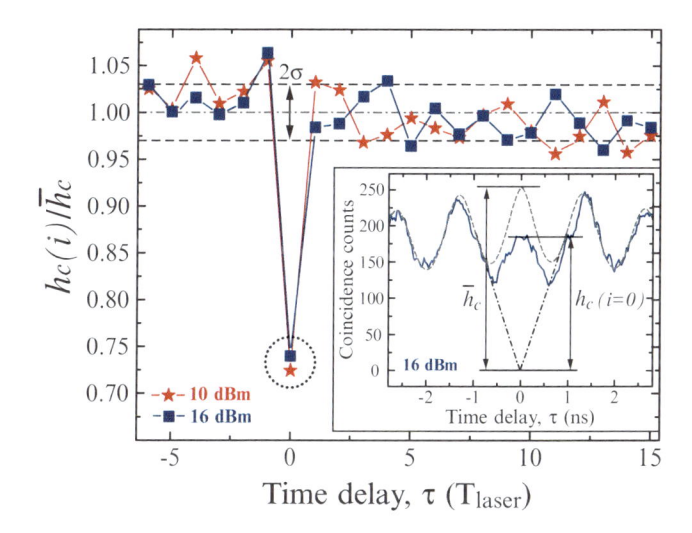

Fig. 9.16 Normalized intensities of central correlation peaks $h_c(i)$ appearing at times $\tau = iT_{\text{laser}}$ for all laser cycles i ($i = -6, -7, \ldots, 33$) for the data in Fig. 9.15. Symbols at $\tau = 0$ surrounded by a *dotted circle* show the degree of anti-bunching for $P_{\text{rf}} = 10$ dBm (*star*) and 16 dBm (*square*). The *dot-dashed and the two dashed lines* display the average value of $h_c(i)/\bar{h}_c$ ratio and the deviation σ, respectively. *Inset* shows the magnified portion of the 16 dBm-photon correlation histogram displayed in Fig. 9.15. The *dashed envelope* represents the average intensity of the correlation peaks in the absence of anti-bunching. The amplitudes of the central peaks $h_c(i)$ are measured from the resolution-limited background level, which is estimated from the overlap of neighboring side peaks represented by the *dash-dotted lines*

Finally, we conclude this section by briefly discussing the mechanisms leading to non-classical light emission from the acoustically populated QWRs. We have seen in Sect. 9.2.2 that acoustically driven single-photon emission can be achieved by (i) limiting the number of carriers transported per SAW cycle to the recombination area. If only a few carriers reach the QWR, their recombination can lead to the observed levels of anti-bunching. Alternatively, single-photon sources can be obtained (ii) using two-level systems based on discrete QD states to limit the photon emission rate. The relative contribution of these two mechanisms was investigated by measuring the degree of photon anti-bunching for various laser excitation densities at a fixed SAW power P_{rf}. The laser excitation power controls the number of photo-excited carriers and, hence, the average number of carriers transported in each SAW cycle. By keeping P_{rf} constant, we ensure that the efficiency of the acoustic transport as well as the carrier capture and extraction processes in the QD states are not changed. We have found that the degree of anti-bunching remains approximately the same for different laser excitation conditions, therefore suggesting that the non-classical light emission is governed by the carrier capture and recombination in the quantized QD states within the QWR (corresponding to mechanism (ii)). The capture and the recombination of carriers in the discrete QD states, which are controlled by the SAW piezoelectric field, also opens the way for the dynamic tuning of the single photon energy indicated in Fig. 9.13.

9.6 Conclusions and Future Perspectives

Coherent acoustic phonons in the form of acoustic waves propagating on a semiconductor surface provide an efficient mechanism for the controlled transport and manipulation of electron–hole pairs. Full exploitation of this ability require electrically interconnected low-dimensional structures like QWs, QWRs, and QDs on a semiconductor surface. These structures can be fabricated using MBE growth on patterned (311)A GaAs substrates described in Sect. 9.3.1. The preferential growth on the sidewalls of etched mesa patterns is scalable and compatible with existing semiconductor processing and allows for an almost perfect site control defined by lithography.

Optical investigations of the charge transport in GaAs (311)A sidewall QWRs by SAW have shown that carriers can be efficiently transported over several tens of micrometers (cf. Sect. 9.4). As shown in Sect. 9.4.1, potential fluctuations along the QWR axis act as QD states, which capture the carriers during acoustic transport and induce their recombination. It was shown in Sect. 9.5 that by adjusting the acoustic transport conditions (determined by the amplitude of the SAW field), it is possible to selectively populate a single QD within the ensemble giving rise to single-photon emission at the SAW frequency. In addition, the SAW amplitude selects which QD is populated: using this process, we have demonstrated on-demand single-photon source with acoustically tunable emission energy.

While the results of Sect. 9.5 demonstrate the feasibility for high-frequency single-photon sources based on acoustic transport, considerable improvements in the anti-bunching levels and in the photon throughput are required to enable applications. The former can be enhanced by reducing the numbers of the QD-like trapping sites within the QWR. The results from Sect. 9.5.3 show that the emitting QD states are closely located (less than a few hundreds of nanometers), so it is unlikely that their density can be reduced by simply decreasing the length of the QWRs or the size of the photon collection area. Therefore, further investigations of the growth mechanism for QD formation, which may lead to a microscopic arrangement of one-dimensional surface corrugations, are required. In particular, the use of atomic hydrogen to control step-bunching during growth has not been investigated [43].

The efficiency of the acoustic SPSs can be significantly enhanced by embedding the QWRs in an optical microcavity. The microcavity would also allow operation at higher temperatures [19]. Also, higher photon yield SPSs can be achieved by optimizing the charge transfer efficiency. For instance, tailored acoustic fields can be used to maximize the transfer of carriers from the generation areas in the QW to the embedded QDs. Such fields can be generated using focusing acoustic transducers [20]. Another possibility is to pump the short QWRs using dynamic dots formed by the interference of the two orthogonal SAW beams [6]. Finally, electrically driven SPSs can be implemented by using the SAW to extract electrons and holes from intentionally doped n- and p-type regions on the (311)A crystal.

Our results open the door to a variety of quantum information experiments using SAWs. Arrays of synchronized single-photon emitters can be realized by simultaneously populating several QDs using a single SAW beam. Also, the acoustic transport of charge carriers in sidewall QWRs can be integrated with spin devices to produce spin-polarized single photon emission. The combination of ambipolar acoustic transport with the one-dimensional confinement provided by the QWR should suppress spin dephasing through spin–orbit effects [36]. As in Fig. 9.1, the spin vector can be manipulated by magnetic or electric control gates along the QWRs. The wires can direct the spin-polarized carriers to the QDs, leading to the subsequent emission of single-photons with well-defined polarization. Polarized states of single photons have important implications for single qubit (quantum bit) operations in quantum information technology [39]. Despite the efficient charge transport in sidewall QWRs, the spin transport distances reported for the GaAs (311)A QWRs are currently too short (comparable to the SAW wavelength of a few micrometers), probably due to strong spin scattering at potential fluctuations during transport [5]. An important issue yet to be addressed concerning the spin dynamics in lithographically patterned wires is the influence of the etching process on the QWR quality. In particular, the fluctuations induced by inhomogeneities in the QWR thickness can also be related to the roughness of the mesa edges defined by photolithography. It remains to be investigated whether the wire quality and, thus, the spin coherence length can be further enhanced by improving the fabrication procedure.

Acknowledgments This work has been the results of many fruitful collaborations. Our special thanks are addressed to R Hey for the supply of state-of-the-art molecular-beam epitaxy samples, as well as to F. Alsina, F. Iikawa, J. A. H. Stotz, R. Nötzel, and U. Jahn for the collaboration in the field of carrier transport in QWRs. We also thank A. Tahraoui for discussions and comments on the manuscript. Finally, we also acknowledge the technical support from A.-K. Bluhm, M. Höricke, S. Krauß, W. Seidel, H.-P. Schönherr, and E. Wiebicke in the fabrication of the samples. This work was supported by the NanoQUIT consortium, Bundesministerium für Bildung und Forschung (BMBF), Germany.

References

1. Alsina F, Santos PV, Hey R (2002) Spatial-dispersion-induced acoustic anisotropy in semiconductor structures. Phys Rev B 65:193301
2. Alsina F, Santos PV, Hey R, García-Cristóbal A, Cantarero A (2001) Dynamic carrier distribution in quantum wells modulated by surface acoustic waves. Phys Rev B 64:0410304(R)
3. Alsina F, Santos PV, Schönherr HP, Nötzel R, Ploog KH (2003) Real-time dynamics of the acoustically-induced carrier transport in GaAs quantum wires. Phys Rev B 67:161305(R)

4. Alsina F, Santos PV, Schönherr HP, Seidel W, Nötzel R, Ploog KH (2002)Surface-acoustic-wave-induced carrier transport in quantum wires. Phys Rev B 65:165330
5. Alsina F, Stotz JAH, Hey R, Jahn U, Santos PV (2008) Acoustic charge and spin transport in GaAs quantum wires. Phys Stat Sol (c) 9:2907
6. Alsina F, Stotz JAH, Hey R, Santos PV (2004) Acoustically induced potential dots on GaAs quantum wells. Solid State Commun 129:453
7. Alsina F, Stotz JAH, Hey R, Santos PV (2006) Radiative recombination during acoustically induced transport in GaAs quantum wells. J Vac Sci Technol B 24:2029
8. Auld BA (1990) Acoustic fields and waves in solids. Robert E. Krieger Publishing Company, Malabar, Florida
9. Batista PD, Hey R, Santos PV (2008) Efficient electrical detection of ambipolar acoustic transport in GaAs. Appl Phys Lett 93:262108
10. Bennett AJ, Unitt DC, See P, Shields AJ, Atkinson P, Cooper K, Ritchie DA (2005) Electrical control of the uncertainty in the time of single photon emission events. Phys Rev B 72(3):033316
11. Benson O, Santori C, Pelton M, Yamamoto Y (2000) Regulated and entangled photons from a single quantum dot. Phys Rev Lett 84:2513
12. Biasiol G, Reinhardt F, Gustafsson A, Kapon E (1997) Self-limiting OMCVD growth of GaAs on V-grooved substrates with application to InGaAs/GaAs quantum wires. J Electron Mater 26:1194
13. Bouwmeester D, Ekert AK, Zeilinger A (eds) (2000) The physics of quantum information. Springer, Berlin
14. Bödefeld C, Ebbecke J, Toivonen J, Sopanen M, Lipsanen H, Wixforth A (2006) Experimental investigation towards a periodically pumped single-photon source. Phys Rev B 74:035407
15. Brassard G, Lütkenhaus N, Mor T, Sanders BC (2000) Limitations on practical quantum cryptography. Phys Rev Lett 85:1330
16. Couto Jr. ODD, Iikawa F, Rudolph J, Hey R, Santos PV (2007) Anisotropic spin transport in (110) GaAs quantum wells. Phys Rev Lett 98:036603
17. Couto Jr ODD, Lazić S, Iikawa F, Stotz J, Hey R, Santos PV (2009) Photon anti-bunching in acoustically pumped quantum dots. Nat Phot 3:645
18. Craig NJ, Taylor JM, Lester EA, Marcus CM, Hanson MP, Gossard AC (2004) Tunable nonlocal spin control in a coupled-quantum dot system. Science 304:565
19. de Lima Jr. MM, Hey R, Stotz JAH, Santos PV (2004) Acoustic manipulation of electron–hole pairs in GaAs at room temperature. Appl Phys Lett 84:2569
20. de Lima Jr. MM, Santos PV (2005) Modulation of photonic structures by surface acoustic waves. Rep Prog Phys 68:1639
21. Eberl K, Petroff PM, Demeester P (eds) (1995) Low dimensional structures prepared by epitaxial growth or regrowth on patterned substrates, NATO advanced science institute series E, vol 298. Kluwer Academic, Dordrecht
22. Foden CL, Talyanskii VI, Milburn GJ, Leadbeater ML, Pepper M (2000) High-frequency acousto-electric single-photon source. Phys Rev A 62:011803(R)
23. Fricke J, Notzel R, Jahn U, Niu Z, Schönherr HP, Ramsteiner M, Ploog KH (1999) Patterned growth on GaAs (311)A substrates: Engineering of growth selectivity for lateral semiconductor nanostructures. J Appl Phys 86:2896
24. Geddes CD, Lakowicz JR (eds) (2005) Reviews in fluorescence 2005. Springer Science + Business Media, New York
25. Gell JR, Atkinson P, Bremner SP, Sfigakis F, Kataoka M, Anderson D, Jones GAC, Barnes CHW, Ritchie DA, Ward MB, Norman CE, Shields AJ (2006) Surface-acoustic-wave-driven luminescence from a lateral p-n junction. Appl Phys Lett 89:243505
26. Gell JR, Ward MB, Shields AJ, Atkinson P, Bremner SP, Anderson D, Kataoka M, Barnes CHW, Jones GAC, Ritchie DA (2007) Temporal characteristics of surface-acoustic-wave-driven luminescence from a lateral p-n junction. Appl Phys Lett 91:013506

27. Gell JR, Ward MB, Young RJ, Stevenson RM, Atkinson P, Anderson D, Jones GAC, Ritchie DA, Shields AJ (2008) Modulation of single quantum dot energy levels by a surface-acoustic-waves. Appl Phys Lett 93:081115
28. Gisin N, Ribordy G, Tittel W, Zbinden H (2002) Quantum cryptography. Rev Mod Phys 74:145
29. Hey R, Friedland KJ, Kostial H, Ploog KH (2004) Conductance anisotropy of high-mobility two-dimensional hole gas at GaAs/(Al,Ga)As (113)A single heterojunctions. Phys E 21:737
30. Hosey T, Talyanskii V, Vijendran S, Jones GAC, Ward MB, Unitt DC, Norman CE, Shields AJ (2004) Lateral n-p junction for acoustoelectric nanocircuits. Appl Phys Lett 85:491
31. Hoskins MJ, Morkoç H, Hunsinger BJ (1982) Charge transport by surface acoustic waves in GaAs. Appl Phys Lett 41:332
32. Imamoglu A (1992) Nonclassical light generation by coulomb blockade of resonant tunneling. Phys Rev B 46:15982
33. Intonti F, Emiliani V, Lienau C, Elsaesser T, Nötzel R, Ploog KH (2001) Near-field optical spectroscopy of localized and delocalized excitons in a single GaAs quantum wire. Phys Rev B 63:75313
34. Jiao SJ, Batista PD, Biermann K, Hey R, Santos PV (2009) Electrical detection of ambipolar acoustic carrier transport by surface acoustic waves. J Appl Phys 106:053708
35. Kim J, Benson O, Kan H, Yamamoto Y (1999) A single-photon turnstile device. Nature 397:500
36. Kiselev AA, Kim KW (2000) Progressive suppression of spin relaxation in two-dimensional channels of finite width. Phys Rev B 61:13115
37. Korpel A (1997) Acousto-optics. Marcel Dekker, New York
38. Lienau C, Richter A, Behme G, Süplitz M, Heinrich D, Elsaesser T, Ramsteiner M, Nötzel R, Ploog KH (1998) Nanoscale mapping of confinement potentials in single semiconductor quantum wires by near-field optical spectroscopy. Phys Rev B 58:2045
39. Lounis B, Orrit M (2005) Single-photon sources. Rep Prog Phys 68:1129
40. Michler P, Kiraz A, Becher C, Schoenfeld WV, Petroff PM, Zhang L, Hu E, Imamoğlu A (2000) A quantum dot single-photon turnstile device. Science 290:2282
41. Miskinis R, Rutkowski O, Urba E (1996) Surface acoustic waves on the (11n) cuts of gallium arsenide. J Appl Phys 80:4867
42. Nötzel R, Menniger J, Ramsteiner M, Ruiz A, Schönherr HP, Ploog KH (1996) Selectivity of growth on patterned GaAs (311)A substrates. Appl Phys Lett 68:1132
43. Nötzel R, Niu ZC, Ramsteiner M, Schönherr HP, Trampert A, Däweritz L, Ploog KH (1998) Uniform quantum-dot arrays formed by natural self-faceting on patterned substrates. Nature (London) 392:56
44. Nötzel R, Ploog KH (2000) Patterned growth on high-index GaAs (311)A substrates. Appl Surf Sci 166:406
45. Nötzel R, Ramsteiner M, Menniger J, Trampert A, Schönherr HP, Däweritz L, Ploog KH (1996) Micro-photoluminescence study at room temperature of sidewall quantum wires formed on patterned GaAs (311)A substrates by molecular beam epitaxy. J Appl Phys 35:L297
46. Nötzel R, Ramsteiner M, Niu Z, Schönherr HP, Däweritz L, Ploog KH (1997) Enhancement of optical nonlinearity in strained (InGa)As sidewall quantum wires on patterned GaAs (311)A substrates. Appl Phys Lett 70:1578
47. Pochung Chen CP, Sham LJ (2001) Control of exciton dynamics in nanodots for quantum operations. Phys Rev Lett 87:067401
48. Rayleigh L (1885) On waves propagated along the plane surface of an elastic solid. Proc Lond Math Soc s1-17(1):4
49. Richter A, Behme G, Süptitz M, Lienau C, Elsaesser T, Nötzel R, Ploog KH (1997) Real-space transfer and trapping of carriers into single GaAs quantum wires studied by near-field optical spectroscopy. Phys Rev Lett 79:2145
50. Richter A, Süptitz M, Heinrich D, Lienau C, Elsaesser T, Nötzel R, Ploog KH (1998) Exciton transport into a single GaAs quantum wire studied by picosecond near-field optical spectroscopy. Appl Phys Lett 73:2176

51. Rocke C, Zimmermann S, Wixforth A, Kotthaus JP, Böhm G, Weimann G (1997) Acoustically driven storage of light in a quantum well. Phys Rev Lett 78:4099
52. Royer D, Dieulesaint E (2000) Elastic waves in solids. Springer, Heidelberg
53. Santori C, Pelton M, Solomon G, Dale Y, Yamamoto Y (2001) Triggered single photons from a quantum dot. Phys Rev Lett 86:1502
54. Santos PV, Nötzel R, Ploog KH (1999) Polarization anisotropy in quasi-planar sidewall quantum wires on patterned GaAs (311)A substrates. J Appl Phys 85:8228
55. Santos PV, Stotz JAH, Hey R (2005) Control of photogenerated carriers and spins using surface acoustic waves. In: Takayanagi H, Nitta J (eds) Realizing controllable quantum states: Proc. of the Int. Symp. on Mesoscopic Superconductivity and Spintronics - In the light of quantum computation, p 357. World Scientific, Singapore
56. Shilton JM, Talyanskii VI, Pepper M, Ritchie DA, Frost JEF, Ford CJB, Smith CG, Jones GAC (1996) High-frequency single-electron transport in a quasi-one-dimensional GaAs channel induced by surface acoustic waves. J Phys Condens Matter 8:L531
57. Sogawa T, Gotoh H, Hiyarama Y, Santos P, Ploog K (2007) Dimensional oscillation in GaAs/AlAs quantum wells by 2-dimensional standing surface acoustic waves. Appl Phys Lett 91:141917
58. Sogawa T, Santos PV, Zhang SK, Eshlaghi S, Wieck AD, Ploog KH (2001) Dynamic band structure modulation of quantum wells by surface acoustic waves. Phys Rev B 63:121307(R)
59. Sogawa T, Santos PV, Zhang SK, Eshlaghi S, Wieck AD, Ploog KH (2001) Transport and lifetime enhancement of photoexcited spins in GaAs by surface acoustic waves. Phys Rev Lett 87:276601
60. Stotz JAH, Hey R, Santos PV, Ploog KH (2005) Coherent spin transport via dynamic quantum dots. Nat Mater 4:585
61. Talyanskii VI, Milburn GJ, Stotz JAH, Santos PV (2007) Acoustoelectric single-photon detector. Semicond Sci Technol 22:209
62. Tanski WJ, Merritt SW, Sacks RN, Cullen DE, Branciforte EJ, Carroll RD, Eschrich TC (1987) Heterojunction acoustic charge transport devices on GaAs. Appl Phys Lett 52:18
63. Tsai CS (1990) Guided-wave acousto-optics. Springer, Berlin
64. Wang XL, Voliotis V (2006) Epitaxial growth and optical properties of semiconductor quantum wires. J Appl Phys 99:121301
65. White RM (1970) Surface elastic waves. In: Proc. of the IEEE, vol 58. IEEE, New York, p 1238
66. White RM, Vollmer FW (1965) Direct piezoelectric coupling to surface elastic waves. Appl Phys Lett 7(12):314
67. Wiele C, Haake F, Rocke C, Wixforth A (1998) Photon trains and lasing: The periodically pumped quantum dot. Phys Rev A 58:R2680
68. Wixforth A, Scriba J, Wassermeier M, Kotthaus J, Weimann G, Schlapp W (1989) Surface acoustic waves on GaAs/Al$_x$Ga$_{1-x}$As heterostructures. Phys Rev B 40:7874
69. Yamamoto Y, Santori C, Solomon G, Vuckovic J, Fattal D, Waks E, Diamanti E (2005) Single photons for quantum information systems. Progr Informat 1:5
70. Zhang V, Lefebvre HE, Gryba T (1997) Theoretical study of surface acoustic waves in (n11) GaAs-cuts. IEEE Trans Sonics Ultrason SU-44:406

Index

A

Ab initio method, 3, 6–10, 15, 17, 19, 20, 22, 23, 25–28, 30, 31, 82, 88, 97, 137–170
Acoustic, 8, 42, 88, 125, 153, 178, 208, 259
Acoustic detection, 220
Acoustic generation, 214, 215, 220, 223
Acoustic mismatch theory, 246
Acoustic phonons, 30, 42, 100, 108, 128, 163, 183, 185, 186, 209, 213, 214, 216–219, 221–223, 229–231, 250, 288
Adiabatic bond charge model, 3, 4, 8, 9, 26, 30, 35, 88–91, 108
Alloys, 4, 111, 116, 137–139, 142, 147, 160, 165–166
Amplification, 227–238, 251–254
Anharmonicity, 85, 86, 106, 116, 119, 169, 181
Anharmonic scattering, 97–100
Antisymmetric mode, 73, 74

B

Bicrystal, 184, 187–192
Boltzmann transport equation, 85, 121, 128, 131, 134, 138–140, 169, 177
Bragg, 240–246, 249, 261
Brillouin, 5, 6, 8, 15, 23, 28, 31, 34, 88, 89, 91–93, 98, 99, 101, 116, 120, 122–125, 127, 128, 134, 140, 141, 144–147, 149–151, 153, 160, 164, 177, 181, 185, 208, 211–213, 216, 217, 223, 240, 248, 250
Bulk semiconductors, 82, 88, 93, 102–104, 116

C

C(001)(2×1), 17, 22–28, 30–32
Callaway theory, 87, 106
Carbon nanotubes (CNTs), 41, 42, 49–51, 53, 58–61, 63, 77
Carrier-phonon scattering, 48, 49, 67
Cavity, 229, 231, 239, 240, 244–247, 249–251, 254, 262, 288
Clean surfaces, 189
CNTs. *See* Carbon nanotubes (CNTs)
Coherent, 231, 232, 249, 250, 253, 254, 286, 288
Confined phonon modes, 73
Continuum simulation, 200–202

D

Debye frequency, 60, 93, 181, 210
Density functional, 4–6, 14, 22, 116, 117, 122–124, 147–154, 168, 169
Density functional perturbation theory, 22, 116, 117, 122–124, 149, 153, 169
Density functional theory (DFT), 5, 6, 14, 124, 147, 153, 154, 169
Density of states, 6–15, 18, 21, 30, 36, 38, 88, 89, 91, 94, 96, 98, 108, 143, 144, 165, 182–184, 208, 210
DFB. *See* Distributed feedback (DFB)
DFT. *See* Density functional theory (DFT)
Diamond, 4–9, 89, 138, 155, 162, 163, 167, 169, 175, 190, 209
Dielectric continuum model, 46–50, 72
Differential measurement, 194
Diffusive mismatch theory, 183
Dimensionally-confined structures, 77

Dispersion, 1, 2, 4–10, 15, 17, 18, 21–23, 26, 28, 30, 31, 34, 35, 37, 51, 53, 58–61, 63–66, 68, 69, 73, 82, 88–92, 101, 108, 112, 122, 123, 128, 138, 147–161, 164–166, 169, 177–180, 185, 186, 209–213, 217, 240, 241, 247, 248
  relation, 2, 4–6, 37, 51, 53, 58–61, 63, 65, 68, 73, 82, 88, 90, 91, 108, 112, 177, 179, 185, 209, 211, 213
Distributed feedback (DFB), 245, 247–251

E
Effective medium approach, 182
Effective thermal conductivity, 196–200
Elastic continuum model, 42–46, 51, 53, 63
Electron
  backscatter diffraction, 189, 190
  energy loss spectroscopy, 2, 17, 188
Electron-phonon
  interaction, 13–16, 69, 234
  scattering, 105
Exciton, 250, 261, 273, 274, 277

F
First principles, 2, 4–6, 84, 97, 115–135, 138, 147, 148, 152, 168, 170
  calculations, 2, 129, 224
  methods, 4–6
Force constants, 2, 4, 5, 33, 116, 117, 120, 122, 123, 125, 134, 138, 141, 142, 147, 149–160, 169, 179
Frozen phonon approach, 5

G
GaAs suspended nanobeams, 104–105
Gaussian smearing, 144, 145, 147
Ge(001)(2×1), 17, 20–23, 30–32
Germanium, 115–135, 138, 160, 161
GHz-THz phonons, 208
Grain Boundarie(s), 60, 116, 175–177, 182, 184, 185, 203
Graphene nanoribbons, 41
Graphene quantum dot, 41
Graphite, 42, 60, 64, 68–69, 77
Green's function, 84, 138, 143, 166

H
Heterostructure, 73, 78, 231, 261, 268
High resolution transmission electron microscopy, 188

I
Inelastic X-ray scattering, 186
Inner-atomic bonding energy, 180
Interatomic force constants, 2, 116, 117, 120, 122, 123, 125, 138, 141, 147–158, 179
Interface
  phonon modes, 79
  scattering, 109, 112, 176, 183, 189
  thermal resistance, 175
Isotope scattering, 167

K
Kapitza resistance, 108, 182–185, 187, 188, 192, 193

L
Lanthanum, 6, 8–13, 37
Linear response, 5, 17, 22, 82, 83, 149, 184
Linear response approach, 5, 82, 83

M
Matthiessen's rule, 180, 210
MgCNi3, 6, 13–16, 37
Miniband, 232, 236
Mirror, 19, 23–25, 27–29, 155, 189, 218, 239–247, 249, 253
Molecular dynamics simulation, 85, 128, 168, 184, 194
Multiple-wall nanotube, 53–58

N
Nanostructures, 1–37, 42, 43, 49, 81, 82, 88, 90–92, 111–113, 138, 152, 182, 186, 203, 211, 250, 263, 268–273
Nanowire(s), 3, 33–36, 91, 92, 95, 106–107, 109, 111–113, 138, 153, 155, 156, 158, 166–168, 176, 184, 186, 193, 203, 208, 211, 259–289
Neutron scattering, 1, 2, 15, 185, 187, 213
Noncontinuum simulation, 194–195
Normalization condition, 48, 49, 52, 55, 67, 72
Normal scattering, 179, 181, 185

O
3-omega method, 186, 187, 192
Optical phonons, 46, 47, 63–67, 77, 124, 128, 129, 150, 186, 209

Index 295

P

Phonon
 anomalies, 10
 boltzmann transport equation, 85, 134
 density of states, 8–10, 12, 13, 15, 18, 21, 36, 88, 89, 91, 108, 143, 144, 165, 183, 208, 210
 dispersion relation, 1, 2, 4–6, 61, 68, 82, 108, 112, 185, 211, 213
 distribution, 82, 86, 178, 181
 emission, 59, 71, 72, 212, 229, 230, 233, 247
 engineering, 110–113
 interactions, 61, 97, 138, 150, 170, 180, 193, 200, 202, 209–211, 228
 lifetimes, 82, 112, 115–135, 140, 146, 170
 mean-free-path, 222
 scattering, 48, 49, 60, 63, 67, 85, 87, 88, 93, 95, 96, 99, 104, 105, 108, 112, 116, 118, 119, 123, 139, 141, 160, 163–166, 176, 180–183, 185, 186, 194, 197, 202, 209–211, 233, 245, 250
 spectral specific heat, 60
 transport, 81–113, 138, 154, 176, 178, 182–185, 187, 202, 207–223
 transport theories, 82–83
Planar
 force constant approach, 5
Propagating modes, 72

Q

Quantum
 dots, 59, 69, 74–76, 111, 176, 265, 268, 278–288
 wells, 69, 72–74, 232, 234, 262
 wires, 268–270, 274, 276, 278–280

R

Relaxation time, 84–85, 87, 88, 97, 98, 101, 103, 104, 106, 112, 116, 121–124, 126, 127, 129, 131, 138, 139, 146, 162, 169, 179–183, 195, 209–211
Relaxation time approximation, 116, 121, 122, 124, 127, 131, 146, 169, 179, 180, 209
Ruby, 228–230

S

Saser, 227, 231, 232, 239, 244–245, 247–251, 253, 254
Scanning thermal microscopy, 186
Scattering
 alloy, 142, 165, 166
 boundary, 60, 62, 93–94, 104, 106, 109, 166, 169, 175, 182, 195, 203
 impurity, 142–144, 194, 195
 3-phonon, 87, 99, 104, 119, 123, 139, 141–142, 160, 163, 164, 166, 181
 Scattering rates, 71, 72, 76, 77, 85, 87, 93–96, 105, 108, 112, 116, 119–121, 140, 142, 143, 147–160, 165–167, 179, 185, 208
Seeding, 249–253
Semiconductors, 3–8, 17, 37, 46, 49, 82, 88, 93, 100–104, 111–113, 116, 128, 147, 160, 176, 188, 208, 215, 218, 227–254, 261–263, 265, 267, 268, 288
Si(001)(2×1), 17–20, 22, 23, 30, 31
SiGe, 90, 111, 165–166
Si/Ge superlattices, 33, 34, 107–110, 112, 113
Silicon, 34, 35, 115–135, 138, 164, 179, 189, 191, 195, 200, 209, 211, 236
Si nanostructures, 34–37, 90, 91
Single-wall nanotubes, 51–53
$\alpha-$Sn(001)(2×1), 28–31
Specularity parameter, 104, 182, 193–196, 199, 200, 202, 203
Spontaneous, 229, 233, 249
Stark, 176, 232, 233, 235–239, 247, 249, 251–254
Stimulated, 213, 227, 229–231, 233, 234, 236, 237, 239, 249, 252
Superconductivity, 8–13, 37
Superlattices, 4, 5, 33, 34, 90, 91, 95–100, 107–113, 116, 176, 185, 203, 208, 218, 227–254
Surface phonons, 2, 17–33, 76, 77, 216
Symmetric mode, 73, 74
Symmetrization, 154–160

T

Thermal conductivity, 51, 60–62, 68, 69, 81–83, 85–113, 117–124, 126–134, 137–139, 145–148, 150, 153, 158, 160–166, 168, 169, 176–182, 184–186, 193, 195–202, 208, 209, 215, 222
Thermal transport, 60, 106, 116, 128, 129, 135, 137–170, 175, 176, 178, 182, 184–189, 192, 193, 200, 207–211
Thermoelectric materials, 82, 110–112, 165, 176
Thermoreflectance methods, 187
Thin films, 33, 111, 112, 128, 176, 184, 186, 187, 189, 193, 203, 218
Three phonon scattering, 87, 99, 104, 119, 123, 139, 141–142, 160, 163, 164, 166, 181

Threshold, 229, 230, 239–240, 245, 247, 248, 253, 254
Time-of-flight, 208, 222
Time resolved thermal wave microscopy (TRTWM), 187–189, 203
Tin, 10, 14
Transport, 49, 82, 121, 137, 175, 207, 232, 260
TRTWM. *See* Time resolved thermal wave microscopy (TRTWM)
Tunable acoustic source, 207

U
Ultrafast acoustic spectroscopy, 208
Umklapp scattering, 179, 181, 182, 185, 210
Uniaxial materials, 47, 48, 70

V
Virtual crystal approximation (VCA), 47, 142, 143

W
Wurtzite structures, 42, 47, 69–77

Printed by Printforce, the Netherlands